Lecture Notes in Statistics 101

Edited by P. Diggle, S. Fienberg, K. Krickeberg, I. Olkin, N. Wermuth

H.H. Andersen, M. Højbjerre, D. Sørensen,
and P.S. Eriksen

Linear and Graphical Models

for the Multivariate Complex Normal Distribution

Springer-Verlag
New York Berlin Heidelberg London Paris
Tokyo Hong Kong Barcelona Budapest

H.H. Andersen, M. Højbjerre, D. Sørensen, and P.S. Eriksen
Department of Mathematics and Computer Science
Aalborg University
Fredrik Bajers Vej 7 E
9220 Aalborg 0
Denmark

Library of Congress Cataloging-in-Publication Data Available
Printed on acid-free paper.

Camera ready copy provided by the author.
Printed and bound by Braun-Brumfield, Ann Arbor, MI.
Printed in the United States of America.

9 8 7 6 5 4 3 2 1

ISBN 0-387-94521-0 Springer-Verlag New York Berlin Heidelberg

Preface

The multivariate complex normal distribution has earlier been used indirectly in the work of a few physicists, but Wooding (1956) was the first who explicitly introduced it. The work in this area was highly motivated by Goodman (1963) and this is also the main reference in the area. The interest of other multivariate complex distributions also began with Goodman (1963).

Within the latest decades graphical models have become increasingly popular. Dempster (1972) introduced the covariance selection models and Wermuth (1976) demonstrated the relation between contingency tables and covariance selection models. Graphical models for contingency tables were introduced by Darroch, Lauritzen & Speed (1980) which nourished the interest in graphical models.

Until now graphical models for the multivariate complex normal distribution have never been studied in the literature. First of all this book is intended as a comprehensive introduction to the undirected graphical models for the multivariate complex normal distribution. Distributional results of multivariate complex distributions are important tools for such a study. As it is hard to find literature which systematically treats these distributions we have included the material here. The theory of linear models for the multivariate complex normal distribution is very easily obtained from the study of multivariate complex distributions, whereby a study of these models is included. We also consider other prerequisites for the study of graphical models for the multivariate complex normal distribution. These prerequisites are graph theoretical results and properties concerning conditional independence.

A brief summary of the book is given below.

Chapter 1. Prerequisites: First we introduce the notation used. Then we define a complex random variable, vector and matrix and we consider the associated expectation, variance and covariance operators. The characteristic function of a complex random variable and of a complex random vector is also defined and we consider mutual independence of complex random vectors.

Chapter 2. Multivariate Complex Normal Distribution: The multivariate complex normal distribution is presented starting with the univariate case and continuing through to the multivariate case. We also treat marginal and conditional distributions of multivariate complex normally distributed random vectors and matrices.

Chapter 3. Complex Wishart Distribution and Complex U-distribution: The focus is on the complex Wishart distribution and the complex U-distribution. Important results on these distributions are stated and we consider their respective relation to the chi-square distribution and the beta distribution.

Chapter 4. Multivariate Linear Complex Normal Models: Complex MANOVA models are treated, including definition of the model, maximum likelihood estimation of the parameters and hypothesis testing. The hypothesis tests cover testing of the mean structure and a test concerning independence. This presentation of complex MANOVA models is highly based on vector space considerations and matrix algebra.

Chapter 5. Simple Undirected Graphs: This chapter includes graph theoretical definitions and results.

Chapter 6. Conditional Independence and Markov Properties: Without using measure theory we consider a definition of conditional independence and discuss associated properties, first generally and then in a more specific form. Besides the pairwise, local and global Markov properties are defined and equivalence among them is discussed. We also investigate the factorization criterion.

Chapter 7. Complex Normal Graphical Models: We investigate undirected graphical models for the multivariate complex normal distribution without using exponential families. It includes definition of the model, maximum likelihood estimation of the concentration matrix and problems of hypothesis testing concerning conditional independence.

Appendix A. Complex Matrices: This appendix collects elementary but useful results from complex matrix algebra.

Appendix B. Orthogonal Projections: We present results about orthogonal projections.

The basic prerequisite for reading this book is an acquaintance with the multivariate real normal distribution. Additionally we presume some basic knowledge of complex number theory, matrix algebra, mathematical analysis, basic probability theory and fundamental statistical inference. It is also an advantage, but not a necessity, to be familiar with linear models for the univariate or the multivariate real normal distribution. Furthermore acquaintance with graphical models for contingency tables or covariance selection models is helpful.

For background material we refer to the bibliography, which can be found on page 163. References to the bibliography are given by the surname of the author and the year published.

Sufficient detail has been provided with the intention that the book constitutes a self-contained whole. We have attempted to be extremely consistent in the notation and all definitions, theorems etc. are clearly and precisely stated with all assumptions included. This is done to make it easier to look up e.g. a particular theorem for a reference and for readers who do not read the book from one end to the other. Most of the proofs are given in detail or otherwise the main ideas are stated as hints. Finally if none of the previous is fulfilled the result is well known and a reference for further information is given.

We use the word iff in place of the phrase if and only if. Definitions, theorems etc. are written in italic typeface and the symbol ■ is used both to indicate the end of a proof and the end of an example. When we consider sets consisting of only one element we omit the braces. Moreover we untraditionally let $A \times A$, where A is an arbitrary set, denote the set of all unordered pairs of elements from A. A list of the symbols used is included on page 181. This list also contains a short explanation for each symbol. On page 177 an index is provided.

We have tried to achieve a level of presentation appropriate for graduate students in mathematical statistics. Furthermore statisticians with interest in linear or especially graphical models may also be possible readers.

Many of the chapters are based on results from earlier chapters, therefore we recommend that they be read in the order they are presented. However, readers who are only interested in graphical models for the multivariate complex normal distribution may omit reading Chapter 4. Conversely readers who are not particularly interested in graphical models need not read Chapter 5 to 7. Readers who are not familiar with complex matrix algebra are recommended to look at the appendices before reading the book. However, the appendices are primarily meant for references, therefore they are short and compact and the results are stated without proof. For a detailed treatment of the subjects we suggest Halmos (1974) and MacDuffee (1956).

Our access to the subjects treated in the book has primarily been theoretical and we have chosen not to attach great importance to practical applications. However, we are convinced that useful applications of the theory in question will appear eventually. The multivariate complex distributions have already shown their usefulness in multiple time series in the area of meteorological forecasts and signal detection. In the book we briefly consider some examples illustrating how the multivariate complex distributions can be used within multiple time series.

The book has what we believe to be a number of distinctive features. First of all it is relatively easily read and it constitute a self-contained whole on linear and graphical models for the multivariate complex normal distribution. In the real case such a systematic and complete presentation does not yet exist. In addition the book is the first complete presentation of the multivariate complex distributions and the first published presentation of the complex normal graphical models at all. Moreover the book contains a good and wide-ranging systematic presentation of the topics conditional independence and Markov properties. In addition general measure theory and the theory of exponential families are not prerequisites for reading the book.

Through our work we have experienced that the results obtained are very similar to the corresponding results in the real case, but often the results are expressed in a more clean form, which we find appealing. Concerning graphical models we have seen that a covariance selection model with a certain structure on the concentration matrix transforms into a complex normal decomposable model. Using this transformation the analysis becomes much easier according to simpler calculations in the complex case. We find the theory treated in the book very exciting and comprehensive. We feel that we by this work have provided a solid base for studying and understanding the multivariate complex distributions and various associated subjects. In addition the book also establishes a fine and instructive insight to the real counterpart.

We are grateful to Professor Steffen L. Lauritzen for critically reading the book and giving helpful comments. We also thank the Danish Research Councils which have supported the work in part through the PIFT programme.

Aalborg University, January , 1995.

Heidi H. Andersen, Malene Højbjerre, Dorte Sørensen & Poul Svante Eriksen.

Contents

1

Prerequisites

This chapter presents some definitions and results frequently used throughout this book. The purpose is first of all to establish the notation, therefore the presentation is at times short and compact. Apart from notation this chapter also contains an important interpretation of the direct product of two complex matrices, and we define the direct product of two complex vector spaces. Further we consider a real isomorphism between a p-dimensional complex vector space and a $2p$-dimensional real vector space. This isomorphism is based on considering a complex vector space as a real vector space. Next we study the concepts of complex random variables, vectors and matrices. Besides definitions of those we regard the expectation, variance and covariance operators individually for each case. Both definitions and properties for the operators are considered. We also define the characteristic function of a complex random variable and of a complex random vector. We see that a characteristic function determines a distribution uniquely. Finally mutual independence of complex random vectors is defined.

1.1 Complex Matrix Algebra

Let $\mathbb{C}$ denote the field of complex numbers. Any element $c \in \mathbb{C}$ has a unique representation as $c = a + ib$, where $a, b \in \mathbb{R}$. Here a is the real part, b the imaginary part and i the imaginary unit, that is $i^2 = -1$. The letter i is reserved for this purpose. We also denote the real and imaginary parts of c by $\mathrm{Re}\,(c)$ and $\mathrm{Im}\,(c)$, respectively. The conjugate of a complex number is given by $\bar{c} = a - ib$ and the absolute value, also called modulus, is given as $|c| = (c\bar{c})^{\frac{1}{2}} = (a^2 + b^2)^{\frac{1}{2}}$.

We let $\mathbb{C}^p$ be the p-dimensional complex vector space of p-tuples of complex numbers. These tuples are represented as columns and are denoted as $\mathbf{c} = (c_k)$, where $c_k \in \mathbb{C}$ for $k = 1, 2, \ldots, p$. The number c_k is called the k'th element of $\mathbf{c}$ and $\mathbf{c}$ is called a p-dimensional complex vector. A p-dimensional complex vector can uniquely be written as $\boldsymbol{c} = \boldsymbol{a} + i\boldsymbol{b}$, where $\boldsymbol{a}, \boldsymbol{b} \in \mathbb{R}^p$ contain the real and imaginary parts of $\mathbf{c}$, respectively. We define the inner product on $\mathbb{C}^p$ as follows.

Definition 1.1 Inner product on $\mathbb{C}^p$
The inner product of $\mathbf{c} = (c_k)$ and $\mathbf{d} = (d_k)$ in $\mathbb{C}^p$ is defined as

$$\langle \mathbf{c}, \mathbf{d} \rangle = \sum_{k=1}^{p} c_k \bar{d}_k \, .$$

Remark that $\langle \cdot \, , \cdot \rangle : \mathbb{C}^p \times \mathbb{C}^p \mapsto \mathbb{C}$ is a conjugate bilinear operator.

An $n \times p$ array $\boldsymbol{C} = (c_{jk})$, where $c_{jk} \in \mathbb{C}$ for $j = 1, 2, \ldots, n$ and $k = 1, 2, \ldots, p$, is called an $n \times p$ complex matrix. It can uniquely be written as $\boldsymbol{C} = \boldsymbol{A} + i\boldsymbol{B}$, where $\boldsymbol{A}$ and $\boldsymbol{B}$ are real matrices of same dimension as $\boldsymbol{C}$. Here $\boldsymbol{A}$ contains the real parts of $\boldsymbol{C}$, whereas $\boldsymbol{B}$ contains the imaginary parts. The set of all $n \times p$ complex matrices is denoted by $\mathbb{C}^{n\times p}$. This set is a complex vector space of dimension np. Note that a complex matrix $\boldsymbol{C} \in \mathbb{C}^{n\times p}$ can be identified with a linear transformation from $\mathbb{C}^p$ to $\mathbb{C}^n$.

The conjugate transpose of an $n \times p$ complex matrix $\boldsymbol{C} = (c_{jk})$ is the $p \times n$ complex matrix given by $\boldsymbol{C}^* = (\bar{c}_{kj})$ for $k = 1, 2, \ldots, p$ and $j = 1, 2, \ldots, n$. We define the inner product on $\mathbb{C}^{n\times p}$ as below.

Definition 1.2 Inner product on $\mathbb{C}^{n\times p}$
The inner product of $\boldsymbol{C}$ and $\boldsymbol{D}$, where $\boldsymbol{C}, \boldsymbol{D} \in \mathbb{C}^{n\times p}$, is defined as

$$\langle \boldsymbol{C}, \boldsymbol{D} \rangle = \operatorname{tr}(\boldsymbol{C}\boldsymbol{D}^*) .$$

Observe that the inner product on $\mathbb{C}^{n\times p}$ is defined in agreement with the inner product on $\mathbb{C}^p$ and that $\langle \cdot, \cdot \rangle : \mathbb{C}^{n\times p} \times \mathbb{C}^{n\times p} \mapsto \mathbb{C}$ is a conjugate bilinear operator.

The direct product of matrices, also called the Kronecker product or tensor product, is an important concept.

Definition 1.3 The direct product of complex matrices
Let $\boldsymbol{C} = (c_{jk}) \in \mathbb{C}^{n\times p}$ and $\boldsymbol{D} = (d_{rs}) \in \mathbb{C}^{m\times q}$. The direct product of $\boldsymbol{C}$ and $\boldsymbol{D}$ is the $nm \times pq$ complex matrix $\boldsymbol{C} \otimes \boldsymbol{D}$ with elements given as

$$(\boldsymbol{C} \otimes \boldsymbol{D})_{jr,ks} = c_{jk}\bar{d}_{rs} .$$

When displaying $\boldsymbol{C} \otimes \boldsymbol{D}$ we use the representation of the block matrix with $m \times q$ blocks of size $n \times p$, i.e.

$$\boldsymbol{C} \otimes \boldsymbol{D} = \begin{pmatrix} \boldsymbol{C}\bar{d}_{11} & \cdots & \boldsymbol{C}\bar{d}_{1q} \\ \vdots & & \vdots \\ \boldsymbol{C}\bar{d}_{m1} & \cdots & \boldsymbol{C}\bar{d}_{mq} \end{pmatrix} .$$

The direct product $\boldsymbol{C} \otimes \boldsymbol{D}$ can be identified with a linear transformation from $\mathbb{C}^{p\times q}$ to $\mathbb{C}^{n\times m}$. This is stated in the following theorem.

Theorem 1.1
Let $\boldsymbol{C} \in \mathbb{C}^{n\times p}$ and $\boldsymbol{D} \in \mathbb{C}^{m\times q}$. The direct product of $\boldsymbol{C}$ and $\boldsymbol{D}$ can be identified with the linear transformation $\boldsymbol{C} \otimes \boldsymbol{D} : \mathbb{C}^{p\times q} \mapsto \mathbb{C}^{n\times m}$ given by

$$(\boldsymbol{C} \otimes \boldsymbol{D})(\boldsymbol{E}) = \boldsymbol{C}\boldsymbol{E}\boldsymbol{D}^* ,$$

where $\boldsymbol{E} \in \mathbb{C}^{p\times q}$.

Proof:
Let $C = (c_{jk}) \in \mathbb{C}^{n\times p}, D = (d_{rs}) \in \mathbb{C}^{m\times q}$ and $E = (e_{ks}) \in \mathbb{C}^{p\times q}$. To show that the identification is correct we consider

$$\begin{aligned}(CED^*)_{jr} &= \sum_{k=1}^{p}\sum_{s=1}^{q} c_{jk}e_{ks}\bar{d}_{rs} \\ &= \sum_{k=1}^{p}\sum_{s=1}^{q} c_{jk}\bar{d}_{rs}e_{ks} \\ &= \sum_{k=1}^{p}\sum_{s=1}^{q} (C\otimes D)_{jr,ks}\, e_{ks}\,,\end{aligned}$$

which completes the proof. ∎

For $C \in \mathbb{C}^{n\times p}$ and $D \in \mathbb{C}^{m\times q}$ the complex matrix $C\otimes D$ has dimension $nm \times pq$, hence in $(C\otimes D)(E)$ the complex matrix $E \in \mathbb{C}^{p\times q}$ is considered as a $pq \times 1$ complex vector and $(C\otimes D)(E)$ is considered as an $nm \times 1$ complex vector. In CED^* we consider C, D and E as the complex matrices they originally are, and CED^* has dimension $m \times n$. We have no notational distinction between these representations and according to Theorem 1.1 they are equal. This means that we sometimes interpret a complex matrix as a complex vector and vice versa.

In this book we use the direct product of the two complex vector spaces N and $\mathbb{C}^p$, where N is a subspace of $\mathbb{C}^n$. We define this complex vector space as

$$N\otimes\mathbb{C}^p = \left\{ C \in \mathbb{C}^{n\times p} \;\middle|\; C = \sum_j n_j \otimes d_j^*, \text{ where } n_j \in N, d_j \in \mathbb{C}^p \right\}.$$

1.2 A Vector Space Isomorphism

We know that each element in $\mathbb{C}$ can be identified with a unique element in $\mathbb{R}^2$. Therefore the p-dimensional complex vector space, $\mathbb{C}^p$, can be regarded as a real vector space. Using this we are able to define a real isomorphism between $\mathbb{C}^p$ and $\mathbb{R}^{2p}$.

Definition 1.4 Isomorphism between $\mathbb{C}^p$ and $\mathbb{R}^{2p}$
Let $c = a + ib \in \mathbb{C}^p$, *where* $a, b \in \mathbb{R}^p$. *The isomorphism* $[\cdot] : \mathbb{C}^p \mapsto \mathbb{R}^{2p}$ *is given by*

$$[c] = \begin{pmatrix} a \\ b \end{pmatrix}.$$

By regarding $\mathbb{C}^p$ as a real vector space we see that $[\cdot]$ is linear, since

$$[\alpha c_1 + \beta c_2] = \alpha [c_1] + \beta [c_2] \,,$$

where $c_1, c_2 \in \mathbb{C}^p$ and $\alpha, \beta \in \mathbb{R}$. Moreover $[\cdot]$ is bijective, since a complex vector $c \in \mathbb{C}^p$ has a unique representation as $c = a + ib$, where $a, b \in \mathbb{R}^p$. Hence we conclude that $\mathbb{C}^p$ and $\mathbb{R}^{2p}$ are isomorphic.

Let $C \in \mathbb{C}^{n \times p}$ and $c \in \mathbb{C}^p$ be written as $C = A + iB$ and $c = a + ib$, where $A, B \in \mathbb{R}^{n \times p}$ and $a, b \in \mathbb{R}^p$, then

$$Cc = (A + iB)(a + ib) = Aa - Bb + i(Ab + Ba) \ .$$

Using the isomorphism $[\cdot]$ on Cc we get

$$(1.1) \qquad [Cc] = \begin{pmatrix} Aa - Bb \\ Ab + Ba \end{pmatrix} = \begin{pmatrix} A & -B \\ B & A \end{pmatrix} \begin{pmatrix} a \\ b \end{pmatrix} .$$

This leads to the following definition.

Definition 1.5
Let $C = A + iB \in \mathbb{C}^{n \times p}$, where $A, B \in \mathbb{R}^{n \times p}$. The partitioned matrix $\{C\} \in \mathbb{R}^{2n \times 2p}$ is given by

$$\{C\} = \begin{pmatrix} A & -B \\ B & A \end{pmatrix} .$$

From (1.1) and Definition 1.5 we see that

$$(1.2) \qquad [Cc] = \{C\}[c] \ .$$

Hence multiplication of a complex matrix $C \in \mathbb{C}^{n \times p}$ and a complex vector $c \in \mathbb{C}^p$ corresponds to multiplication of the real matrix $\{C\} \in \mathbb{R}^{2n \times 2p}$ and the real vector $[c] \in \mathbb{R}^{2p}$. Notice that the following theorem holds.

Theorem 1.2
For $C \in \mathbb{C}^{n \times p}$ the following properties hold.

1. $\{C^*\} = \{C\}^{\mathsf{T}}$, *where* ^{T} *denotes the transpose of a real matrix.*
2. $\{\alpha C + \beta D\} = \alpha\{C\} + \beta\{D\}$, *where* $D \in \mathbb{C}^{n \times p}$ *and* $\alpha, \beta \in \mathbb{R}$.
3. $\{CD\} = \{C\}\{D\}$, *where* $D \in \mathbb{C}^{p \times q}$.

1.3 Complex Random Variables

We define a complex random variable and associated operators in this section.

Definition 1.6 Complex random variable
Let U and V be real random variables. The unique random variable given by $X = U + iV$ is a complex random variable.

From Definition 1.6 we find that a complex random variable is a random variable taking values in $\mathbb{C}$.

When we consider complex random variables, three operators are important. These operators are defined on the complex vector space given by

$$\mathcal{L}_2(\mathbb{C}) = \left\{X \mid X \text{ is a complex random variable and } \mathbb{E}\left(X\overline{X}\right) < \infty\right\},$$

where $\mathbb{E}$ denotes the expectation operator of a real random variable. Writing $X = U + iV$ we see that $\mathbb{E}\left(X\overline{X}\right) = \mathbb{E}(U^2 + V^2)$, thus $\mathcal{L}_2(\mathbb{C})$ is the vector space of complex random variables with real and imaginary parts each having finite second moment. The subscript 2 in $\mathcal{L}_2(\mathbb{C})$ arises from the fact that $\mathbb{E}\left(X\overline{X}\right) = \mathbb{E}(|X|^2)$, which means that $\mathcal{L}_2(\mathbb{C})$ also is the vector space of complex random variables having finite square length.

Definition 1.7 The expectation operator of a complex random variable
Let $X = U+iV$ be a complex random variable. The expectation operator of X, $\mathbb{E} : \mathcal{L}_2(\mathbb{C}) \mapsto \mathbb{C}$, is defined as

$$\mathbb{E}(X) = \mathbb{E}(U) + i\,\mathbb{E}(V) .$$

Note that we use $\mathbb{E}$ both as the symbol for the expectation operator of a real and complex random variable. The value of the expectation operator evaluated at X is referred to as the mean of X. Observe from Definition 1.7 that

$$\mathbb{E}\left(X^2\right) = \mathbb{E}\left((U+iV)^2\right) = \mathbb{E}\left(U^2 - V^2\right) + 2i\mathbb{E}(UV) . \tag{1.3}$$

The covariance operator of two complex random variables is defined below.

Definition 1.8 The covariance operator of complex random variables
Let X and Y be complex random variables. The covariance operator of X and Y, $\mathbb{C} : \mathcal{L}_2(\mathbb{C}) \times \mathcal{L}_2(\mathbb{C}) \mapsto \mathbb{C}$, is defined as

$$\mathbb{C}(X,Y) = \mathbb{E}\left((X - \mathbb{E}(X))\,\overline{(Y - \mathbb{E}(Y))}\right) .$$

We refer to the value of $\mathbb{C}(X,Y)$ as the covariance of X and Y (in that order). Note that the symbol $\mathbb{C}$ is used both for the field of complex numbers and the covariance operator. As these quantities are quite different it should not confuse the reader.

In the special case where $X = Y$ the covariance operator is called the variance operator. This provides us with the following definition.

Definition 1.9 The variance operator of a complex random variable
Let X be a complex random variable. The variance operator of X, $\mathbb{V}\colon \mathcal{L}_2(\mathbb{C}) \mapsto \overline{\mathbb{R}}_+$, is defined as

$$\begin{aligned}\mathbb{V}(X) &= \mathbb{C}(X,X)\\ &= \mathbb{E}\left((X-\mathbb{E}(X))\,\overline{(X-\mathbb{E}(X))}\right).\end{aligned}$$

The value of $\mathbb{V}(X)$ is called the variance of X. Remark that the complex conjugate is necessary in the variance operator as we require the variance of a complex random variable to be a nonnegative real number.

The following theorems hold for the expectation, covariance and variance operators, respectively. They are stated without proof as they follow immediately from the definitions above or each other.

Theorem 1.3 Rules for the expectation operator
Let X and Y be complex random variables and let $c, d \in \mathbb{C}$. The following rules hold for the expectation operator.

1. $\mathbb{E}(cX+d) = c\mathbb{E}(X)+d$.
2. $\mathbb{E}(X+Y) = \mathbb{E}(X)+\mathbb{E}(Y)$.
3. $\mathbb{E}\left(\overline{X}\right) = \overline{\mathbb{E}(X)}$.

This indicates that the expectation operator is a linear operator on $\mathcal{L}_2(\mathbb{C})$.

Theorem 1.4 Rules for the covariance operator
Let X, Y and Z be complex random variables and let $c_1, c_2, d_1, d_2 \in \mathbb{C}$. The following rules hold for the covariance operator.

1. $\mathbb{C}(X,Y) = \mathbb{E}\left(X\overline{Y}\right) - \mathbb{E}(X)\overline{\mathbb{E}(Y)}$.
2. $\mathbb{C}(X,Y) = \overline{\mathbb{C}(Y,X)}$.
3. $\mathbb{C}(X,Y+Z) = \mathbb{C}(X,Y)+\mathbb{C}(X,Z)$.
4. $\mathbb{C}(c_1X+d_1, c_2Y+d_2) = c_1\overline{c}_2\mathbb{C}(X,Y)$.

According to Theorem 1.4 we see that the covariance operator is a conjugate bilinear operator on $\mathcal{L}_2(\mathbb{C}) \times \mathcal{L}_2(\mathbb{C})$.

Theorem 1.5 Rules for the variance operator
Let $X = U + iV$ and Y be complex random variables and let $c, d \in \mathbb{C}$. The following rules hold for the variance operator.

1. $\mathbb{V}(X) = \mathbb{E}\left(X\overline{X}\right) - \mathbb{E}(X)\overline{\mathbb{E}(X)}$.
2. $\mathbb{V}(X) = \mathbb{V}(U) + \mathbb{V}(V)$.
3. $\mathbb{V}(cX + d) = c\overline{c}\mathbb{V}(X)$.
4. $\mathbb{V}(X + Y) = \mathbb{V}(X) + \mathbb{V}(Y) + 2\,\mathrm{Re}\,(\mathbb{C}(X, Y))$.

Our aim is now to define the characteristic function of a complex random variable. Let $X = U + i\boldsymbol{V}$ be a complex random variable, then

$$[X] = \begin{pmatrix} U \\ V \end{pmatrix},$$

where $[\cdot]$ is the real isomorphism between $\mathbb{C}$ and $\mathbb{R}^2$. This isomorphism is defined in Definition 1.4 page 3. The characteristic function of $[X]$ is known from the real case as

$$\varphi_{[X]}([\xi]) = \mathbb{E}\left(\exp\left(i[\xi]^{\top}[X]\right)\right), \quad [\xi] \in \mathbb{R}^2 .$$

Since $[\cdot]$ is an isomorphism, there is a one-to-one correspondence between the distributions of X and $[X]$. Therefore we define the characteristic function of X, such that it is identical to the characteristic function of $[X]$.

Definition 1.10 The characteristic function of a complex random variable
Let X be a complex random variable. The characteristic function of X is defined as

$$\varphi_X(\xi) = \mathbb{E}\left(\exp\left(i\,\mathrm{Re}\left(\overline{\xi}X\right)\right)\right), \quad \xi \in \mathbb{C} .$$

1.4 Complex Random Vectors and Matrices

Later we study a situation where we have several complex random variables. It can be an advantage to arrange these complex random variables in vectors or matrices. First we arrange p complex random variables in a vector. Let $X_1, X_2, \ldots, X_p$ be complex random variables. The p-dimensional vector $\boldsymbol{X}$ given by

$$\boldsymbol{X} = \begin{pmatrix} X_1 \\ X_2 \\ \vdots \\ X_p \end{pmatrix}$$

is called a p-dimensional complex random vector. For typographical convenience we often write $\boldsymbol{X} = (X_k)$. Formally we define a complex random vector as follows.

Definition 1.11 Complex random vector
Let $\boldsymbol{X} = (X_k)$ be a p-dimensional vector, where X_k for $k = 1, 2, \ldots, p$ is a complex random variable. Then $\boldsymbol{X}$ is called a p-dimensional complex random vector.

Let X_{jk} for $j = 1, 2, \ldots, n$ and $k = 1, 2, \ldots, p$ be complex random variables, and let these be arranged in an $n \times p$ matrix as

$$\boldsymbol{X} = \begin{pmatrix} X_{11} & X_{12} & \cdots & X_{1p} \\ X_{21} & X_{22} & \cdots & X_{2p} \\ \vdots & \vdots & & \vdots \\ X_{n1} & X_{n2} & \cdots & X_{np} \end{pmatrix} .$$

The matrix $\boldsymbol{X}$ is called an $n \times p$ complex random matrix and is formally defined below.

Definition 1.12 Complex random matrix
Let $\boldsymbol{X} = (X_{jk})$ be an $n \times p$ matrix, where X_{jk} for $j = 1, 2, \ldots, n$ and $k = 1, 2, \ldots, p$ is a complex random variable. Then $\boldsymbol{X}$ is called an $n \times p$ complex random matrix.

We define expectation, covariance and variance operator on complex random matrices. The operators are considered on the complex vector space given by

$$\mathcal{L}_2\left(\mathbb{C}^{n \times p}\right) = \{\boldsymbol{X} = (X_{jk}) | \, \boldsymbol{X} \text{ is an } n \times p \text{ complex random matrix and } \mathbb{E}\left(\operatorname{tr}\left(\boldsymbol{X}\boldsymbol{X}^*\right)\right) < \infty\} .$$

The same operator notation is used both for variables and matrices (and later on for vectors).

Definition 1.13 The expectation operator of a complex random matrix
Let $\boldsymbol{X} = (X_{jk})$ be an $n \times p$ complex random matrix. The expectation operator of $\boldsymbol{X}$, $\mathbb{E} : \mathcal{L}_2\left(\mathbb{C}^{n \times p}\right) \mapsto \mathbb{C}^{n \times p}$, is defined as

$$\mathbb{E}\left(\boldsymbol{X}\right) = \left(\mathbb{E}\left(X_{jk}\right)\right) .$$

The expectation operator evaluated in the $n \times p$ complex random matrix $\boldsymbol{X}$ is an $n \times p$ complex matrix which we call the mean of $\boldsymbol{X}$.

In agreement with Definition 1.13 we define the expectation operator on a complex random vector. In this case the operator is defined on the complex vector space given by

$$\mathcal{L}_2\left(\mathbb{C}^{p}\right) = \{\boldsymbol{X} = (X_k) | \text{ is a } p\text{-dimensional complex random vector and } \mathbb{E}\left(\boldsymbol{X}^*\boldsymbol{X}\right) < \infty\} .$$

Definition 1.14 The expectation operator of a complex random vector
Let $\boldsymbol{X} = (X_k)$ be a p-dimensional complex random vector. The expectation operator of $\boldsymbol{X}$, $\mathbb{E} : \mathcal{L}_2\left(\mathbb{C}^{p}\right) \mapsto \mathbb{C}^{p}$, is defined as

$$\mathbb{E}\left(\boldsymbol{X}\right) = \left(\mathbb{E}\left(X_k\right)\right) .$$

The covariance operator of two complex random matrices is defined next.

Definition 1.15 The covariance operator of complex random matrices
Let $\boldsymbol{X} = (X_{jk})$ and $\boldsymbol{Y} = (Y_{rs})$ be complex random matrices of dimensions $n \times p$ and $m \times q$, respectively. The covariance operator of $\boldsymbol{X}$ and $\boldsymbol{Y}$, $\mathbb{C} : \mathcal{L}_2(\mathbb{C}^{n\times p}) \times \mathcal{L}_2(\mathbb{C}^{m\times q}) \mapsto \mathbb{C}^{np\times mq}$, is defined as

$$\mathbb{C}(\boldsymbol{X}, \boldsymbol{Y}) = (\mathbb{C}(X_{jk}, Y_{rs})) \ .$$

The covariance operator of the complex random matrices $\boldsymbol{X}$ and $\boldsymbol{Y}$ of dimensions $n \times p$ and $m \times q$, respectively, is an $np \times mq$ complex matrix. This matrix is referred to as the covariance matrix of $\boldsymbol{X}$ and $\boldsymbol{Y}$ (in that order).

We define the variance operator of a complex random matrix in the same way as it is defined for a complex random variable.

Definition 1.16 The variance operator of a complex random matrix
Let $\boldsymbol{X}$ be an $n \times p$ complex random matrix. The variance operator of $\boldsymbol{X}$, $\mathbb{V} : \mathcal{L}_2(\mathbb{C}^{n\times p}) \mapsto \mathbb{C}^{np\times np}$, is defined as

$$\mathbb{V}(\boldsymbol{X}) = \mathbb{C}(\boldsymbol{X}, \boldsymbol{X}) \ .$$

We refer to the value of $\mathbb{V}(\boldsymbol{X})$ as the variance matrix of $\boldsymbol{X}$.

Again the definitions can be transformed to complex random vectors. From Definition 1.15 we have the following definition of the covariance operator of complex random vectors.

Definition 1.17 The covariance operator of complex random vectors
Let $\boldsymbol{X} = (X_k)$ and $\boldsymbol{Y} = (Y_s)$ be complex random vectors of dimensions p and q, respectively. The covariance operator of $\boldsymbol{X}$ and $\boldsymbol{Y}$, $\mathbb{C} : \mathcal{L}_2(\mathbb{C}^{p}) \times \mathcal{L}_2(\mathbb{C}^{q}) \mapsto \mathbb{C}^{p\times q}$, is defined as

$$\mathbb{C}(\boldsymbol{X}, \boldsymbol{Y}) = (\mathbb{C}(X_k, Y_s)) \ .$$

The definition of the variance operator of a complex random vector becomes as follows.

Definition 1.18 The variance operator of a complex random vector
Let $\boldsymbol{X}$ be a p-dimensional complex random vector. The variance operator of $\boldsymbol{X}$, $\mathbb{V} : \mathcal{L}_2(\mathbb{C}^{p}) \mapsto \mathbb{C}^{p\times p}$, is defined as

$$\mathbb{V}(\boldsymbol{X}) = \mathbb{C}(\boldsymbol{X}, \boldsymbol{X}) \ .$$

When $\boldsymbol{X} = (X_k)$ and $\boldsymbol{Y} = (Y_s)$ are complex random vectors of dimensions p and q, respectively, the $p \times q$ complex covariance matrix of $\boldsymbol{X}$ and $\boldsymbol{Y}$ can be interpreted as a $pq \times 1$ complex vector given as

$$\begin{aligned}
\mathbb{C}(\boldsymbol{X},\boldsymbol{Y}) &= (\mathbb{C}(X_k, Y_s)) \\
&= \left(\mathbb{E}\left((X_k - \mathbb{E}(X_k))\,\overline{(Y_s - \mathbb{E}(Y_s))}\right)\right) \\
&= \mathbb{E}((\boldsymbol{X} - \mathbb{E}(\boldsymbol{X}))\,(\boldsymbol{Y} - \mathbb{E}(\boldsymbol{Y}))^*) \\
&= \mathbb{E}((\boldsymbol{X} - \mathbb{E}(\boldsymbol{X})) \otimes (\boldsymbol{Y} - \mathbb{E}(\boldsymbol{Y}))) \,. \qquad (1.4)
\end{aligned}$$

This interpretation of $\mathbb{C}(\boldsymbol{X},\boldsymbol{Y})$ is useful in the proof of Theorem 1.7 below. The following theorem holds for the expectation operator of a complex random matrix. It is also necessary in the proof of Theorem 1.7.

Theorem 1.6 Rule for the expectation operator
Let $\boldsymbol{X}$ be an $n \times p$ complex random matrix and let $\boldsymbol{C} \in \mathbb{C}^{m\times n}$, $\boldsymbol{D} \in \mathbb{C}^{p\times q}$ and $\boldsymbol{E} \in \mathbb{C}^{m\times q}$. It holds that

$$\mathbb{E}(\boldsymbol{C}\boldsymbol{X}\boldsymbol{D} + \boldsymbol{E}) = \boldsymbol{C}\mathbb{E}(\boldsymbol{X})\,\boldsymbol{D} + \boldsymbol{E} \,.$$

Proof:
The theorem follows directly from Definition 1.13 page 8 and Theorem 1.3 page 6. ∎

The following theorem states the covariance of linear transformations of complex random vectors.

Theorem 1.7 Rule for the covariance operator
Let $\boldsymbol{X}$ and $\boldsymbol{Y}$ be complex random vectors of dimensions p and q, respectively, and let $\boldsymbol{C} \in \mathbb{C}^{n\times p}$ and $\boldsymbol{D} \in \mathbb{C}^{m\times q}$. It holds that

$$\mathbb{C}(\boldsymbol{C}\boldsymbol{X}, \boldsymbol{D}\boldsymbol{Y}) = \boldsymbol{C}\mathbb{C}(\boldsymbol{X},\boldsymbol{Y})\,\boldsymbol{D}^* \,.$$

Proof:
Let $\boldsymbol{X}$ and $\boldsymbol{Y}$ be complex random vectors of dimensions p and q, respectively, and let $\boldsymbol{C} \in \mathbb{C}^{n\times p}$ and $\boldsymbol{D} \in \mathbb{C}^{m\times q}$.

The interpretation of the covariance operator stated in (1.4) and Theorem 1.6 gives us

$$\begin{aligned}
\mathbb{C}(\boldsymbol{C}\boldsymbol{X}, \boldsymbol{D}\boldsymbol{Y}) &= \mathbb{E}((\boldsymbol{C}\boldsymbol{X} - \mathbb{E}(\boldsymbol{C}\boldsymbol{X})) \otimes (\boldsymbol{D}\boldsymbol{Y} - \mathbb{E}(\boldsymbol{D}\boldsymbol{Y}))) \\
&= \mathbb{E}((\boldsymbol{C} \otimes \boldsymbol{D})\,((\boldsymbol{X} - \mathbb{E}(\boldsymbol{X})) \otimes (\boldsymbol{Y} - \mathbb{E}(\boldsymbol{Y})))) \\
&= (\boldsymbol{C} \otimes \boldsymbol{D})\,\mathbb{E}((\boldsymbol{X} - \mathbb{E}(\boldsymbol{X})) \otimes (\boldsymbol{Y} - \mathbb{E}(\boldsymbol{Y}))) \\
&= \boldsymbol{C}\mathbb{C}(\boldsymbol{X},\boldsymbol{Y})\,\boldsymbol{D}^* \,.
\end{aligned}$$

∎

The corollary below follows from Theorem 1.7 and it states the corresponding rule for the variance operator.

Corollary 1.1 Rule for the variance operator
Let $\boldsymbol{X}$ be a p-dimensional complex random vector and let $\boldsymbol{C} \in \mathbb{C}^{n \times p}$. It holds that

$$\mathbb{V}(\boldsymbol{C}\boldsymbol{X}) = \boldsymbol{C}\mathbb{V}(\boldsymbol{X})\,\boldsymbol{C}^* .$$

Let $\boldsymbol{X}$ be a p-dimensional complex random vector and consider the $p \times p$ variance matrix $\mathbb{V}(\boldsymbol{X})$. According to Theorem 1.4 page 6 we observe that the ks'th element of $\mathbb{V}(\boldsymbol{X})$ is given as

$$(\mathbb{V}(\boldsymbol{X}))_{ks} = \mathbb{C}(X_k, X_s) = \overline{\mathbb{C}(X_s, X_k)} = \overline{(\mathbb{V}(\boldsymbol{X}))_{sk}} ,$$

thus $\mathbb{V}(\boldsymbol{X})$ is Hermitian. Further for all $\boldsymbol{c} \in \mathbb{C}^p$ we have from Corollary 1.1 that

$$\begin{aligned} \mathbb{V}(\boldsymbol{c}^*\boldsymbol{X}) &= \boldsymbol{c}^*\mathbb{V}(\boldsymbol{X})\,\boldsymbol{c} \\ &\geq 0 , \end{aligned}$$

as the variance of a complex random variable is nonnegative. Thus $\mathbb{V}(\boldsymbol{X}) \geq \boldsymbol{O}$, which means that $\mathbb{V}(\boldsymbol{X}) \in \mathbb{C}_S^{p \times p}$. This leads to the following theorem.

Theorem 1.8
Let $\boldsymbol{X}$ be a p-dimensional complex random vector with variance matrix $\mathbb{V}(\boldsymbol{X})$. Then it holds that $\mathbb{V}(\boldsymbol{X}) \in \mathbb{C}_S^{p \times p}$.

Concerning the $np \times np$ variance matrix $\mathbb{V}(\boldsymbol{X})$ of an $n \times p$ complex random matrix $\boldsymbol{X}$ we deduce from Theorem 1.8, by regarding $\boldsymbol{X}$ as an $np \times 1$ complex random vector, that $\mathbb{V}(\boldsymbol{X}) \in \mathbb{C}_S^{np \times np}$.

When we later define the univariate complex normal distribution, we need a special type of covariance structure.

Definition 1.19 Complex covariance structure
Let $\boldsymbol{X}$ be a p-dimensional complex random vector. The 2p-dimensional real random vector $[\boldsymbol{X}]$ is said to have a complex covariance structure if

$$\mathbb{V}([\boldsymbol{X}]) = \begin{pmatrix} \Sigma & -\boldsymbol{A} \\ \boldsymbol{A} & \Sigma \end{pmatrix} ,$$

where $\Sigma, \boldsymbol{A} \in \mathbb{R}^{p \times p}$.

Notice that Σ is symmetric and $\boldsymbol{A}$ is skew symmetric, since we know from the real case that $\mathbb{V}([\boldsymbol{X}])$ is symmetric.

Let $\boldsymbol{X}$ be a p-dimensional complex random vector. Using the one-to-one correspondence between the distributions of $\boldsymbol{X}$ and $[\boldsymbol{X}]$ we define the characteristic function of $\boldsymbol{X}$. From the real case the characteristic function of $[\boldsymbol{X}]$ is given by

$$\varphi_{[\boldsymbol{X}]}([\boldsymbol{\xi}]) = \mathbb{E}\left(\exp\left(i[\boldsymbol{\xi}]^{\top}[\boldsymbol{X}]\right)\right) \ , \ [\boldsymbol{\xi}] \in \mathbb{R}^{2p} \ ,$$

whereby the following definition is natural.

Definition 1.20 The characteristic function of a complex random vector
Let $\boldsymbol{X}$ be a p-dimensional complex random vector. The characteristic function of $\boldsymbol{X}$ is defined as

$$\varphi_{\boldsymbol{X}}(\boldsymbol{\xi}) = \mathbb{E}(\exp(i \operatorname{Re}(\boldsymbol{\xi}^{*}\boldsymbol{X}))) \ , \ \boldsymbol{\xi} \in \mathbb{C}^{p} \ .$$

Combining Definition 1.10 page 7 and Definition 1.20 implies for a p-dimensional complex random vector $\boldsymbol{X}$ that

$$\varphi_{\boldsymbol{\xi}^{*}\boldsymbol{X}}(1) = \varphi_{\boldsymbol{X}}(\boldsymbol{\xi}) \ , \ \boldsymbol{\xi} \in \mathbb{C}^{p} \ . \tag{1.5}$$

A basic property of a characteristic function is that it determines a distribution uniquely.

Theorem 1.9
Let $\boldsymbol{X}$ and $\boldsymbol{Y}$ be p-dimensional complex random vectors with characteristic functions $\varphi_{\boldsymbol{X}}(\boldsymbol{\xi})$ and $\varphi_{\boldsymbol{Y}}(\boldsymbol{\xi})$, where $\boldsymbol{\xi} \in \mathbb{C}^{p}$, respectively. If

$$\varphi_{\boldsymbol{X}}(\boldsymbol{\xi}) = \varphi_{\boldsymbol{Y}}(\boldsymbol{\xi}) \quad \forall \boldsymbol{\xi} \in \mathbb{C}^{p} \ ,$$

then

$$\mathcal{L}(\boldsymbol{X}) = \mathcal{L}(\boldsymbol{Y}) \ .$$

Proof:
Using the one-to-one correspondence between the distributions of $[\boldsymbol{X}]$ and $\boldsymbol{X}$ and similarly between $[\boldsymbol{Y}]$ and $\boldsymbol{Y}$ together with the uniqueness theorem in the real case (Cramér 1945, p. 101) the theorem is deduced. ■

An interesting consequence of this uniqueness theorem follows, if we let $\boldsymbol{X}$ be a p-dimensional complex random vector and assume that the distribution of $\boldsymbol{c}^{*}\boldsymbol{X}$ is known for all $\boldsymbol{c} \in \mathbb{C}^{p}$. Then $\varphi_{\boldsymbol{c}^{*}\boldsymbol{X}}(\xi)$, where $\xi \in \mathbb{C}$, is known for all $\boldsymbol{c} \in \mathbb{C}^{p}$. Since $\varphi_{\boldsymbol{c}^{*}\boldsymbol{X}}(1) = \varphi_{\boldsymbol{X}}(\boldsymbol{c})$, the characteristic function of $\boldsymbol{X}$ is known for all $\boldsymbol{c} \in \mathbb{C}^{p}$. Therefore, according to Theorem 1.9, we also know the distribution of $\boldsymbol{X}$.

Finally we define mutual independence of complex random vectors as follows.

Definition 1.21 Mutual independence
Let $X_1, X_2, \ldots, X_n$ be n complex random vectors. If, for all measurable sets $A_1, A_2, \ldots, A_n$ in the sample spaces of $X_1, X_2, \ldots X_n$, respectively, it holds that

$$P(X_1 \in A_1, X_2 \in A_2, \ldots, X_n \in A_n) = P(X_1 \in A_1)\, P(X_2 \in A_2) \cdots P(X_n \in A_n) \ ,$$

then $X_1, X_2, \ldots, X_n$ are said to be mutually independent.

If X and Y are independent complex random vectors we denote this by writing $X \perp\!\!\!\perp Y$. Remark that the complex random vectors are said to be dependent if they are not independent. Furthermore Definition 1.21 also includes definition of mutual independence of complex random variables and matrices, as a vector can consist of only one element and a matrix can be interpreted as a vector.

Theorem 1.10
Let X and Y be independent complex random vectors of dimensions p and q with mean $\mathbb{E}(X)$ and $\mathbb{E}(Y)$, respectively. Then $\mathbb{E}(XY^*)$ exists and is given by

$$\mathbb{E}(XY^*) = \mathbb{E}(X)\,\mathbb{E}(Y)^* \ .$$

Proof:
The theorem is shown by considering the real counterpart of this relation. ■

The converse of this theorem does not necessarily hold. Theorem 1.10 leads to the following corollary.

Corollary 1.2
Let X and Y be independent complex random vectors of dimensions p and q with mean $\mathbb{E}(X)$ and $\mathbb{E}(Y)$, respectively. Then

$$\mathbb{C}(X, Y) = O \ .$$

Proof:
Using $\mathbb{C}(X, Y) = \mathbb{E}(XY^*) - \mathbb{E}(X)\,\mathbb{E}(Y)^*$ and Theorem 1.10 implies the corollary. ■

Again the converse is not necessarily true.

2

The Multivariate Complex Normal Distribution

This chapter presents the multivariate complex normal distribution. It is introduced by Wooding (1956), but it is Goodman (1963) who initiates a more thorough study of this area. Furthermore Eaton (1983) describes the distribution by using vector space approach. In this book we have also used vector space approach and the book is the first to give a systematic and wide-ranging presentation of the multivariate complex normal distribution. The results presented are known from the literature or from the real case. First the univariate case is considered. We define the standard normal distribution on $\mathbb{C}$ and by means of this an arbitrary normal distribution on $\mathbb{C}$ is defined. For the univariate standard complex normal distribution we study the rotation invariance, which says that the univariate standard complex normal distribution is invariant under multiplication by a complex unit. For an arbitrary complex normal distribution the property of reproductivity is examined. This is the characteristic that the sum of complex numbers and independent complex normally distributed random variables multiplied by complex numbers still is complex normally distributed. The normal distribution on $\mathbb{C}^p$ is defined and also the reproductivity property for it is studied. For all the distributions the relation to the real normal distribution is determined and the density function and the characteristic function are stated. We also specify independence results in the multivariate complex normal distribution and furthermore marginal and conditional distributions are examined. We investigate some of the results for the complex normal distribution on $\mathbb{C}^p$ in matrix form, i.e. for the complex normal distribution on $\mathbb{C}^{n \times p}$.

2.1 The Univariate Complex Normal Distribution

First we turn to reflections on the univariate complex normal distribution. We begin by considering the standard complex normal distribution and carry on with an arbitrary complex normal distribution.

2.1.1 The Standard Complex Normal Distribution

Recall that a complex random variable X can be written as

$$X = U + iV ,$$

where U and V are the uniquely determined real random variables corresponding to the real and the imaginary parts of X, respectively. Moreover we have

$$[X] = \begin{pmatrix} U \\ V \end{pmatrix},$$

where $[\cdot]$ is the real isomorphism between $\mathbb{C}$ and $\mathbb{R}^2$. This isomorphism is defined in Definition 1.4 page 3. The distribution of X on $\mathbb{C}$ determines the joint distribution of U and V on $\mathbb{R}^2$ and conversely, since $[\cdot]$ is an isomorphism. Thus the properties defining the distribution of X can be stated on $[X]$.

We say that X has a univariate complex normal distribution with mean zero and variance one, also called the *univariate standard complex normal distribution*, iff

1. $[X]$ has a bivariate normal distribution on $\mathbb{R}^2$.
2. $[X]$ has a complex covariance structure.
3. $\mathbb{E}(X) = 0$ and $\mathbb{V}(X) = 1$.

Condition 1 is the requirement that a normally distributed random variable in $\mathbb{C}$ is transformed into a normally distributed random variable in $\mathbb{R}^2$ under $[\cdot]$. Complex covariance structure mentioned in condition 2 is stated in Definition 1.19 page 11, and since a real variance matrix is symmetric it follows that $\mathbb{C}(U, V) = 0$. Therefore condition 2 implies that U and V are independent with $\mathbb{V}(U) = \mathbb{V}(V)$ and furthermore it is the requirement that the contour levels of the density function must be circular. Condition 3 standardizes the mean and the variance.

Using $\mathbb{E}(X) = \mathbb{E}(U) + i\,\mathbb{E}(V)$, $\mathbb{V}(X) = \mathbb{V}(U) + \mathbb{V}(V)$ and $\mathbb{V}(U) = \mathbb{V}(V)$ we hereby get

$$\begin{aligned} \mathbb{E}(U) &= \mathbb{E}(V) = 0 \\ \mathbb{V}(U) &= \mathbb{V}(V) = \frac{1}{2}. \end{aligned} \tag{2.1}$$

From condition 1 and 2 we require that $[X]$ is normally distributed on $\mathbb{R}^2$ with $\mathbb{C}(U, V) = 0$. Together with (2.1) this gives the conditions to put on $[X]$ to ensure that X is univariate standard complex normally distributed, namely

$$\mathcal{L}([X]) = \mathcal{N}_2\left(\mathbf{0}, \frac{1}{2}\boldsymbol{I}_2\right).$$

The above leads to the following definition.

Definition 2.1 The univariate standard complex normal distribution
A complex random variable X has a univariate complex normal distribution with mean zero and variance one if

$$\mathcal{L}([X]) = \mathcal{N}_2\left(\mathbf{0}, \frac{1}{2}\boldsymbol{I}_2\right).$$

This is denoted by $\mathcal{L}(X) = \mathbb{C}\mathcal{N}(0, 1)$.

When it is obvious from the context that we consider a complex random variable the word univariate can be omitted.

Using (2.1) page 16 together with the independence of U and V we observe from (1.3) page 5 that

$$\begin{aligned}
\mathbb{E}\left(X^2\right) &= \mathbb{E}\left(U^2 - V^2\right) + 2i\mathbb{E}(UV) \\
&= \mathbb{V}(U) - \mathbb{V}(V) + 2i\mathbb{E}(U)\,\mathbb{E}(V) \\
&= 0\,.
\end{aligned}$$

This tells us that a univariate standard complex normally distributed random variable have second moment equal to zero.

The following theorem states that a univariate standard complex normally distributed random variable remains a univariate standard complex normally distributed random variable when it is multiplied by a complex number with absolute value one.

Theorem 2.1 Rotation invariance
Let X be a complex random variable with $\mathcal{L}(X) = \mathbb{C}\mathcal{N}(0,1)$ and let $c \in \mathbb{C}$ with $|c| = 1$. It holds that

$$\mathcal{L}(cX) = \mathbb{C}\mathcal{N}(0,1)\ .$$

Proof:
Let X be a complex random variable with $\mathcal{L}(X) = \mathbb{C}\mathcal{N}(0,1)$. According to Definition 2.1 page 16 this is equivalent to

$$\mathcal{L}([X]) = \mathcal{N}_2\left(\mathbf{0}, \frac{1}{2}\boldsymbol{I}_2\right)\ .$$

Let $c \in \mathbb{C}$ with $|c| = 1$ be written as $c = a + ib$, where $a, b \in \mathbb{R}$. Then by (1.2) page 4 we get

$$\begin{aligned}
[cX] &= \{c\}\,[X] \\
&= \begin{pmatrix} a & -b \\ b & a \end{pmatrix} [X]\ .
\end{aligned}$$

Notice that the matrix $\{c\}$ is orthogonal, since $(a,b)\,(-b,a)^{\top} = 0$ and since $a^2 + b^2 = |c| = 1$.

From the multivariate real normal distribution it is well known that given a p-dimensional random vector $\boldsymbol{Y}$ with $\mathcal{L}(\boldsymbol{Y}) = \mathcal{N}_p(\boldsymbol{\theta}, \Sigma)$, a $q \times p$ real matrix $\boldsymbol{B}$ and a q-dimensional real vector $\boldsymbol{\beta}$, then $\mathcal{L}(\boldsymbol{BY} + \boldsymbol{\beta}) = \mathcal{N}_q\left(\boldsymbol{B\theta} + \boldsymbol{\beta}, \boldsymbol{B}\Sigma\boldsymbol{B}^{\top}\right)$. Using this result and the orthogonality of $\{c\}$ we get

$$\begin{aligned}
\mathcal{L}([cX]) &= \mathcal{L}(\{c\}\,[X]) \\
&= \mathcal{N}_2\left(\mathbf{0}, \{c\}\,\frac{1}{2}\boldsymbol{I}_2\,\{c\}^{\top}\right) \\
&= \mathcal{N}_2\left(\mathbf{0}, \frac{1}{2}\boldsymbol{I}_2\right)\,,
\end{aligned}$$

which says that

$$\mathcal{L}(cX) = \mathbb{C}\mathcal{N}(0,1) .$$

■

This theorem characterizes the univariate standard complex normal distribution to be invariant when multiplying by a complex unit. The proof shows that this is equivalent to the rotation invariance of $\mathcal{N}_2\left(0, \frac{1}{2}\boldsymbol{I}_2\right)$. Furthermore note that the theorem is a direct consequence of the later proven Theorem 2.6 page 21.

From the multivariate real normal distribution we know that, when $\mathcal{L}([X]) = \mathcal{N}_2\left(0, \frac{1}{2}\boldsymbol{I}_2\right)$, then the density of $[X]$ w.r.t. Lebesgue measure on $\mathbb{R}^2$ is given as

$$\begin{aligned} f_{[X]}([x]) &= (2\pi)^{-1}\det\left(\frac{1}{2}\boldsymbol{I}_2\right)^{-\frac{1}{2}}\exp\left(-\frac{1}{2}[x]^{\mathrm{T}}\left(\frac{1}{2}\boldsymbol{I}_2\right)^{-1}[x]\right) \\ &= \frac{1}{\pi}\exp\left(-[x]^{\mathrm{T}}[x]\right), \quad [x] \in \mathbb{R}^2 . \end{aligned}$$

From Definition 2.1 page 16 there is a one-to-one correspondence between the univariate complex standard normal distribution and this bivariate normal distribution established by the isomorphism $[\cdot]$. Thus the density function of X w.r.t. Lebesgue measure on $\mathbb{C}$ is identical to the density function of $[X]$ w.r.t. Lebesgue measure on $\mathbb{R}^2$. This leads to the following theorem.

Theorem 2.2 The density function of $\mathbb{C}\mathcal{N}(0,1)$
Let X be a complex random variable with $\mathcal{L}(X) = \mathbb{C}\mathcal{N}(0,1)$. The density function of X w.r.t. Lebesgue measure on $\mathbb{C}$ is given as

$$f_X(x) = \frac{1}{\pi}\exp(-\overline{x}x), \quad x \in \mathbb{C} .$$

2.1.2 The Complex Normal Distribution

From the definition of the standard normal distribution on $\mathbb{C}$ we are able to define an arbitrary normal distribution on $\mathbb{C}$. Let Z be a complex random variable with $\mathcal{L}(Z) = \mathbb{C}\mathcal{N}(0,1)$. Further let X be the complex random variable given by

$$X = \theta + cZ ,$$

where $\theta, c \in \mathbb{C}$. Then we say that X is complex normally distributed with mean and variance given from Theorem 1.3 page 6 and Theorem 1.5 page 7 as

$$\mathbb{E}(X) = \theta \text{ and } \mathbb{V}(X) = c\overline{c} = |c|^2 .$$

Note that both $|c|$ and $|c|^2$ are nonnegative and real. Thus $c = |c|\,d$, where d is a complex unit. Hereby we deduce by Theorem 2.1 page 17 that $\mathcal{L}(\theta + cZ) = \mathcal{L}(\theta + |c|\,Z)$.

The above leads to the following definition of an arbitrary univariate normal distribution on $\mathbb{C}$.

Definition 2.2 The univariate complex normal distribution
Let Z be a complex random variable with $\mathcal{L}(Z) = \mathbb{C}\mathcal{N}(0,1)$ and let $\theta \in \mathbb{C}$ and $\sigma \in \overline{\mathbb{R}}_+$. The complex random variable $X = \theta + \sigma Z$ has a univariate complex normal distribution with mean θ and variance σ^2. This is denoted by $\mathcal{L}(X) = \mathbb{C}\mathcal{N}(\theta, \sigma^2)$.

Again the word univariate can be omitted.

There is a one-to-one correspondence between the univariate complex normal distribution and the bivariate normal distribution established by the isomorphism $[\cdot]$. The following theorem establishes this.

Theorem 2.3 The relation to the multivariate real normal distribution
For a complex random variable X it holds that

$$\mathcal{L}(X) = \mathbb{C}\mathcal{N}\left(\theta, \sigma^2\right)$$

iff

$$\mathcal{L}([X]) = \mathcal{N}_2\left([\theta], \frac{\sigma^2}{2}\boldsymbol{I}_2\right) ,$$

where $\theta \in \mathbb{C}$ and $\sigma^2 \in \overline{\mathbb{R}}_+$.

Proof:
Let X be a complex random variable with

$$\mathcal{L}(X) = \mathbb{C}\mathcal{N}\left(\theta, \sigma^2\right) ,$$

where $\theta \in \mathbb{C}$ and $\sigma^2 \in \overline{\mathbb{R}}_+$. Further let Z be a complex random variable with $\mathcal{L}(Z) = \mathbb{C}\mathcal{N}(0,1)$. Note according to Definition 2.1 page 16 that

$$\mathcal{L}(Z) = \mathbb{C}\mathcal{N}(0,1) \quad \Leftrightarrow \quad \mathcal{L}([Z]) = \mathcal{N}_2\left(\boldsymbol{0}, \frac{1}{2}\boldsymbol{I}_2\right) . \tag{2.2}$$

Using Definition 2.2 we get that $\mathcal{L}(X) = \mathcal{L}(\theta + \sigma Z)$. Since $[\cdot]$ an isomorphism this holds iff

$$\begin{aligned} \mathcal{L}([X]) &= \mathcal{L}([\theta + \sigma Z]) \\ &= \mathcal{L}([\theta] + \sigma [Z]) . \end{aligned}$$

Using the result from the multivariate real normal distribution noted on page 17 and the result in (2.2) page 19 we find

$$\mathcal{L}([X]) = \mathcal{N}_2\left([\theta], \frac{\sigma^2}{2}\boldsymbol{I}_2\right) .$$

■

Theorem 2.3 tells us that there is a one-to-one correspondence between $\mathcal{L}(X) = \mathbb{C}\mathcal{N}(\theta, \sigma^2)$ and $\mathcal{L}([X]) = \mathcal{N}_2\left([\theta], \frac{\sigma^2}{2}\boldsymbol{I}_2\right)$, which enables us to find the density function of X w.r.t. Lebesgue measure on $\mathbb{C}$. Whenever $\sigma^2 \in \mathbb{R}_+$ we recall from the multivariate real normal distribution that the density function of $[X]$ w.r.t. Lebesgue measure on $\mathbb{R}^2$ is given as

$$\begin{aligned} f_{[X]}([x]) &= (2\pi)^{-1}\det\left(\frac{\sigma^2}{2}\boldsymbol{I}_2\right)^{-\frac{1}{2}}\exp\left(-\frac{1}{2}[x-\theta]^{\mathsf{T}}\left(\frac{\sigma^2}{2}\boldsymbol{I}_2\right)^{-1}[x-\theta]\right) \\ &= \frac{1}{\pi\sigma^2}\exp\left(-\frac{1}{\sigma^2}[x-\theta]^{\mathsf{T}}[x-\theta]\right), \quad [x] \in \mathbb{R}^2 . \end{aligned}$$

The one-to-one correspondence established by the isomorphism $[\cdot]$ infers that the density function of X w.r.t. Lebesgue measure on $\mathbb{C}$ is identical to the density function of $[X]$ w.r.t. Lebesgue measure on $\mathbb{R}^2$. This gives us the following theorem.

Theorem 2.4 The density function of $\mathbb{C}\mathcal{N}(\theta, \sigma^2)$
Let X be a complex random variable with $\mathcal{L}(X) = \mathbb{C}\mathcal{N}(\theta, \sigma^2)$, where $\theta \in \mathbb{C}$ and $\sigma^2 \in \mathbb{R}_+$. The density function of X w.r.t. Lebesgue measure on $\mathbb{C}$ is given as

$$f_X(x) = \frac{1}{\pi\sigma^2}\exp\left(-\frac{1}{\sigma^2}\left(\overline{x-\theta}\right)(x-\theta)\right), x \in \mathbb{C} .$$

From the multivariate real normal distribution we recall that the characteristic function of $[X]$ with $\mathcal{L}([X]) = \mathcal{N}_2\left([\theta], \frac{\sigma^2}{2}\boldsymbol{I}_2\right)$ is given as

$$\begin{aligned} \varphi_{[X]}([\xi]) &= \exp\left(i[\xi]^{\mathsf{T}}[\theta] - \frac{1}{2}[\xi]^{\mathsf{T}}\frac{\sigma^2}{2}\boldsymbol{I}_2[\xi]\right) \\ &= \exp\left(i[\xi]^{\mathsf{T}}[\theta] - \frac{\sigma^2}{4}[\xi]^{\mathsf{T}}[\xi]\right), \quad [\xi] \in \mathbb{R}^2 . \end{aligned}$$

By the same argument as above the characteristic functions of X and $[X]$ are identical, whereby the following theorem is obtained.

Theorem 2.5 The characteristic function of $\mathbb{C}\mathcal{N}(\theta, \sigma^2)$
Let X be a complex random variable with $\mathcal{L}(X) = \mathbb{C}\mathcal{N}(\theta, \sigma^2)$, where $\theta \in \mathbb{C}$ and $\sigma^2 \in \overline{\mathbb{R}}_+$. The characteristic function of X is given as

$$\varphi_X(\xi) = \exp\left(i\,\mathrm{Re}\left(\bar{\xi}\theta\right) - \frac{\sigma^2}{4}\bar{\xi}\xi\right), \ \xi \in \mathbb{C} .$$

Note that the characteristic function only depends on θ and σ^2. Hence a univariate complex normal distribution is completely specified by the mean θ and the variance σ^2. This is equivalent to the real case. Another characteristic which holds for the complex normal distribution as for the real normal distribution is the reproductivity property. This is shown in the following theorem by means of the characteristic function.

Theorem 2.6 The reproductivity property of $\mathbb{C}\mathcal{N}$
Let $X_1, X_2, \ldots, X_p$ be mutually independent complex random variables with $\mathcal{L}(X_k) = \mathbb{C}\mathcal{N}(\theta_k, \sigma_k^2)$, where $\theta_k \in \mathbb{C}$ and $\sigma_k^2 \in \overline{\mathbb{R}}_+$, and let $c_k, d_k \in \mathbb{C}$ for $k = 1, 2, \ldots, p$. It holds that

$$\mathcal{L}\left(\sum_{k=1}^{p}(c_k X_k + d_k)\right) = \mathbb{C}\mathcal{N}\left(\sum_{k=1}^{p}(c_k\theta_k + d_k), \sum_{k=1}^{p} c_k\overline{c}_k\sigma_k^2\right).$$

Proof:
Let $X_1, X_2, \ldots, X_p$ be mutually independent complex random variables with $\mathcal{L}(X_k) = \mathbb{C}\mathcal{N}(\theta_k, \sigma_k^2)$, where $\theta_k \in \mathbb{C}$ and $\sigma_k^2 \in \overline{\mathbb{R}}_+$, and let $c_k, d_k \in \mathbb{C}$ for $k = 1, 2, \ldots, p$.

In order to find the distribution of $\sum_{k=1}^{p}(c_k X_k + d_k)$ we consider the characteristic function of it. By Definition 1.10 page 7 this function is for $\xi \in \mathbb{C}$ given as

$$\begin{aligned}
\varphi_{\sum_{k=1}^{p}(c_k X_k + d_k)}(\xi) &= \mathbb{E}\left(\exp\left(i\,\mathrm{Re}\left(\overline{\xi}\sum_{k=1}^{p}(c_k X_k + d_k)\right)\right)\right) \\
&= \mathbb{E}\left(\prod_{k=1}^{p}\exp\left(i\,\mathrm{Re}\left(\overline{\xi}\,(c_k X_k + d_k)\right)\right)\right) \\
&= \mathbb{E}\left(\prod_{k=1}^{p}\exp\left(i\,\mathrm{Re}\left(\overline{\xi}d_k\right)\right)\exp\left(i\,\mathrm{Re}\left(\overline{\xi}c_k X_k\right)\right)\right).
\end{aligned}$$

Using the mutual independence of the X_k's we find for $\xi \in \mathbb{C}$ that

$$\begin{aligned}
\varphi_{\sum_{k=1}^{p}(c_k X_k + d_k)}(\xi) &= \prod_{k=1}^{p}\exp\left(i\,\mathrm{Re}(\overline{\xi}d_k)\right)\mathbb{E}\left(\exp\left(i\,\mathrm{Re}(\overline{\xi}c_k X_k)\right)\right) \\
&= \prod_{k=1}^{p}\exp\left(i\,\mathrm{Re}(\overline{\xi}d_k)\right)\varphi_{X_k}(\xi\overline{c}_k) \\
&= \prod_{k=1}^{p}\exp\left(i\,\mathrm{Re}(\overline{\xi}d_k)\right)\exp\left(i\,\mathrm{Re}(\overline{\xi}c_k\theta_k) - \frac{\sigma_k^2}{4}\overline{\xi}c_k\xi\overline{c}_k\right) \\
&= \prod_{k=1}^{p}\exp\left(i\,\mathrm{Re}\left(\overline{\xi}\,(c_k\theta_k + d_k)\right) - \frac{c_k\overline{c}_k\sigma_k^2}{4}\overline{\xi}\xi\right) \\
&= \exp\left(i\,\mathrm{Re}\left(\overline{\xi}\sum_{k=1}^{p}(c_k\theta_k + d_k)\right) - \frac{\sum_{k=1}^{p}c_k\overline{c}_k\sigma_k^2}{4}\overline{\xi}\xi\right).
\end{aligned}$$

This is the characteristic function of a complex normal distribution with mean $\sum_{k=1}^{p}(c_k\theta_k + d_k)$ and variance $\sum_{k=1}^{p}c_k\overline{c}_k\sigma_k^2$. According to Theorem 1.9 page 12 a characteristic function determines a distribution uniquely, whereby we deduce

$$\mathcal{L}\left(\sum_{k=1}^{p}(c_k X_k + d_k)\right) = \mathbb{C}\mathcal{N}\left(\sum_{k=1}^{p}(c_k\theta_k + d_k), \sum_{k=1}^{p} c_k\overline{c}_k\sigma_k^2\right).$$

∎

From the reproductivity property of the univariate complex normal distribution we see for $p = 1$, that the class of univariate complex normal distributions is invariant under addition and multiplication of complex constants.

2.2 The Multivariate Complex Normal Distribution

After having studied the univariate complex normal distribution we are ready to consider the complex normal distribution on the p-dimensional complex vector space $\mathbb{C}^p$.

Similar to the multivariate normal distribution on $\mathbb{R}^p$ we say that a p-dimensional complex random vector $\boldsymbol{X}$ has a p-variate complex normal distribution iff for all $\boldsymbol{c} \in \mathbb{C}^p$ the inner product $\langle \boldsymbol{X}, \boldsymbol{c} \rangle$ has a univariate complex normal distribution. This is stated in the following definition.

Definition 2.3 The multivariate complex normal distribution
A p-dimensional complex random vector $\boldsymbol{X} = (X_k)$ *has a p-variate complex normal distribution, if for all* $\mathbf{c} = (c_k) \in \mathbb{C}^p$ *the inner product*

$$\langle \boldsymbol{X}, \boldsymbol{c} \rangle = \sum_{k=1}^{p} \bar{c}_k X_k = \boldsymbol{c}^* \boldsymbol{X}$$

has a univariate complex normal distribution.

When it is obvious from the context that we consider a p-dimensional complex random vector, the words multivariate and p-variate can be omitted. According to the considerations on page 12 the distribution of $\langle \boldsymbol{X}, \boldsymbol{c} \rangle$ for all $\boldsymbol{c} \in \mathbb{C}^p$ is sufficient to determine the distribution of $\boldsymbol{X}$.

If we let $\boldsymbol{X}$ be a p-variate complex normally distributed random vector, we deduce from Definition 2.3 that each X_k, $k = 1, 2, \ldots, p$, has a univariate complex normal distribution. This implies that $\mathbb{E}(X_k)$ and $\mathbb{V}(X_k)$ exist and are finite, and further for $k \neq l$ we see that $\mathbb{C}(X_k, X_l)$ exists and is finite. Thus $\mathbb{E}(\boldsymbol{X})$ and $\mathbb{V}(\boldsymbol{X})$ exist and we denote these by $\boldsymbol{\theta}$ and $\boldsymbol{H}$, respectively. It holds that $\boldsymbol{\theta} \in \mathbb{C}^p$ and from Theorem 1.8 page 11 that $\boldsymbol{H} \in \mathbb{C}_S^{p \times p}$. The notation used for $\boldsymbol{X}$ having a p-variate complex normal distribution with mean $\boldsymbol{\theta}$ and variance matrix $\boldsymbol{H}$ is $\mathcal{L}(\boldsymbol{X}) = \mathbb{C}\mathcal{N}_p(\boldsymbol{\theta}, \boldsymbol{H})$.

For all $\boldsymbol{c} \in \mathbb{C}^p$ it holds by Theorem 1.6 page 10 that

$$\begin{aligned} \mathbb{E}(\boldsymbol{c}^* \boldsymbol{X}) &= \boldsymbol{c}^* \mathbb{E}(\boldsymbol{X}) \\ &= \boldsymbol{c}^* \boldsymbol{\theta} \end{aligned}$$

and by Corollary 1.1 page 11 that

$$\begin{aligned} \mathbb{V}(\boldsymbol{c}^* \boldsymbol{X}) &= \boldsymbol{c}^* \mathbb{V}(\boldsymbol{X}) \boldsymbol{c} \\ &= \boldsymbol{c}^* \boldsymbol{H} \boldsymbol{c}\,. \end{aligned}$$

Together with Definition 2.3 page 22 this means that, when $\mathcal{L}(\boldsymbol{X}) = \mathbb{C}\mathcal{N}_p(\boldsymbol{\theta}, \boldsymbol{H})$, then

$$\mathcal{L}(\boldsymbol{c}^*\boldsymbol{X}) = \mathbb{C}\mathcal{N}(\boldsymbol{c}^*\boldsymbol{\theta}, \boldsymbol{c}^*\boldsymbol{H}\boldsymbol{c}) \tag{2.3}$$

for all $\boldsymbol{c} \in \mathbb{C}^p$. Using this result we are able to find the characteristic function of $\boldsymbol{X}$ with $\mathcal{L}(\boldsymbol{X}) = \mathbb{C}\mathcal{N}_p(\boldsymbol{\theta}, \boldsymbol{H})$. From Theorem 2.5 page 20 the characteristic function of $\boldsymbol{c}^*\boldsymbol{X}$ with $\mathcal{L}(\boldsymbol{c}^*\boldsymbol{X}) = \mathbb{C}\mathcal{N}(\boldsymbol{c}^*\boldsymbol{\theta}, \boldsymbol{c}^*\boldsymbol{H}\boldsymbol{c})$ is given as

$$\varphi_{\boldsymbol{c}^*\boldsymbol{X}}(\xi) = \exp\left(i\,\mathrm{Re}\left(\bar{\xi}\boldsymbol{c}^*\boldsymbol{\theta}\right) - \frac{\boldsymbol{c}^*\boldsymbol{H}\boldsymbol{c}}{4}\bar{\xi}\xi\right), \ \xi \in \mathbb{C} .$$

From (1.5) page 12 we know that $\varphi_{\boldsymbol{c}^*\boldsymbol{X}}(1) = \varphi_{\boldsymbol{X}}(\boldsymbol{c})$, whereby the characteristic function of $\boldsymbol{X}$ is

$$\varphi_{\boldsymbol{X}}(\boldsymbol{c}) = \exp\left(i\,\mathrm{Re}\left(\boldsymbol{c}^*\boldsymbol{\theta}\right) - \frac{\boldsymbol{c}^*\boldsymbol{H}\boldsymbol{c}}{4}\right), \ \boldsymbol{c} \in \mathbb{C}^p.$$

This leads to the following theorem.

Theorem 2.7 The characteristic function of $\mathbb{C}\mathcal{N}_p(\boldsymbol{\theta}, \boldsymbol{H})$
Let $\boldsymbol{X}$ be a p-dimensional complex random vector with $\mathcal{L}(\boldsymbol{X}) = \mathbb{C}\mathcal{N}_p(\boldsymbol{\theta}, \boldsymbol{H})$, where $\boldsymbol{\theta} \in \mathbb{C}^p$ and $\boldsymbol{H} \in \mathbb{C}_S^{p\times p}$. The characteristic function of $\boldsymbol{X}$ is given as

$$\varphi_{\boldsymbol{X}}(\boldsymbol{\xi}) = \exp\left(i\,\mathrm{Re}\left(\boldsymbol{\xi}^*\boldsymbol{\theta}\right) - \frac{\boldsymbol{\xi}^*\boldsymbol{H}\boldsymbol{\xi}}{4}\right), \ \boldsymbol{\xi} \in \mathbb{C}^p.$$

Notice that the characteristic function only depends on the mean $\boldsymbol{\theta}$ and the variance matrix $\boldsymbol{H}$, which implies that the multivariate complex normal distribution is completely determined by the mean and the variance matrix.

The next theorem states some rules of the multivariate complex normal distribution.

Theorem 2.8 Properties of the multivariate complex normal distribution
Let $\boldsymbol{X}$ be a p-dimensional complex random vector.

1. *If there exist $\boldsymbol{\theta} \in \mathbb{C}^p$ and $\boldsymbol{H} \in \mathbb{C}_S^{p\times p}$ such that for all $\boldsymbol{c} \in \mathbb{C}^p$ it holds that*

$$\mathcal{L}(\boldsymbol{c}^*\boldsymbol{X}) = \mathbb{C}\mathcal{N}(\boldsymbol{c}^*\boldsymbol{\theta}, \boldsymbol{c}^*\boldsymbol{H}\boldsymbol{c}) ,$$

   then

$$\mathcal{L}(\boldsymbol{X}) = \mathbb{C}\mathcal{N}_p(\boldsymbol{\theta}, \boldsymbol{H}) .$$

2. *Let $\boldsymbol{D} \in \mathbb{C}^{q\times p}$ and $\boldsymbol{d} \in \mathbb{C}^q$. If $\mathcal{L}(\boldsymbol{X}) = \mathbb{C}\mathcal{N}_p(\boldsymbol{\theta}, \boldsymbol{H})$, where $\boldsymbol{\theta} \in \mathbb{C}^p$ and $\boldsymbol{H} \in \mathbb{C}_S^{p\times p}$, then*

$$\mathcal{L}(\boldsymbol{D}\boldsymbol{X} + \boldsymbol{d}) = \mathbb{C}\mathcal{N}_q(\boldsymbol{D}\boldsymbol{\theta} + \boldsymbol{d}, \boldsymbol{D}\boldsymbol{H}\boldsymbol{D}^*) .$$

Proof:
Let $\boldsymbol{X}$ be a p-dimensional complex random vector.

Re 1:
Assume that $\boldsymbol{\theta} \in \mathbb{C}^p$ and $\boldsymbol{H} \in \mathbb{C}_S^{p\times p}$ exist such that $\mathcal{L}(\boldsymbol{c}^*\boldsymbol{X}) = \mathbb{C}\mathcal{N}(\boldsymbol{c}^*\boldsymbol{\theta}, \boldsymbol{c}^*\boldsymbol{H}\boldsymbol{c})$ for all $\boldsymbol{c} \in \mathbb{C}^p$. From the distribution of $\boldsymbol{c}^*\boldsymbol{X}$ and the fact that $\varphi_{\boldsymbol{c}^*\boldsymbol{X}}(1) = \varphi_{\boldsymbol{X}}(\boldsymbol{c})$ we find as on page 23 the characteristic function of $\boldsymbol{X}$ to be

$$\varphi_{\boldsymbol{X}}(\boldsymbol{c}) = \exp\left(i\,\mathrm{Re}\,(\boldsymbol{c}^*\boldsymbol{\theta}) - \frac{\boldsymbol{c}^*\boldsymbol{H}\boldsymbol{c}}{4}\right),\ \boldsymbol{c} \in \mathbb{C}^p.$$

By Theorem 2.7 page 23 this is the characteristic function of a p-variate complex normal distribution with mean $\boldsymbol{\theta}$ and variance matrix $\boldsymbol{H}$. According to Theorem 1.9 page 12 a characteristic function determines a distribution uniquely, therefore we conclude that

$$\mathcal{L}(\boldsymbol{X}) = \mathbb{C}\mathcal{N}_p(\boldsymbol{\theta}, \boldsymbol{H})\ .$$

Re 2:
Let $\boldsymbol{D} \in \mathbb{C}^{q\times p}$, $\boldsymbol{d} \in \mathbb{C}^q$ and let $\mathcal{L}(\boldsymbol{X}) = \mathbb{C}\mathcal{N}_p(\boldsymbol{\theta}, \boldsymbol{H})$, where $\boldsymbol{\theta} \in \mathbb{C}^p$ and $\boldsymbol{H} \in \mathbb{C}_S^{p\times p}$. According to (2.3) page 23 this means for all $\boldsymbol{c} \in \mathbb{C}^p$ that

$$\mathcal{L}(\boldsymbol{c}^*\boldsymbol{X}) = \mathbb{C}\mathcal{N}(\boldsymbol{c}^*\boldsymbol{\theta}, \boldsymbol{c}^*\boldsymbol{H}\boldsymbol{c})\ . \tag{2.4}$$

Consider the q-dimensional complex random vector given by $\boldsymbol{DX} + \boldsymbol{d}$. Let $\boldsymbol{e} \in \mathbb{C}^q$, then we observe that

$$\begin{aligned} \boldsymbol{e}^*(\boldsymbol{DX} + \boldsymbol{d}) &= \boldsymbol{e}^*\boldsymbol{DX} + \boldsymbol{e}^*\boldsymbol{d} \\ &= (\boldsymbol{D}^*\boldsymbol{e})^*\boldsymbol{X} + \boldsymbol{e}^*\boldsymbol{d}\,. \end{aligned} \tag{2.5}$$

Since we know that (2.4) holds for all $\boldsymbol{c} \in \mathbb{C}^p$ it holds especially for $\boldsymbol{D}^*\boldsymbol{e} \in \mathbb{C}^p$, which means that $\mathcal{L}(\boldsymbol{e}^*\boldsymbol{DX}) = \mathbb{C}\mathcal{N}(\boldsymbol{e}^*\boldsymbol{D\theta}, \boldsymbol{e}^*\boldsymbol{DHD}^*\boldsymbol{e})$ for all $\boldsymbol{e} \in \mathbb{C}^q$. As the class of univariate complex normal distributions is invariant under addition of a complex constant we deduce from (2.5) that for all $\boldsymbol{e} \in \mathbb{C}^q$

$$\mathcal{L}(\boldsymbol{e}^*(\boldsymbol{DX} + \boldsymbol{d})) = \mathbb{C}\mathcal{N}(\boldsymbol{e}^*(\boldsymbol{D\theta} + \boldsymbol{d}), \boldsymbol{e}^*\boldsymbol{DHD}^*\boldsymbol{e})\ .$$

Using part 1 we hereby conclude that

$$\mathcal{L}(\boldsymbol{DX} + \boldsymbol{d}) = \mathbb{C}\mathcal{N}_q(\boldsymbol{D\theta} + \boldsymbol{d}, \boldsymbol{DHD}^*)\ .$$

■

Note that the converse of Theorem 2.8 part 1 holds and it has been shown on page 22. Furthermore part 2 is the complex equivalent to the real version stated on page 17.

In the following we concentrate on how to create an arbitrary p-dimensional complex normally distributed random vector from p mutually independent standard complex normally distributed random variables. Let $Z_1, Z_2, \ldots, Z_p$ be p mutually independent complex random variables with $\mathcal{L}(Z_k) = \mathbb{C}\mathcal{N}(0,1)$ for $k = 1, 2, \ldots, p$ and let $\boldsymbol{Z}$ be the p-dimensional random vector defined by $\boldsymbol{Z} = (Z_k)$. Moreover let $\boldsymbol{c} = (c_k) \in \mathbb{C}^p$. From the reproductivity property of the univariate complex normal distribution (Theorem 2.6 page 21) we have that

$$\begin{aligned} \mathcal{L}(\boldsymbol{c}^*\boldsymbol{Z}) &= \mathbb{C}\mathcal{N}\left(0, \sum_{k=1}^{p} \bar{c}_k c_k\right) \\ &= \mathbb{C}\mathcal{N}(\boldsymbol{c}^*\boldsymbol{0}, \boldsymbol{c}^*\boldsymbol{I}_p\boldsymbol{c}) \ . \end{aligned}$$

Thus by Theorem 2.8 page 23

$$\mathcal{L}(\boldsymbol{Z}) = \mathbb{C}\mathcal{N}_p(\boldsymbol{0}, \boldsymbol{I}_p) \ .$$

Let $\boldsymbol{C} \in \mathbb{C}^{p \times p}$, $\boldsymbol{\theta} \in \mathbb{C}^p$ and let $\boldsymbol{X}$ be the complex random vector defined by

$$\boldsymbol{X} = \boldsymbol{\theta} + \boldsymbol{C}\boldsymbol{Z} \ .$$

From Theorem 2.8 we observe that

$$\mathcal{L}(\boldsymbol{X}) = \mathbb{C}\mathcal{N}_p(\boldsymbol{\theta}, \boldsymbol{C}\boldsymbol{C}^*) \ .$$

Note that the variance matrix of $\boldsymbol{X}$ is a positive semidefinite matrix as it must be. We deduce that we are able to create a complex normal distribution on $\mathbb{C}^p$ with an arbitrary mean $\boldsymbol{\theta} \in \mathbb{C}^p$ and a positive semidefinite variance matrix $\boldsymbol{C}\boldsymbol{C}^* \in \mathbb{C}_S^{p \times p}$.

The theorem below states likewise the univariate case (Theorem 2.3 page 19) the one-to-one correspondence between the p-variate complex normal distribution and the $2p$-variate normal distribution established by the isomorphism $[\cdot]$.

Theorem 2.9 The relation to the multivariate real normal distribution
For a p-dimensional complex random vector $\boldsymbol{X}$ it holds that

$$\mathcal{L}(\boldsymbol{X}) = \mathbb{C}\mathcal{N}_p(\boldsymbol{\theta}, \boldsymbol{H})$$

iff

$$\mathcal{L}([\boldsymbol{X}]) = \mathcal{N}_{2p}\left([\boldsymbol{\theta}], \frac{1}{2}\{\boldsymbol{H}\}\right) ,$$

where $\boldsymbol{\theta} \in \mathbb{C}^p$ and $\boldsymbol{H} \in \mathbb{C}_S^{p \times p}$.

Proof:
Let $\boldsymbol{X}$ be a p-dimensional complex random vector and let $\boldsymbol{\theta} \in \mathbb{C}^p$ and $\boldsymbol{H} \in \mathbb{C}_S^{p \times p}$.

Assume that $\mathcal{L}(\boldsymbol{X}) = \mathbb{C}\mathcal{N}_p(\boldsymbol{\theta}, \boldsymbol{H})$. From the considerations stated before the theorem we know that

$$\mathcal{L}(\boldsymbol{X}) = \mathcal{L}(\boldsymbol{\theta} + \boldsymbol{C}\boldsymbol{Z}) \ , \tag{2.6}$$

where $\boldsymbol{Z}$ is a p-dimensional complex random vector with $\mathcal{L}(\boldsymbol{Z}) = \mathbb{C}\mathcal{N}_p(0, \boldsymbol{I}_p)$ and $\boldsymbol{C} \in \mathbb{C}^{p \times p}$ such that $\boldsymbol{H} = \boldsymbol{C}\boldsymbol{C}^*$. First we easily deduce that

$$\mathcal{L}(\boldsymbol{Z}) = \mathbb{C}\mathcal{N}_p(0, \boldsymbol{I}_p) \Leftrightarrow \mathcal{L}([\boldsymbol{Z}]) = \mathcal{N}_{2p}\left(0, \frac{1}{2}\boldsymbol{I}_{2p}\right) . \tag{2.7}$$

Since $[\cdot]$ is an isomorphism (2.6) holds iff

$$\begin{aligned} \mathcal{L}([\boldsymbol{X}]) &= \mathcal{L}([\boldsymbol{\theta} + \boldsymbol{C}\boldsymbol{Z}]) \\ &= \mathcal{L}([\boldsymbol{\theta}] + \{\boldsymbol{C}\}[\boldsymbol{Z}]) . \end{aligned}$$

From (2.7) this is equivalent to

$$\mathcal{L}([\boldsymbol{X}]) = \mathcal{N}_{2p}\left([\boldsymbol{\theta}], \{\boldsymbol{C}\}\frac{1}{2}\boldsymbol{I}_{2p}\{\boldsymbol{C}\}^{\mathsf{T}}\right) .$$

Because $\boldsymbol{H} = \boldsymbol{C}\boldsymbol{C}^* \Leftrightarrow \{\boldsymbol{H}\} = \{\boldsymbol{C}\}\{\boldsymbol{C}\}^{\mathsf{T}}$ we have

$$\mathcal{L}([\boldsymbol{X}]) = \mathcal{N}_{2p}\left([\boldsymbol{\theta}], \frac{1}{2}\{\boldsymbol{H}\}\right) .$$

■

Let $\boldsymbol{X}$ be a p-dimensional complex random vector with $\mathcal{L}(\boldsymbol{X}) = \mathbb{C}\mathcal{N}_p(\boldsymbol{\theta}, \boldsymbol{H})$. We know the distribution of $[\boldsymbol{X}]$ on $\mathbb{R}^{2p}$ and therefore we are able to specify the density function of $\boldsymbol{X}$ w.r.t. Lebesgue measure on $\mathbb{C}^p$. When $\mathcal{L}([\boldsymbol{X}]) = \mathcal{N}_{2p}\left([\boldsymbol{\theta}], \frac{1}{2}\{\boldsymbol{H}\}\right)$ and $\{\boldsymbol{H}\} > \boldsymbol{O}$, the density function of $[\boldsymbol{X}]$ w.r.t. Lebesgue measure on $\mathbb{R}^{2p}$ is given by

$$\begin{aligned} f_{[\boldsymbol{X}]}([\boldsymbol{x}]) &= (2\pi)^{-p} \det\left(\frac{1}{2}\{\boldsymbol{H}\}\right)^{-\frac{1}{2}} \exp\left(-\frac{1}{2}\left([\boldsymbol{x}-\boldsymbol{\theta}]^{\mathsf{T}}\left(\frac{1}{2}\{\boldsymbol{H}\}\right)^{-1}[\boldsymbol{x}-\boldsymbol{\theta}]\right)\right) \\ &= \pi^{-p} \det(\{\boldsymbol{H}\})^{-\frac{1}{2}} \exp\left(-[\boldsymbol{x}-\boldsymbol{\theta}]^{\mathsf{T}}\{\boldsymbol{H}\}^{-1}[\boldsymbol{x}-\boldsymbol{\theta}]\right) , \quad [\boldsymbol{x}] \in \mathbb{R}^{2p} . \end{aligned}$$

Theorem 2.9 page 25 points out the one-to-one correspondence between $\mathcal{L}(\boldsymbol{X}) = \mathbb{C}\mathcal{N}_p(\boldsymbol{\theta}, \boldsymbol{H})$ and $\mathcal{L}([\boldsymbol{X}]) = \mathcal{N}_{2p}\left([\boldsymbol{\theta}], \frac{1}{2}\{\boldsymbol{H}\}\right)$ established by the isomorphism $[\cdot]$. Hereby the density function of $\boldsymbol{X}$ w.r.t. Lebesgue measure on $\mathbb{C}^p$ is equal to the density function of $[\boldsymbol{X}]$ w.r.t. Lebesgue measure on $\mathbb{R}^{2p}$. By observing that $\det(\{\boldsymbol{H}\}) = \det(\boldsymbol{H})^2$, as $\boldsymbol{H}$ is Hermitian, we get the theorem below.

Theorem 2.10 The density function of $\mathbb{C}\mathcal{N}_p(\boldsymbol{\theta}, \boldsymbol{H})$
Let $\boldsymbol{X}$ be a p-dimensional complex random vector with $\mathcal{L}(\boldsymbol{X}) = \mathbb{C}\mathcal{N}_p(\boldsymbol{\theta}, \boldsymbol{H})$, where $\boldsymbol{\theta} \in \mathbb{C}^p$ and $\boldsymbol{H} \in \mathbb{C}_+^{p \times p}$. The density function of $\boldsymbol{X}$ w.r.t. Lebesgue measure on $\mathbb{C}^p$ is given as

$$f_{\boldsymbol{X}}(\boldsymbol{x}) = \pi^{-p} \det(\boldsymbol{H})^{-1} \exp\left(-(\boldsymbol{x}-\boldsymbol{\theta})^* \boldsymbol{H}^{-1}(\boldsymbol{x}-\boldsymbol{\theta})\right) , \quad \boldsymbol{x} \in \mathbb{C}^p .$$

The reproductivity property holds for the multivariate complex normal distribution as it does in the univariate case (Theorem 2.6 page 21).

Theorem 2.11 The reproductivity property of $\mathbb{C}\mathcal{N}_p$

Let $\boldsymbol{X}_1, \boldsymbol{X}_2, \ldots, \boldsymbol{X}_n$ be mutually independent p-dimensional complex random vectors with $\mathcal{L}(\boldsymbol{X}_j) = \mathbb{C}\mathcal{N}_p(\boldsymbol{\theta}_j, \boldsymbol{H}_j)$, where $\boldsymbol{\theta}_j \in \mathbb{C}^p$ and $\boldsymbol{H}_j \in \mathbb{C}_S^{p\times p}$, and let $c_j \in \mathbb{C}$ and $\boldsymbol{d}_j \in \mathbb{C}^p$ for $j = 1, 2, \ldots, n$. It holds that

$$\mathcal{L}\left(\sum_{j=1}^{n} (c_j \boldsymbol{X}_j + \boldsymbol{d}_j)\right) = \mathbb{C}\mathcal{N}_p\left(\sum_{j=1}^{n} (c_j \boldsymbol{\theta}_j + \boldsymbol{d}_j), \sum_{j=1}^{n} c_j \bar{c}_j \boldsymbol{H}_j\right) .$$

Proof:

Let $\boldsymbol{X}_1, \boldsymbol{X}_2, \ldots, \boldsymbol{X}_n$ be mutually independent p-dimensional complex random vectors with $\mathcal{L}(\boldsymbol{X}_j) = \mathbb{C}\mathcal{N}_p(\boldsymbol{\theta}_j, \boldsymbol{H}_j)$, where $\boldsymbol{\theta}_j \in \mathbb{C}^p$ and $\boldsymbol{H}_j \in \mathbb{C}_S^{p\times p}$, and let $c_j \in \mathbb{C}$ and $\boldsymbol{d}_j \in \mathbb{C}^p$ for $j = 1, 2, \ldots, n$. Define the p-dimensional complex random vector $\boldsymbol{Y}$ as

$$\boldsymbol{Y} = \sum_{j=1}^{n} (c_j \boldsymbol{X}_j + \boldsymbol{d}_j) .$$

For all $\boldsymbol{e} \in \mathbb{C}^p$ we consider

$$\boldsymbol{e}^* \boldsymbol{Y} = \sum_{j=1}^{n} (c_j \boldsymbol{e}^* \boldsymbol{X}_j + \boldsymbol{e}^* \boldsymbol{d}_j) .$$

From (2.3) page 23 we know that $\mathcal{L}(\boldsymbol{e}^*\boldsymbol{X}_j) = \mathbb{C}\mathcal{N}(\boldsymbol{e}^*\boldsymbol{\theta}_j, \boldsymbol{e}^*\boldsymbol{H}_j\boldsymbol{e})$ for $j = 1, 2, \ldots, n$. Furthermore $\boldsymbol{e}^*\boldsymbol{X}_1, \boldsymbol{e}^*\boldsymbol{X}_2, \ldots, \boldsymbol{e}^*\boldsymbol{X}_n$ are mutually independent complex random variables. Thus by Theorem 2.6 page 21 we get for all $\boldsymbol{e} \in \mathbb{C}^p$ that

$$\begin{aligned}\mathcal{L}(\boldsymbol{e}^*\boldsymbol{Y}) &= \mathbb{C}\mathcal{N}\left(\sum_{j=1}^{n} (c_j \boldsymbol{e}^* \boldsymbol{\theta}_j + \boldsymbol{e}^* \boldsymbol{d}_j), \sum_{j=1}^{n} c_j \bar{c}_j \boldsymbol{e}^* \boldsymbol{H}_j \boldsymbol{e}\right) \\ &= \mathbb{C}\mathcal{N}\left(\boldsymbol{e}^* \sum_{j=1}^{n} (c_j \boldsymbol{\theta}_j + \boldsymbol{d}_j), \boldsymbol{e}^* \left(\sum_{j=1}^{n} c_j \bar{c}_j \boldsymbol{H}_j\right) \boldsymbol{e}\right) ,\end{aligned}$$

Use of Theorem 2.8 page 23 completes the proof. ∎

As in the univariate case we see, by letting $n = 1$ in the reproductivity property of the multivariate complex normal distribution, that the class of multivariate complex normal distributions is invariant under addition of a complex vector and multiplication of a complex constant.

2.3 Independence, Marginal and Conditional Distributions

In this section we discuss results concerning partitioning of a multivariate complex normally distributed random vector. We state a necessary and sufficient condition for independence of two parts of a complex normally distributed random vector. Furthermore a necessary and sufficient condition for independence of transformations of a complex normally distributed random vector is given. We also consider marginal and conditional distributions of the parts of a partitioned complex normally distributed random vector.

Theorem 2.12
Let $\boldsymbol{X}$ be a p-dimensional complex random vector with $\mathcal{L}(\boldsymbol{X}) = \mathbb{C}\mathcal{N}_p(\boldsymbol{\theta}, \boldsymbol{H})$, where $\boldsymbol{\theta} \in \mathbb{C}^p$ and $\boldsymbol{H} \in \mathbb{C}_S^{p \times p}$. Let $\boldsymbol{X}$, $\boldsymbol{\theta}$ and $\boldsymbol{H}$ be partitioned as

$$\boldsymbol{X} = \begin{pmatrix} \boldsymbol{X}_1 \\ \boldsymbol{X}_2 \end{pmatrix}, \ \boldsymbol{\theta} = \begin{pmatrix} \boldsymbol{\theta}_1 \\ \boldsymbol{\theta}_2 \end{pmatrix} \ \text{and} \ \boldsymbol{H} = \begin{pmatrix} \boldsymbol{H}_{11} & \boldsymbol{H}_{12} \\ \boldsymbol{H}_{21} & \boldsymbol{H}_{22} \end{pmatrix},$$

where $\boldsymbol{X}_j$ and $\boldsymbol{\theta}_j$ are $p_j \times 1$ and $\boldsymbol{H}_{jk}$ is $p_j \times p_k$ for $j,k = 1,2$ and $p = p_1 + p_2$. It holds that $\boldsymbol{X}_1$ and $\boldsymbol{X}_2$ are independent iff $\boldsymbol{H}_{12} = \boldsymbol{O}$.

Proof:
Let $\boldsymbol{X}$ be a p-dimensional complex random vector with $\mathcal{L}(\boldsymbol{X}) = \mathbb{C}\mathcal{N}_p(\boldsymbol{\theta}, \boldsymbol{H})$, where $\boldsymbol{\theta} \in \mathbb{C}^p$ and $\boldsymbol{H} \in \mathbb{C}_S^{p \times p}$. Further let $\boldsymbol{X}, \boldsymbol{\theta}$ and $\boldsymbol{H}$ be partitioned as in the theorem.

The matrix $\boldsymbol{H}_{12}$ is the matrix of the covariance between the components of $\boldsymbol{X}_1$ and the components of $\boldsymbol{X}_2$, so independence of $\boldsymbol{X}_1$ and $\boldsymbol{X}_2$ implies $\boldsymbol{H}_{12} = \boldsymbol{O}$.

Conversely suppose that $\boldsymbol{H}_{12} = \boldsymbol{O}$. Let $\widetilde{\boldsymbol{X}}_1$ and $\widetilde{\boldsymbol{X}}_2$ be independent complex random vectors of dimensions p_1 and p_2 and distributed as

$$\mathcal{L}\left(\widetilde{\boldsymbol{X}}_1\right) = \mathbb{C}\mathcal{N}_{p_1}(\boldsymbol{\theta}_1, \boldsymbol{H}_{11}) \ \text{and} \ \mathcal{L}\left(\widetilde{\boldsymbol{X}}_2\right) = \mathbb{C}\mathcal{N}_{p_2}(\boldsymbol{\theta}_2, \boldsymbol{H}_{22}) \ ,$$

respectively. This implies for all $\boldsymbol{c}_1 \in \mathbb{C}^{p_1}$ and all $\boldsymbol{c}_2 \in \mathbb{C}^{p_2}$ that

$$\mathcal{L}\left(\boldsymbol{c}_1^*\widetilde{\boldsymbol{X}}_1\right) = \mathbb{C}\mathcal{N}(\boldsymbol{c}_1^*\boldsymbol{\theta}_1, \boldsymbol{c}_1^*\boldsymbol{H}_{11}\boldsymbol{c}_1) \ \text{and} \ \mathcal{L}\left(\boldsymbol{c}_2^*\widetilde{\boldsymbol{X}}_2\right) = \mathbb{C}\mathcal{N}(\boldsymbol{c}_2^*\boldsymbol{\theta}_2, \boldsymbol{c}_2^*\boldsymbol{H}_{22}\boldsymbol{c}_2) \ .$$

As $\widetilde{\boldsymbol{X}}_1$ and $\widetilde{\boldsymbol{X}}_2$ are independent we deduce that $\boldsymbol{c}_1^*\widetilde{\boldsymbol{X}}_1$ and $\boldsymbol{c}_2^*\widetilde{\boldsymbol{X}}_2$ are independent. Then by use of the reproductivity property for the univariate complex normal distribution (Theorem 2.6 page 21) we get

$$\mathcal{L}\left(\boldsymbol{c}_1^*\widetilde{\boldsymbol{X}}_1 + \boldsymbol{c}_2^*\widetilde{\boldsymbol{X}}_2\right) = \mathbb{C}\mathcal{N}(\boldsymbol{c}_1^*\boldsymbol{\theta}_1 + \boldsymbol{c}_2^*\boldsymbol{\theta}_2, \boldsymbol{c}_1^*\boldsymbol{H}_{11}\boldsymbol{c}_1 + \boldsymbol{c}_2^*\boldsymbol{H}_{22}\boldsymbol{c}_2) \ .$$

Letting

$$\widetilde{\boldsymbol{X}} = \begin{pmatrix} \widetilde{\boldsymbol{X}}_1 \\ \widetilde{\boldsymbol{X}}_2 \end{pmatrix} \ \text{and} \ \boldsymbol{c} = \begin{pmatrix} \boldsymbol{c}_1 \\ \boldsymbol{c}_2 \end{pmatrix}$$

we observe that

$$\mathcal{L}\left(\boldsymbol{c}^*\widetilde{\boldsymbol{X}}\right) = \mathbb{C}\mathcal{N}(\boldsymbol{c}^*\boldsymbol{\theta}, \boldsymbol{c}^*\boldsymbol{H}\boldsymbol{c}) \ ,$$

where

$$\boldsymbol{H} = \begin{pmatrix} \boldsymbol{H}_{11} & \boldsymbol{O} \\ \boldsymbol{O} & \boldsymbol{H}_{22} \end{pmatrix} .$$

Hence by Theorem 2.8 page 23 we have that $\mathcal{L}(\boldsymbol{X}) = \mathbb{C}\mathcal{N}_p(\boldsymbol{\theta}, \boldsymbol{H})$ and $\mathcal{L}\left(\widetilde{\boldsymbol{X}}\right) = \mathbb{C}\mathcal{N}_p(\boldsymbol{\theta}, \boldsymbol{H})$, i.e. $\boldsymbol{X}$ and $\widetilde{\boldsymbol{X}}$ are identically distributed. Hence $\boldsymbol{X}_1$ and $\boldsymbol{X}_2$ are independent. ∎

Remark that Theorem 2.12 also can be established by using the fact that the characteristic function of $\boldsymbol{X}$ factorizes into the product of the characteristic functions of $\boldsymbol{X}_1$ and $\boldsymbol{X}_2$, when $\boldsymbol{H}_{12} = \boldsymbol{O}$.

A useful consequence of Theorem 2.12 is shown in Theorem 2.13 below. It gives a necessary and sufficient condition for independence of two transformations of a complex normally distributed random vector.

Theorem 2.13
Let $\boldsymbol{X}$ be a p-dimensional complex random vector with $\mathcal{L}(\boldsymbol{X}) = \mathbb{C}\mathcal{N}_p(\boldsymbol{\theta}, \boldsymbol{H})$, where $\boldsymbol{\theta} \in \mathbb{C}^p$ and $\boldsymbol{H} \in \mathbb{C}_S^{p \times p}$. Further let $\boldsymbol{C} \in \mathbb{C}^{m \times p}$ and $\boldsymbol{D} \in \mathbb{C}^{q \times p}$. It holds that $\boldsymbol{CX}$ and $\boldsymbol{DX}$ are independent iff $\boldsymbol{CHD}^* = \boldsymbol{O}$.

Proof:
Let $\boldsymbol{X}$ be a p-dimensional complex random vector with $\mathcal{L}(\boldsymbol{X}) = \mathbb{C}\mathcal{N}_p(\boldsymbol{\theta}, \boldsymbol{H})$, where $\boldsymbol{\theta} \in \mathbb{C}^p$ and $\boldsymbol{H} \in \mathbb{C}_S^{p \times p}$. Further let $\boldsymbol{C} \in \mathbb{C}^{m \times p}$ and $\boldsymbol{D} \in \mathbb{C}^{q \times p}$. By Theorem 2.8 page 23 we get

$$\mathcal{L}\left(\begin{pmatrix} \boldsymbol{C} \\ \boldsymbol{D} \end{pmatrix} \boldsymbol{X}\right) = \mathbb{C}\mathcal{N}_{m+q}\left(\begin{pmatrix} \boldsymbol{C\theta} \\ \boldsymbol{D\theta} \end{pmatrix}, \begin{pmatrix} \boldsymbol{CHC}^* & \boldsymbol{CHD}^* \\ \boldsymbol{DHC}^* & \boldsymbol{DHD}^* \end{pmatrix}\right) .$$

Using Theorem 2.12 page 28 we observe that $\boldsymbol{CX}$ and $\boldsymbol{DX}$ are independent iff $\boldsymbol{CHD}^* = \boldsymbol{O}$.
■

The marginal distribution of a part of a complex normally distributed random vector is given in the following theorem. It is useful when one is interested in the joint distribution of some of the complex random variables contained in the complex random vector.

Theorem 2.14 The marginal distribution
Let $\boldsymbol{X}$ be a p-dimensional complex random vector with $\mathcal{L}(\boldsymbol{X}) = \mathbb{C}\mathcal{N}_p(\boldsymbol{\theta}, \boldsymbol{H})$, where $\boldsymbol{\theta} \in \mathbb{C}^p$ and $\boldsymbol{H} \in \mathbb{C}_S^{p \times p}$. Furthermore let $\boldsymbol{X}$, $\boldsymbol{\theta}$ and $\boldsymbol{H}$ be partitioned as

$$\boldsymbol{X} = \begin{pmatrix} \boldsymbol{X}_1 \\ \boldsymbol{X}_2 \end{pmatrix}, \quad \boldsymbol{\theta} = \begin{pmatrix} \boldsymbol{\theta}_1 \\ \boldsymbol{\theta}_2 \end{pmatrix} \quad \text{and } \boldsymbol{H} = \begin{pmatrix} \boldsymbol{H}_{11} & \boldsymbol{H}_{12} \\ \boldsymbol{H}_{21} & \boldsymbol{H}_{22} \end{pmatrix},$$

where $\boldsymbol{X}_j$ and $\boldsymbol{\theta}_j$ are $p_j \times 1$ and $\boldsymbol{H}_{jk}$ is $p_j \times p_k$ for $j, k = 1, 2$ and $p = p_1 + p_2$. For $j = 1, 2$ it holds that

$$\mathcal{L}(\boldsymbol{X}_j) = \mathbb{C}\mathcal{N}_{p_j}(\boldsymbol{\theta}_j, \boldsymbol{H}_{jj}) .$$

Proof:
Let $\boldsymbol{X}$ be a p-dimensional complex random vector with $\mathcal{L}(\boldsymbol{X}) = \mathbb{C}\mathcal{N}_p(\boldsymbol{\theta}, \boldsymbol{H})$, where $\boldsymbol{\theta} \in \mathbb{C}^p$ and $\boldsymbol{H} \in \mathbb{C}_S^{p \times p}$, and let $\boldsymbol{X}$, $\boldsymbol{\theta}$ and $\boldsymbol{H}$ be partitioned as in the theorem. Further let $\boldsymbol{D}$ be the $p_1 \times p$ complex matrix given by $\boldsymbol{D} = (\boldsymbol{I}_{p_1}\ ,\ \boldsymbol{O})$, then $\boldsymbol{DX} = \boldsymbol{X}_1$. Using Theorem 2.8 page 23 we get

$$\mathcal{L}(\boldsymbol{X}_1) = \mathbb{C}\mathcal{N}_{p_1}(\boldsymbol{\theta}_1, \boldsymbol{H}_{11}) .$$

Similarly we obtain the marginal distribution of $\boldsymbol{X}_2$ by letting $\boldsymbol{D}$ be the $p_2 \times p$ complex matrix given by $\boldsymbol{D} = (\boldsymbol{O}\ ,\ \boldsymbol{I}_{p_2})$. ■

The next theorem states the distribution of a special transformation of a partitioned complex random vector. When we later consider the conditional distribution of a part of a complex random vector given the remaining variables in the vector, this theorem turns out to be useful.

Theorem 2.15
Let $\boldsymbol{X}$ be a p-dimensional complex random vector with $\mathcal{L}(\boldsymbol{X}) = \mathbb{C}\mathcal{N}_p(\boldsymbol{\theta}, \boldsymbol{H})$, where $\boldsymbol{\theta} \in \mathbb{C}^p$ and $\boldsymbol{H} \in \mathbb{C}_S^{p\times p}$. Furthermore let $\boldsymbol{X}$, $\boldsymbol{\theta}$ and $\boldsymbol{H}$ be partitioned as

$$\boldsymbol{X} = \begin{pmatrix} \boldsymbol{X}_1 \\ \boldsymbol{X}_2 \end{pmatrix},\ \boldsymbol{\theta} = \begin{pmatrix} \boldsymbol{\theta}_1 \\ \boldsymbol{\theta}_2 \end{pmatrix} \text{ and } \boldsymbol{H} = \begin{pmatrix} \boldsymbol{H}_{11} & \boldsymbol{H}_{12} \\ \boldsymbol{H}_{21} & \boldsymbol{H}_{22} \end{pmatrix},$$

where $\boldsymbol{X}_j$ and $\boldsymbol{\theta}_j$ are $p_j \times 1$ and $\boldsymbol{H}_{jk}$ is $p_j \times p_k$ for $j, k = 1, 2$ and $p = p_1 + p_2$. If $\boldsymbol{H}_{22} > \boldsymbol{O}$, then it holds that

$$\mathcal{L}\left(\boldsymbol{X}_1 - \boldsymbol{H}_{12}\boldsymbol{H}_{22}^{-1}\boldsymbol{X}_2\right) = \mathbb{C}\mathcal{N}_{p_1}\left(\boldsymbol{\theta}_1 - \boldsymbol{H}_{12}\boldsymbol{H}_{22}^{-1}\boldsymbol{\theta}_2, \boldsymbol{H}_{11} - \boldsymbol{H}_{12}\boldsymbol{H}_{22}^{-1}\boldsymbol{H}_{21}\right)$$

and

$$\boldsymbol{X}_1 - \boldsymbol{H}_{12}\boldsymbol{H}_{22}^{-1}\boldsymbol{X}_2 \perp\!\!\!\perp \boldsymbol{X}_2\,.$$

Proof:
Let $\boldsymbol{X}$ be a p-dimensional complex random vector with $\mathcal{L}(\boldsymbol{X}) = \mathbb{C}\mathcal{N}_p(\boldsymbol{\theta}, \boldsymbol{H})$, where $\boldsymbol{\theta} \in \mathbb{C}^p$ and $\boldsymbol{H} \in \mathbb{C}_S^{p\times p}$. Further let $\boldsymbol{X}$, $\boldsymbol{\theta}$ and $\boldsymbol{H}$ be partitioned as in the theorem and assume that $\boldsymbol{H}_{22} > \boldsymbol{O}$. Hereby $\boldsymbol{H}_{22}^{-1}$ exists. Let $\boldsymbol{C} \in \mathbb{C}^{p\times p}$ be given as

$$\boldsymbol{C} = \begin{pmatrix} \boldsymbol{I}_{p_1} & -\boldsymbol{H}_{12}\boldsymbol{H}_{22}^{-1} \\ \boldsymbol{O} & \boldsymbol{I}_{p_2} \end{pmatrix},$$

which yields that

$$\boldsymbol{C}\boldsymbol{X} = \begin{pmatrix} \boldsymbol{X}_1 - \boldsymbol{H}_{12}\boldsymbol{H}_{22}^{-1}\boldsymbol{X}_2 \\ \boldsymbol{X}_2 \end{pmatrix}.$$

Since $\mathcal{L}(\boldsymbol{X}) = \mathbb{C}\mathcal{N}_p(\boldsymbol{\theta}, \boldsymbol{H})$ we have from Theorem 2.8 page 23 that

$$\mathcal{L}(\boldsymbol{C}\boldsymbol{X}) = \mathbb{C}\mathcal{N}_p\left(\begin{pmatrix} \boldsymbol{\theta}_1 - \boldsymbol{H}_{12}\boldsymbol{H}_{22}^{-1}\boldsymbol{\theta}_2 \\ \boldsymbol{\theta}_2 \end{pmatrix}, \boldsymbol{C}\boldsymbol{H}\boldsymbol{C}^*\right),$$

where

$$\begin{aligned} \boldsymbol{C}\boldsymbol{H}\boldsymbol{C}^* &= \begin{pmatrix} \boldsymbol{I}_{p_1} & -\boldsymbol{H}_{12}\boldsymbol{H}_{22}^{-1} \\ \boldsymbol{O} & \boldsymbol{I}_{p_2} \end{pmatrix} \begin{pmatrix} \boldsymbol{H}_{11} & \boldsymbol{H}_{12} \\ \boldsymbol{H}_{21} & \boldsymbol{H}_{22} \end{pmatrix} \begin{pmatrix} \boldsymbol{I}_{p_1} & -\boldsymbol{H}_{12}\boldsymbol{H}_{22}^{-1} \\ \boldsymbol{O} & \boldsymbol{I}_{p_2} \end{pmatrix}^* \\ &= \begin{pmatrix} \boldsymbol{H}_{11} - \boldsymbol{H}_{12}\boldsymbol{H}_{22}^{-1}\boldsymbol{H}_{21} & \boldsymbol{O} \\ \boldsymbol{H}_{21} & \boldsymbol{H}_{22} \end{pmatrix} \begin{pmatrix} \boldsymbol{I}_{p_1} & \boldsymbol{O} \\ -\boldsymbol{H}_{22}^{-1}\boldsymbol{H}_{21} & \boldsymbol{I}_{p_2} \end{pmatrix} \\ &= \begin{pmatrix} \boldsymbol{H}_{11} - \boldsymbol{H}_{12}\boldsymbol{H}_{22}^{-1}\boldsymbol{H}_{21} & \boldsymbol{O} \\ \boldsymbol{O} & \boldsymbol{H}_{22} \end{pmatrix}. \end{aligned}$$

From Theorem 2.12 page 28 we hereby see that $\boldsymbol{X}_1 - \boldsymbol{H}_{12}\boldsymbol{H}_{22}^{-1}\boldsymbol{X}_2$ and $\boldsymbol{X}_2$ are independent. Furthermore we get the marginal distribution of $\boldsymbol{X}_1 - \boldsymbol{H}_{12}\boldsymbol{H}_{22}^{-1}\boldsymbol{X}_2$ by Theorem 2.14 page 29 as

$$\mathcal{L}\left(\boldsymbol{X}_1 - \boldsymbol{H}_{12}\boldsymbol{H}_{22}^{-1}\boldsymbol{X}_2\right) = \mathbb{C}\mathcal{N}_{p_1}\left(\boldsymbol{\theta}_1 - \boldsymbol{H}_{12}\boldsymbol{H}_{22}^{-1}\boldsymbol{\theta}_2, \boldsymbol{H}_{11} - \boldsymbol{H}_{12}\boldsymbol{H}_{22}^{-1}\boldsymbol{H}_{21}\right) .$$

■

For a complex normally distributed random vector the conditional distribution of a part of the variables given the remaining ones is determined in the subsequent theorem.

Theorem 2.16 The conditional distribution
Let $\boldsymbol{X}$ be a p-dimensional complex random vector with $\mathcal{L}(\boldsymbol{X}) = \mathbb{C}\mathcal{N}_p(\boldsymbol{\theta}, \boldsymbol{H})$, where $\boldsymbol{\theta} \in \mathbb{C}^p$ and $\boldsymbol{H} \in \mathbb{C}_S^{p\times p}$. Furthermore let $\boldsymbol{X}$, $\boldsymbol{\theta}$ and $\boldsymbol{H}$ be partitioned as

$$\boldsymbol{X} = \begin{pmatrix} \boldsymbol{X}_1 \\ \boldsymbol{X}_2 \end{pmatrix}, \ \boldsymbol{\theta} = \begin{pmatrix} \boldsymbol{\theta}_1 \\ \boldsymbol{\theta}_2 \end{pmatrix} \ \text{and}\ \boldsymbol{H} = \begin{pmatrix} \boldsymbol{H}_{11} & \boldsymbol{H}_{12} \\ \boldsymbol{H}_{21} & \boldsymbol{H}_{22} \end{pmatrix},$$

where $\boldsymbol{X}_j$ and $\boldsymbol{\theta}_j$ are $p_j \times 1$ and $\boldsymbol{H}_{jk}$ is $p_j \times p_k$ for $j, k = 1, 2$ and $p = p_1 + p_2$. If $\boldsymbol{H}_{22} > \boldsymbol{O}$, then the conditional distribution of $\boldsymbol{X}_1$ given $\boldsymbol{X}_2$ is

$$\mathcal{L}(\boldsymbol{X}_1|\boldsymbol{X}_2) = \mathbb{C}\mathcal{N}_{p_1}\left(\boldsymbol{\theta}_1 + \boldsymbol{H}_{12}\boldsymbol{H}_{22}^{-1}(\boldsymbol{X}_2 - \boldsymbol{\theta}_2), \boldsymbol{H}_{11} - \boldsymbol{H}_{12}\boldsymbol{H}_{22}^{-1}\boldsymbol{H}_{21}\right) .$$

Proof:
Let $\boldsymbol{X}$ be a p-dimensional complex random vector with $\mathcal{L}(\boldsymbol{X}) = \mathbb{C}\mathcal{N}_p(\boldsymbol{\theta}, \boldsymbol{H})$, where $\boldsymbol{\theta} \in \mathbb{C}^p$ and $\boldsymbol{H} \in \mathbb{C}_S^{p\times p}$. Further let $\boldsymbol{X}$, $\boldsymbol{\theta}$ and $\boldsymbol{H}$ be partitioned as in the theorem and assume that $\boldsymbol{H}_{22} > \boldsymbol{O}$.

According to Theorem 2.15 page 30 we observe that

$$\mathcal{L}\left(\boldsymbol{X}_1 - \boldsymbol{H}_{12}\boldsymbol{H}_{22}^{-1}\boldsymbol{X}_2\right) = \mathbb{C}\mathcal{N}_{p_1}\left(\boldsymbol{\theta}_1 - \boldsymbol{H}_{12}\boldsymbol{H}_{22}^{-1}\boldsymbol{\theta}_2, \boldsymbol{H}_{11} - \boldsymbol{H}_{12}\boldsymbol{H}_{22}^{-1}\boldsymbol{H}_{21}\right)$$

and that $\boldsymbol{X}_1 - \boldsymbol{H}_{12}\boldsymbol{H}_{22}^{-1}\boldsymbol{X}_2$ and $\boldsymbol{X}_2$ are independent. Thus

$$\begin{aligned} \mathcal{L}\left(\boldsymbol{X}_1 - \boldsymbol{H}_{12}\boldsymbol{H}_{22}^{-1}\boldsymbol{X}_2|\boldsymbol{X}_2\right) &= \mathcal{L}\left(\boldsymbol{X}_1 - \boldsymbol{H}_{12}\boldsymbol{H}_{22}^{-1}\boldsymbol{X}_2\right) \\ &= \mathbb{C}\mathcal{N}_{p_1}\left(\boldsymbol{\theta}_1 - \boldsymbol{H}_{12}\boldsymbol{H}_{22}^{-1}\boldsymbol{\theta}_2, \boldsymbol{H}_{11} - \boldsymbol{H}_{12}\boldsymbol{H}_{22}^{-1}\boldsymbol{H}_{21}\right). \end{aligned}$$

By using Theorem 2.8 page 23 with $\boldsymbol{D} = \boldsymbol{I}_{p_1}$ and $\boldsymbol{d} = \boldsymbol{H}_{12}\boldsymbol{H}_{22}^{-1}\boldsymbol{X}_2$ we get

$$\mathcal{L}(\boldsymbol{X}_1|\boldsymbol{X}_2) = \mathbb{C}\mathcal{N}_{p_1}\left(\boldsymbol{\theta}_1 + \boldsymbol{H}_{12}\boldsymbol{H}_{22}^{-1}(\boldsymbol{X}_2 - \boldsymbol{\theta}_2), \boldsymbol{H}_{11} - \boldsymbol{H}_{12}\boldsymbol{H}_{22}^{-1}\boldsymbol{H}_{21}\right) .$$

■

2.4 The Multivariate Complex Normal Distribution in Matrix Notation

Often it can be advantageous to arrange a set of complex normally distributed random vectors of equal dimension in a matrix. This is due to the fact that we frequently have a situation where we on n objects measure p complex random variables. Let X_{jk} for $j = 1, 2, \ldots, n$ and $k = 1, 2, \ldots, p$ be the complex random variables associated with the np measurements. The subscript j refers to the j'th object and the subscript k refers to the k'th variable. Let the complex random variables be arranged in n p-dimensional complex random vectors $\boldsymbol{X}_1, \boldsymbol{X}_2, \ldots, \boldsymbol{X}_n$, where the $\boldsymbol{X}_j$'s are defined by

$$\boldsymbol{X}_j = \begin{pmatrix} X_{j1} \\ X_{j2} \\ \vdots \\ X_{jp} \end{pmatrix} .$$

To obtain a mathematical object which contains the total information of the whole problem we arrange $\boldsymbol{X}_1, \boldsymbol{X}_2, \ldots, \boldsymbol{X}_n$ in the $n \times p$ complex random matrix $\boldsymbol{X}$ given by

$$\boldsymbol{X} = (\boldsymbol{X}_1, \boldsymbol{X}_2, \ldots, \boldsymbol{X}_n)^* = \begin{pmatrix} \boldsymbol{X}_1^* \\ \boldsymbol{X}_2^* \\ \vdots \\ \boldsymbol{X}_n^* \end{pmatrix} .$$

In this section we state some useful results for an $n \times p$ complex random matrix. First we consider the situation where all the complex random vectors $\boldsymbol{X}_1, \boldsymbol{X}_2, \ldots, \boldsymbol{X}_n$ are mutually independent, and where they have equal variance matrices but different means.

Let $\boldsymbol{X}_1, \boldsymbol{X}_2, \ldots, \boldsymbol{X}_n$ be mutually independent p-dimensional complex random vectors with $\mathcal{L}(\boldsymbol{X}_j) = \mathbb{C}\mathcal{N}_p(\boldsymbol{\theta}_j, \boldsymbol{H})$, where $\boldsymbol{\theta}_j \in \mathbb{C}^p$ and $\boldsymbol{H} \in \mathbb{C}_S^{p \times p}$ for $j = 1, 2, \ldots, n$. Further let $\boldsymbol{X}$ be the complex random matrix defined by $\boldsymbol{X} = (\boldsymbol{X}_1, \boldsymbol{X}_2, \ldots, \boldsymbol{X}_n)^*$. Then $\boldsymbol{X}$ has an $(n \times p)$-variate complex normal distribution with mean and variance given by

$$\mathbb{E}(\boldsymbol{X}) = (\boldsymbol{\theta}_1, \boldsymbol{\theta}_2, \ldots, \boldsymbol{\theta}_n)^* = \Theta \text{ and } \mathbb{V}(\boldsymbol{X}) = \boldsymbol{I}_n \otimes \boldsymbol{H} .$$

This is denoted by $\mathcal{L}(\boldsymbol{X}) = \mathbb{C}\mathcal{N}_{n \times p}(\Theta, \boldsymbol{I}_n \otimes \boldsymbol{H})$. The word $(n \times p)$-variate can be omitted if it is obvious from the context.

Remark that for $n = 1$ the matrix $\boldsymbol{X}$ is not a vector similar to the ones considered in Section 2.2 page 22. However it should not be inconvenient as we are only interested in arranging vectors in a matrix for $n > 1$.

Using the mutual independence of the $\boldsymbol{X}_j$'s and Theorem 2.10 page 26 we obtain, when $\boldsymbol{H} > \boldsymbol{O}$, the joint density function of $\boldsymbol{X}_1, \boldsymbol{X}_2, \ldots, \boldsymbol{X}_n$ w.r.t. Lebesgue measure on $\mathbb{C}^{n \times p}$ as

$$\begin{aligned} f_{\boldsymbol{X}}(\boldsymbol{x}) &= \prod_{j=1}^{n} f_{\boldsymbol{X}_j}(\boldsymbol{x}_j) \\ &= \prod_{j=1}^{n} \pi^{-p} \det(\boldsymbol{H})^{-1} \exp\left(-(\boldsymbol{x}_j - \boldsymbol{\theta}_j)^* \boldsymbol{H}^{-1} (\boldsymbol{x}_j - \boldsymbol{\theta}_j)\right) \end{aligned}$$

$$= \pi^{-np} \det(\boldsymbol{H})^{-n} \exp\left(-\sum_{j=1}^{n} (\boldsymbol{x}_j - \boldsymbol{\theta}_j)^* \boldsymbol{H}^{-1} (\boldsymbol{x}_j - \boldsymbol{\theta}_j)\right)$$
$$= \pi^{-np} \det(\boldsymbol{H})^{-n} \exp\left(-\operatorname{tr}\left(\sum_{j=1}^{n} (\boldsymbol{x}_j - \boldsymbol{\theta}_j)(\boldsymbol{x}_j - \boldsymbol{\theta}_j)^* \boldsymbol{H}^{-1}\right)\right)$$
$$= \pi^{-np} \det(\boldsymbol{H})^{-n} \exp\left(-\operatorname{tr}\left((\boldsymbol{x} - \Theta)^* (\boldsymbol{x} - \Theta) \boldsymbol{H}^{-1}\right)\right),$$

where $\Theta = (\boldsymbol{\theta}_1, \boldsymbol{\theta}_2, \ldots, \boldsymbol{\theta}_n)^*$ and $\boldsymbol{x} = (\boldsymbol{x}_1, \boldsymbol{x}_2, \ldots, \boldsymbol{x}_n)^* \in \mathbb{C}^{n \times p}$. This leads to the following theorem.

Theorem 2.17 The density function of $\mathbb{C}\mathcal{N}_{n\times p}(\Theta, \boldsymbol{I}_n \otimes \boldsymbol{H})$
Let $\boldsymbol{X}$ be an $n \times p$ complex random matrix with $\mathcal{L}(\boldsymbol{X}) = \mathbb{C}\mathcal{N}_{n\times p}(\Theta, \boldsymbol{I}_n \otimes \boldsymbol{H})$, where $\Theta \in \mathbb{C}^{n\times p}$ and $\boldsymbol{H} \in \mathbb{C}_+^{p\times p}$. The density function of $\boldsymbol{X}$ w.r.t. Lebesgue measure on $\mathbb{C}^{n\times p}$ is given as

$$f_{\boldsymbol{X}}(\boldsymbol{x}) = \pi^{-np} \det(\boldsymbol{H})^{-n} \exp\left(-\operatorname{tr}\left((\boldsymbol{x} - \Theta) \boldsymbol{H}^{-1} (\boldsymbol{x} - \Theta)^*\right)\right), \ \boldsymbol{x} \in \mathbb{C}^{n\times p}.$$

Let us now consider the characteristic function of $\boldsymbol{X}$ with $\mathcal{L}(\boldsymbol{X}) = \mathbb{C}\mathcal{N}_{n\times p}(\Theta, \boldsymbol{I}_n \otimes \boldsymbol{H})$, where $\Theta = (\boldsymbol{\theta}_1, \boldsymbol{\theta}_2, \ldots, \boldsymbol{\theta}_n)^* \in \mathbb{C}^{n\times p}$ and $\boldsymbol{H} \in \mathbb{C}_S^{p\times p}$. We obtain by using the mutual independence of the $\boldsymbol{X}_j$'s and Theorem 2.7 page 23 for $\boldsymbol{\xi} = (\boldsymbol{\xi}_1, \boldsymbol{\xi}_2, \ldots, \boldsymbol{\xi}_n)^* \in \mathbb{C}^{n\times p}$ that

$$\begin{aligned}
\varphi_{\boldsymbol{X}}(\boldsymbol{\xi}) &= \prod_{j=1}^{n} \varphi_{\boldsymbol{X}_j}(\boldsymbol{\xi}_j) \\
&= \prod_{j=1}^{n} \exp\left(i \operatorname{Re}\left(\boldsymbol{\xi}_j^* \boldsymbol{\theta}_j\right) - \frac{\boldsymbol{\xi}_j^* \boldsymbol{H} \boldsymbol{\xi}_j}{4}\right) \\
&= \exp\left(\sum_{j=1}^{n} i \operatorname{Re}\left(\boldsymbol{\xi}_j^* \boldsymbol{\theta}_j\right) - \frac{\sum_{j=1}^{n} \boldsymbol{\xi}_j^* \boldsymbol{H} \boldsymbol{\xi}_j}{4}\right) \\
&= \exp\left(i \operatorname{Re}\left(\sum_{j=1}^{n} \boldsymbol{\xi}_j^* \boldsymbol{\theta}_j\right) - \frac{\operatorname{tr}\left(\sum_{j=1}^{n} \boldsymbol{\xi}_j \boldsymbol{\xi}_j^* \boldsymbol{H}\right)}{4}\right) \\
&= \exp\left(i \operatorname{Re}\left(\operatorname{tr}\left(\boldsymbol{\xi} \Theta^*\right)\right) - \frac{\operatorname{tr}\left(\boldsymbol{\xi}^* \boldsymbol{\xi} \boldsymbol{H}\right)}{4}\right).
\end{aligned}$$

Hereby the next theorem is obtained.

Theorem 2.18 The characteristic function of $\mathbb{C}\mathcal{N}_{n\times p}(\Theta, \boldsymbol{I}_n \otimes \boldsymbol{H})$
Let $\boldsymbol{X}$ be an $n \times p$ complex random matrix with $\mathcal{L}(\boldsymbol{X}) = \mathbb{C}\mathcal{N}_{n\times p}(\Theta, \boldsymbol{I}_n \otimes \boldsymbol{H})$, where $\Theta \in \mathbb{C}^{n\times p}$ and $\boldsymbol{H} \in \mathbb{C}_S^{p\times p}$. The characteristic function of $\boldsymbol{X}$ is given as

$$\varphi_{\boldsymbol{X}}(\boldsymbol{\xi}) = \exp\left(i \operatorname{Re}\left(\operatorname{tr}\left(\boldsymbol{\xi} \Theta^*\right)\right) - \frac{\operatorname{tr}\left(\boldsymbol{\xi} \boldsymbol{H} \boldsymbol{\xi}^*\right)}{4}\right), \boldsymbol{\xi} \in \mathbb{C}^{n\times p}.$$

Later we consider situations where the complex random vectors $\boldsymbol{X}_1, \boldsymbol{X}_2, \ldots, \boldsymbol{X}_n$ are not necessarily independent. Therefore let a multivariate complex normally distributed random matrix $\boldsymbol{X}$ have a special variance structure of the form $\boldsymbol{J} \otimes \boldsymbol{H}$, where $\boldsymbol{J} \in \mathbb{C}_S^{n \times n}$ and $\boldsymbol{H} \in \mathbb{C}_S^{p \times p}$. In other words let $\mathcal{L}(\boldsymbol{X}) = \mathbb{C}\mathcal{N}_{n \times p}(\boldsymbol{\Theta}, \boldsymbol{J} \otimes \boldsymbol{H})$. In subsequent considerations we need the theorems below. These are stated for such a particular complex random matrix, and they are deduced by using the interpretation of the direct product of two matrices as a linear transformation, i.e. we use that $\boldsymbol{X}$ can be interpreted as an $np \times 1$ complex random vector.

Theorem 2.19 Property for the multivariate complex normal distribution
Let $\boldsymbol{X}$ be an $n \times p$ complex random matrix with $\mathcal{L}(\boldsymbol{X}) = \mathbb{C}\mathcal{N}_{n \times p}(\boldsymbol{\Theta}, \boldsymbol{J} \otimes \boldsymbol{H})$, where $\boldsymbol{\Theta} \in \mathbb{C}^{n \times p}$, $\boldsymbol{J} \in \mathbb{C}_S^{n \times n}$ and $\boldsymbol{H} \in \mathbb{C}_S^{p \times p}$. Furthermore let $\boldsymbol{C} \in \mathbb{C}^{m \times n}$, $\boldsymbol{D} \in \mathbb{C}^{q \times p}$ and $\boldsymbol{E} \in \mathbb{C}^{m \times q}$. It holds that

$$\mathcal{L}(\boldsymbol{C}\boldsymbol{X}\boldsymbol{D}^* + \boldsymbol{E}) = \mathbb{C}\mathcal{N}_{m \times q}(\boldsymbol{C}\boldsymbol{\Theta}\boldsymbol{D}^* + \boldsymbol{E}, \boldsymbol{C}\boldsymbol{J}\boldsymbol{C}^* \otimes \boldsymbol{D}\boldsymbol{H}\boldsymbol{D}^*) .$$

Proof:
Let $\boldsymbol{X}$ be an $n \times p$ complex random matrix with $\mathcal{L}(\boldsymbol{X}) = \mathbb{C}\mathcal{N}_{n \times p}(\boldsymbol{\Theta}, \boldsymbol{J} \otimes \boldsymbol{H})$, where $\boldsymbol{\Theta} \in \mathbb{C}^{n \times p}$, $\boldsymbol{J} \in \mathbb{C}_S^{n \times n}$ and $\boldsymbol{H} \in \mathbb{C}_S^{p \times p}$. Furthermore let $\boldsymbol{C} \in \mathbb{C}^{m \times n}$, $\boldsymbol{D} \in \mathbb{C}^{q \times p}$ and $\boldsymbol{E} \in \mathbb{C}^{m \times q}$. Notice that

$$\boldsymbol{C}\boldsymbol{X}\boldsymbol{D}^* = (\boldsymbol{C} \otimes \boldsymbol{D})(\boldsymbol{X}) ,$$

where $\boldsymbol{X}$ on the left side is an $n \times p$ complex random matrix and on the right side is regarded as an $np \times 1$ complex random vector. If we moreover interpret $\boldsymbol{E}$ as an $mq \times 1$ complex vector and use Theorem 2.8 page 23 we get

$$\begin{aligned}
\mathcal{L}(\boldsymbol{C}\boldsymbol{X}\boldsymbol{D}^* + \boldsymbol{E}) &= \mathbb{C}\mathcal{N}_{m \times q}((\boldsymbol{C} \otimes \boldsymbol{D})(\boldsymbol{\Theta}) + \boldsymbol{E}, (\boldsymbol{C} \otimes \boldsymbol{D})(\boldsymbol{J} \otimes \boldsymbol{H})(\boldsymbol{C} \otimes \boldsymbol{D})^*) \\
&= \mathbb{C}\mathcal{N}_{m \times q}(\boldsymbol{C}\boldsymbol{\Theta}\boldsymbol{D}^* + \boldsymbol{E}, \boldsymbol{C}\boldsymbol{J}\boldsymbol{C}^* \otimes \boldsymbol{D}\boldsymbol{H}\boldsymbol{D}^*) .
\end{aligned}$$

■

We now focus on results, which can be used for determination of independence of two transformations of a complex random matrix. The first result is stated in the theorem below.

Theorem 2.20
Let $\boldsymbol{X}$ be an $n \times p$ complex random matrix with $\mathcal{L}(\boldsymbol{X}) = \mathbb{C}\mathcal{N}_{n \times p}(\boldsymbol{\Theta}, \boldsymbol{J} \otimes \boldsymbol{H})$, where $\boldsymbol{\Theta} \in \mathbb{C}^{n \times p}$, $\boldsymbol{J} \in \mathbb{C}_S^{n \times n}$ and $\boldsymbol{H} \in \mathbb{C}_S^{p \times p}$. Furthermore let $\boldsymbol{C} \in \mathbb{C}^{m \times n}$ and $\boldsymbol{D} \in \mathbb{C}^{q \times n}$. It holds that $\boldsymbol{C}\boldsymbol{X}$ and $\boldsymbol{D}\boldsymbol{X}$ are independent iff $\boldsymbol{C}\boldsymbol{J}\boldsymbol{D}^* \otimes \boldsymbol{H} = \boldsymbol{O}$.

Proof:
Let $\boldsymbol{X}$ be an $n \times p$ complex random matrix with $\mathcal{L}(\boldsymbol{X}) = \mathbb{C}\mathcal{N}_{n \times p}(\boldsymbol{\Theta}, \boldsymbol{J} \otimes \boldsymbol{H})$, where $\boldsymbol{\Theta} \in \mathbb{C}^{n \times p}$, $\boldsymbol{J} \in \mathbb{C}_S^{n \times n}$ and $\boldsymbol{H} \in \mathbb{C}_S^{p \times p}$, and let $\boldsymbol{C} \in \mathbb{C}^{m \times n}$ and $\boldsymbol{D} \in \mathbb{C}^{q \times n}$. Observe that

$$\begin{aligned}
\boldsymbol{C}\boldsymbol{X} &= (\boldsymbol{C} \otimes \boldsymbol{I}_p)(\boldsymbol{X}) \\
\boldsymbol{D}\boldsymbol{X} &= (\boldsymbol{D} \otimes \boldsymbol{I}_p)(\boldsymbol{X}) ,
\end{aligned}$$

where $\boldsymbol{X}$ again is interpreted both as a complex random matrix and as a complex random vector. Using Theorem 2.13 page 29 we see that $\boldsymbol{CX}$ and $\boldsymbol{DX}$ are independent iff $\boldsymbol{CJD}^* \otimes \boldsymbol{H} = \boldsymbol{O}$, since

$$(\boldsymbol{C} \otimes \boldsymbol{I}_p)\,(\boldsymbol{J} \otimes \boldsymbol{H})\,(\boldsymbol{D} \otimes \boldsymbol{I}_p)^* = \boldsymbol{CJD}^* \otimes \boldsymbol{H}\,.$$

■

Note that if $\boldsymbol{H} \neq \boldsymbol{O}$ in Theorem 2.20, then the statement $\boldsymbol{CJD}^* \otimes \boldsymbol{H} = \boldsymbol{O}$ is equivalent to $\boldsymbol{CJD}^* = \boldsymbol{O}$. The following theorem is analogous to Theorem 2.20. It contains other transformations of the complex random matrix.

Theorem 2.21
Let $\boldsymbol{X}$ be an $n \times p$ complex random matrix with $\mathcal{L}(\boldsymbol{X}) = \mathbb{C}\mathcal{N}_{n\times p}(\boldsymbol{\Theta}, \boldsymbol{J} \otimes \boldsymbol{H})$, where $\boldsymbol{\Theta} \in \mathbb{C}^{n\times p}$, $\boldsymbol{J} \in \mathbb{C}_S^{n\times n}$ and $\boldsymbol{H} \in \mathbb{C}_S^{p\times p}$. Furthermore let $\boldsymbol{C} \in \mathbb{C}^{p\times q}$ and $\boldsymbol{D} \in \mathbb{C}^{p\times m}$. It holds that $\boldsymbol{XC}$ and $\boldsymbol{XD}$ are independent iff $\boldsymbol{J} \otimes \boldsymbol{C}^*\boldsymbol{HD} = \boldsymbol{O}$.

Proof:
Analogous to the proof of Theorem 2.20 page 34. ■

Next we consider the partition of a complex random matrix consisting of mutually independent complex random vectors. The theorem below states a necessary and sufficient condition for independence of two parts of a complex random matrix.

Theorem 2.22
Let $\boldsymbol{X}$ be an $n \times p$ complex random matrix with $\mathcal{L}(\boldsymbol{X}) = \mathbb{C}\mathcal{N}_{n\times p}(\boldsymbol{\Theta}, \boldsymbol{I}_n \otimes \boldsymbol{H})$, where $\boldsymbol{\Theta} \in \mathbb{C}^{n\times p}$ and $\boldsymbol{H} \in \mathbb{C}_S^{p\times p}$. Furthermore let $\boldsymbol{X}$, $\boldsymbol{\Theta}$ and $\boldsymbol{H}$ be partitioned as

$$\boldsymbol{X} = (\boldsymbol{X}_1, \boldsymbol{X}_2)\,,\ \boldsymbol{\Theta} = (\boldsymbol{\Theta}_1, \boldsymbol{\Theta}_2) \text{ and } \boldsymbol{H} = \begin{pmatrix} \boldsymbol{H}_{11} & \boldsymbol{H}_{12} \\ \boldsymbol{H}_{21} & \boldsymbol{H}_{22} \end{pmatrix},$$

where $\boldsymbol{X}_j$ and $\boldsymbol{\Theta}_j$ are $n \times p_j$ and $\boldsymbol{H}_{jk}$ is $p_j \times p_k$ for $j = 1, 2$ and $p = p_1 + p_2$. It holds that $\boldsymbol{X}_1$ and $\boldsymbol{X}_2$ are independent iff $\boldsymbol{H}_{12} = \boldsymbol{O}$.

Proof:
Let $\boldsymbol{X}$ be an $n \times p$ complex random matrix with $\mathcal{L}(\boldsymbol{X}) = \mathbb{C}\mathcal{N}_{n\times p}(\boldsymbol{\Theta}, \boldsymbol{I}_n \otimes \boldsymbol{H})$, where $\boldsymbol{\Theta} \in \mathbb{C}^{n\times p}$ and $\boldsymbol{H} \in \mathbb{C}_S^{p\times p}$. Furthermore let $\boldsymbol{X}$, $\boldsymbol{\Theta}$ and $\boldsymbol{H}$ be partitioned as in the theorem. Observe that $\boldsymbol{X}_1$ and $\boldsymbol{X}_2$ can be written as

$$\boldsymbol{X}_1 = \boldsymbol{X}\begin{pmatrix} \boldsymbol{I}_{p_1} \\ \boldsymbol{O} \end{pmatrix} \text{ and } \boldsymbol{X}_2 = \boldsymbol{X}\begin{pmatrix} \boldsymbol{O} \\ \boldsymbol{I}_{p_2} \end{pmatrix}.$$

Theorem 2.21 says that $\boldsymbol{X}_1$ and $\boldsymbol{X}_2$ are independent iff $\boldsymbol{I}_n \otimes (\boldsymbol{I}_{p_1}, \boldsymbol{O})\,\boldsymbol{H}\,(\boldsymbol{O}, \boldsymbol{I}_{p_2})^* = \boldsymbol{I}_n \otimes \boldsymbol{H}_{12} = \boldsymbol{O}$, i.e. $\boldsymbol{H}_{12} = \boldsymbol{O}$. ■

The marginal distribution of a part of a complex random matrix is given in the theorem below. As in the case with complex random vectors this result is useful when only a part of the complex random variables in the matrix is of interest.

Theorem 2.23 The marginal distribution
Let X be an $n \times p$ complex random matrix with $\mathcal{L}(X) = \mathbb{C}\mathcal{N}_{n \times p}(\Theta, I_n \otimes H)$, where $\Theta \in \mathbb{C}^{n \times p}$ and $H \in \mathbb{C}_S^{p \times p}$. Furthermore let X, Θ and H be partitioned as

$$X = (X_1, X_2), \ \Theta = (\Theta_1, \Theta_2) \ \text{and} \ H = \begin{pmatrix} H_{11} & H_{12} \\ H_{21} & H_{22} \end{pmatrix},$$

where X_j and Θ_j are $n \times p_j$ and H_{jk} is $p_j \times p_k$ for $j = 1, 2$ and $p = p_1 + p_2$. For $j = 1, 2$ it holds that

$$\mathcal{L}(X_j) = \mathbb{C}\mathcal{N}_{n \times p_j}(\Theta_j, I_n \otimes H_{jj}) .$$

Proof:
Let X be an $n \times p$ complex random matrix with $\mathcal{L}(X) = \mathbb{C}\mathcal{N}_{n \times p}(\Theta, I_n \otimes H)$, where $\Theta \in \mathbb{C}^{n \times p}$ and $H \in \mathbb{C}_S^{p \times p}$. Furthermore let X, Θ and H be partitioned as in the theorem. Let D be the $p_1 \times p$ complex matrix given by $D = (I_{p_1}, O)$, then $X_1 = XD^*$. Using Theorem 2.19 page 34 we find

$$\mathcal{L}(X_1) = \mathbb{C}\mathcal{N}_{n \times p_1}(\Theta_1, I_n \otimes H_{11}) .$$

Similarly by letting $D = (O, I_{p_2}) \in \mathbb{C}^{p_2 \times p}$ we obtain the marginal distribution of X_2. ∎

As we seek a result on the conditional distribution of a part of the complex random matrix given the remaining complex random matrix, we need the following theorem.

Theorem 2.24
Let X be an $n \times p$ complex random matrix with $\mathcal{L}(X) = \mathbb{C}\mathcal{N}_{n \times p}(\Theta, I_n \otimes H)$, where $\Theta \in \mathbb{C}^{n \times p}$ and $H \in \mathbb{C}_S^{p \times p}$. Furthermore let X, Θ and H be partitioned as

$$X = (X_1, X_2), \ \Theta = (\Theta_1, \Theta_2) \ \text{and} \ H = \begin{pmatrix} H_{11} & H_{12} \\ H_{21} & H_{22} \end{pmatrix},$$

where X_j and Θ_j are $n \times p_j$ and H_{jk} is $p_j \times p_k$ for $j = 1, 2$ and $p = p_1 + p_2$. If $H_{22} > O$, then it holds that

$$\mathcal{L}\left(X_1 - X_2 H_{22}^{-1} H_{21}\right) = \mathbb{C}\mathcal{N}_{n \times p_1}\left(\Theta_1 - \Theta_2 H_{22}^{-1} H_{21}, I_n \otimes \left(H_{11} - H_{12} H_{22}^{-1} H_{21}\right)\right)$$

and

$$X_1 - X_2 H_{22}^{-1} H_{21} \perp\!\!\!\perp X_2 .$$

Proof:
Let $\boldsymbol{X}$ be an $n \times p$ complex random matrix with $\mathcal{L}(\boldsymbol{X}) = \mathbb{C}\mathcal{N}_{n \times p}(\boldsymbol{\Theta}, \boldsymbol{I}_n \otimes \boldsymbol{H})$, where $\boldsymbol{\Theta} \in \mathbb{C}^{n \times p}$ and $\boldsymbol{H} \in \mathbb{C}_S^{p \times p}$. Furthermore let $\boldsymbol{X}$, $\boldsymbol{\Theta}$ and $\boldsymbol{H}$ be partitioned as in the theorem and assume that $\boldsymbol{H}_{22} > \boldsymbol{O}$.

Using Theorem 2.19 page 34 with $\boldsymbol{D} = \left(\boldsymbol{I}_{p_1}, -\boldsymbol{H}_{12}\boldsymbol{H}_{22}^{-1}\right) \in \mathbb{C}^{p_1 \times p}$ we get

$$\mathcal{L}\left(\boldsymbol{X}_1 - \boldsymbol{X}_2\boldsymbol{H}_{22}^{-1}\boldsymbol{H}_{21}\right) = \mathbb{C}\mathcal{N}_{n \times p_1}\left(\boldsymbol{\Theta}_1 - \boldsymbol{\Theta}_2\boldsymbol{H}_{22}^{-1}\boldsymbol{H}_{21}, \boldsymbol{I}_n \otimes \left(\boldsymbol{H}_{11} - \boldsymbol{H}_{12}\boldsymbol{H}_{22}^{-1}\boldsymbol{H}_{21}\right)\right).$$

Writing $\boldsymbol{X}_2 = \boldsymbol{X}\boldsymbol{C}^*$, where $\boldsymbol{C} = (\boldsymbol{O}, \boldsymbol{I}_{p_2}) \in \mathbb{C}^{p_2 \times p}$, we see that $\boldsymbol{I}_n \otimes \boldsymbol{C}\boldsymbol{H}\boldsymbol{D}^* = \boldsymbol{O}$. Then from Theorem 2.21 page 35 we conclude that $\boldsymbol{X}_2$ and $\boldsymbol{X}_1 - \boldsymbol{X}_2\boldsymbol{H}_{22}^{-1}\boldsymbol{H}_{21}$ are independent. ■

Finally the conditional distribution is stated in the theorem below.

Theorem 2.25 The conditional distribution
Let $\boldsymbol{X}$ be an $n \times p$ complex random matrix with $\mathcal{L}(\boldsymbol{X}) = \mathbb{C}\mathcal{N}_{n \times p}(\boldsymbol{\Theta}, \boldsymbol{I}_n \otimes \boldsymbol{H})$, where $\boldsymbol{\Theta} \in \mathbb{C}^{n \times p}$ and $\boldsymbol{H} \in \mathbb{C}_S^{p \times p}$. Furthermore let $\boldsymbol{X}, \boldsymbol{\Theta}$ and $\boldsymbol{H}$ be partitioned as

$$\boldsymbol{X} = (\boldsymbol{X}_1, \boldsymbol{X}_2), \ \boldsymbol{\Theta} = (\boldsymbol{\Theta}_1, \boldsymbol{\Theta}_2) \text{ and } \boldsymbol{H} = \begin{pmatrix} \boldsymbol{H}_{11} & \boldsymbol{H}_{12} \\ \boldsymbol{H}_{21} & \boldsymbol{H}_{22} \end{pmatrix},$$

where $\boldsymbol{X}_j$ and $\boldsymbol{\Theta}_j$ are $n \times p_j$ and $\boldsymbol{H}_{jk}$ is $p_j \times p_k$ for $j = 1, 2$ and $p = p_1 + p_2$. If $\boldsymbol{H}_{22} > \boldsymbol{O}$, then the conditional distribution of $\boldsymbol{X}_1$ given $\boldsymbol{X}_2$ is

$$\mathcal{L}(\boldsymbol{X}_1 | \boldsymbol{X}_2) = \mathbb{C}\mathcal{N}_{n \times p_1}\left(\boldsymbol{\Theta}_1 + (\boldsymbol{X}_2 - \boldsymbol{\Theta}_2)\boldsymbol{H}_{22}^{-1}\boldsymbol{H}_{21}, \boldsymbol{I}_n \otimes \left(\boldsymbol{H}_{11} - \boldsymbol{H}_{12}\boldsymbol{H}_{22}^{-1}\boldsymbol{H}_{21}\right)\right).$$

Proof:
Let $\boldsymbol{X}$ be an $n \times p$ complex random matrix with $\mathcal{L}(\boldsymbol{X}) = \mathbb{C}\mathcal{N}_{n \times p}(\boldsymbol{\Theta}, \boldsymbol{I}_n \otimes \boldsymbol{H})$, where $\boldsymbol{\Theta} \in \mathbb{C}^{n \times p}$ and $\boldsymbol{H} \in \mathbb{C}_S^{p \times p}$. Furthermore let $\boldsymbol{X}$, $\boldsymbol{\Theta}$ and $\boldsymbol{H}$ be partitioned as in the theorem and assume that $\boldsymbol{H}_{22} > \boldsymbol{O}$.

From Theorem 2.24 page 36 we have that $\boldsymbol{X}_1 - \boldsymbol{X}_2\boldsymbol{H}_{22}^{-1}\boldsymbol{H}_{21}$ and $\boldsymbol{X}_2$ are independent and that

$$\mathcal{L}\left(\boldsymbol{X}_1 - \boldsymbol{X}_2\boldsymbol{H}_{22}^{-1}\boldsymbol{H}_{21}\right) = \mathbb{C}\mathcal{N}_{n \times p_1}\left(\boldsymbol{\Theta}_1 - \boldsymbol{\Theta}_2\boldsymbol{H}_{22}^{-1}\boldsymbol{H}_{21}, \boldsymbol{I}_n \otimes \left(\boldsymbol{H}_{11} - \boldsymbol{H}_{12}\boldsymbol{H}_{22}^{-1}\boldsymbol{H}_{21}\right)\right).$$

Thus

$$\begin{aligned} \mathcal{L}\left(\boldsymbol{X}_1 - \boldsymbol{X}_2\boldsymbol{H}_{22}^{-1}\boldsymbol{H}_{21} \middle| \boldsymbol{X}_2\right) &= \mathcal{L}\left(\boldsymbol{X}_1 - \boldsymbol{X}_2\boldsymbol{H}_{22}^{-1}\boldsymbol{H}_{21}\right) \\ &= \mathbb{C}\mathcal{N}_{n \times p_1}\left(\boldsymbol{\Theta}_1 - \boldsymbol{\Theta}_2\boldsymbol{H}_{22}^{-1}\boldsymbol{H}_{21}, \boldsymbol{I}_n \otimes \left(\boldsymbol{H}_{11} - \boldsymbol{H}_{12}\boldsymbol{H}_{22}^{-1}\boldsymbol{H}_{21}\right)\right). \end{aligned}$$

Using Theorem 2.19 page 34 with $\boldsymbol{E} = \boldsymbol{X}_2\boldsymbol{H}_{22}^{-1}\boldsymbol{H}_{21}$ we get

$$\mathcal{L}(\boldsymbol{X}_1 | \boldsymbol{X}_2) = \mathbb{C}\mathcal{N}_{n \times p_1}\left(\boldsymbol{\Theta}_1 + (\boldsymbol{X}_2 - \boldsymbol{\Theta}_2)\boldsymbol{H}_{22}^{-1}\boldsymbol{H}_{21}, \boldsymbol{I}_n \otimes \left(\boldsymbol{H}_{11} - \boldsymbol{H}_{12}\boldsymbol{H}_{22}^{-1}\boldsymbol{H}_{21}\right)\right).$$

■

3

The Complex Wishart Distribution and the Complex *U*-distribution

This chapter contains results on the complex Wishart distribution and the complex *U*-distribution. Goodman (1963) is the first to consider the complex Wishart distribution. Khatri (1965*a*), Khatri (1965*b*) and Giri (1965) make use of Goodmans results in further statistical analysis. The complex *U*-distribution is used in Gupta (1971). In this book the distributions are used in multivariate linear complex normal models in connection with distributional results for the maximum likelihood estimators and the likelihood ratio test statistics, and they are also useful in complex normal graphical models. This presentation of the complex Wishart distribution and the complex *U*-distribution is highly based on matrix algebra and the results are known from the literature or the real case. First the attention is focused on the complex Wishart distribution. We consider definition of the distribution and the mean of a complex Wishart distributed random matrix is determined. The correspondence between the complex Wishart distribution and the chi-square distribution is discussed and then we state under certain assumptions that a complex random quadratic form involving a projection matrix is complex Wishart distributed. Under the restriction of equal variance matrices we find the distribution of the sum of two independent complex Wishart distributed random matrices. We state the very useful result that a complex Wishart distributed random matrix is positive definite with probability one, if the variance matrix is positive definite and the degrees of freedom is greater than or equal to the dimension of the considered matrix. Next we turn to consideration of partitioning a complex Wishart distributed random matrix for which we give the distributions of various complex random matrices associated with this partition. Further the independence of these matrices is established. We also determine the density function of a $p \times p$ complex Wishart distributed random matrix w.r.t. Lebesgue measure on $\mathbb{C}_{+}^{p \times p}$. Finally we show that the distribution of the determinant of a complex Wishart distributed random matrix is proportional to the distribution of a product of mutually independent chi-square distributed random variables. Thereafter we investigate the complex *U*-distribution. To begin with we define the distribution and show that it does not depend on the equal variance matrix of the complex Wishart distributed random matrices involved. We consider some complex *U*-distributed random variables containing projection matrices and we examine independence of such complex random variables. Further it is established that a complex *U*-distribution is equal to the distribution of a product of mutually independent beta distributed random variables and that a complex *U*-distributed random variable is independent of the sum of the complex random matrices which is included in the denominator of the considered random variable. Finally by means of this result we consider the correspondence between the complex *U*-distribution and the beta distribution in a special case.

3.1 The Complex Wishart Distribution

We begin this section by defining the complex Wishart distribution in the same way as it is defined in the real case.

Definition 3.1 The complex Wishart distribution
Let $\boldsymbol{X} = (\boldsymbol{X}_1, \boldsymbol{X}_2, \ldots, \boldsymbol{X}_n)^*$ *be an* $n \times p$ *complex random matrix with* $\mathcal{L}(\boldsymbol{X}) = \mathbb{C}\mathcal{N}_{n\times p}(\boldsymbol{O}, \boldsymbol{I}_n \otimes \boldsymbol{H})$, *where* $\boldsymbol{H} \in \mathbb{C}_S^{p\times p}$. *The distribution of the* $p \times p$ *complex random matrix* $\boldsymbol{W}$ *given by* $\boldsymbol{W} = \boldsymbol{X}^*\boldsymbol{X}$ *is called a complex Wishart distribution with parameters* $\boldsymbol{H}, p$ *and* n. *This is denoted by* $\mathcal{L}(\boldsymbol{W}) = \mathbb{C}\mathcal{W}_p(\boldsymbol{H}, n)$.

The integers p and n are called the dimension and the degrees of freedom, respectively. The mean of a complex Wishart distribution is given in the theorem below.

Theorem 3.1 The mean of the complex Wishart distribution
Let $\boldsymbol{W}$ *be a* $p \times p$ *complex random matrix with* $\mathcal{L}(\boldsymbol{W}) = \mathbb{C}\mathcal{W}_p(\boldsymbol{H}, n)$, *where* $\boldsymbol{H} \in \mathbb{C}_S^{p\times p}$. *It holds that*

$$\mathbb{E}(\boldsymbol{W}) = n\boldsymbol{H} .$$

Proof:
Let $\boldsymbol{W}$ be a $p \times p$ complex random matrix with $\mathcal{L}(\boldsymbol{W}) = \mathbb{C}\mathcal{W}_p(\boldsymbol{H}, n)$, where $\boldsymbol{H} \in \mathbb{C}_S^{p\times p}$. Hence the distribution of $\boldsymbol{W}$ is also the distribution of $\boldsymbol{X}^*\boldsymbol{X}$, where $\boldsymbol{X}$ is an $n \times p$ complex random matrix with $\mathcal{L}(\boldsymbol{X}) = \mathbb{C}\mathcal{N}_{n\times p}(\boldsymbol{O}, \boldsymbol{I}_n \otimes \boldsymbol{H})$. Writing $\boldsymbol{X}$ as $\boldsymbol{X} = (\boldsymbol{X}_1, \boldsymbol{X}_2, \ldots, \boldsymbol{X}_n)^*$, where $\mathcal{L}(\boldsymbol{X}_j) = \mathbb{C}\mathcal{N}_p(\boldsymbol{0}, \boldsymbol{H})$ we get

$$\mathbb{E}(\boldsymbol{W}) = \mathbb{E}(\boldsymbol{X}^*\boldsymbol{X}) = \sum_{j=1}^{n} \mathbb{E}\left(\boldsymbol{X}_j \boldsymbol{X}_j^*\right) = \sum_{j=1}^{n} \mathbb{V}(\boldsymbol{X}_j) = n\boldsymbol{H} .$$

■

In the case where the dimension equals one the complex Wishart distribution becomes a chi-square distribution. This relation is stated below.

Theorem 3.2 The relation to the chi-square distribution
Let W *be a complex random variable with* $\mathcal{L}(W) = \mathbb{C}\mathcal{W}_1(\sigma^2, n)$, *where* $\sigma^2 \in \mathbb{R}_+$. *The distribution of* W *is also given as*

$$\mathcal{L}(W) = \frac{\sigma^2}{2}\chi^2_{2n} .$$

Proof:
Let W be a complex random variable with $\mathcal{L}(W) = \mathbb{C}\mathcal{W}_1(\sigma^2, n)$, where $\sigma^2 \in \mathbb{R}_+$. The

distribution of W is also the distribution of $\boldsymbol{X}^*\boldsymbol{X}$, where $\boldsymbol{X} = (X_j)$ is an $n \times 1$ complex random vector with the X_j's mutually independent identically distributed as $\mathcal{L}(X_j) = \mathbb{C}\mathcal{N}(0, \sigma^2)$ for $j = 1, 2, \ldots, n$. Writing X_j as $U_j + iV_j$ we have from Theorem 2.3 page 19 that

$$\mathcal{L}(U_j) = \mathcal{L}(V_j) = \mathcal{N}\left(0, \frac{\sigma^2}{2}\right)$$

and U_j and V_j are independent. Hereby

$$\mathcal{L}\left(U_j^2 + V_j^2\right) = \frac{\sigma^2}{2}\chi_2^2 .$$

Now since $\sum_{j=1}^n X_j\overline{X}_j = \sum_{j=1}^n \left(U_j^2 + V_j^2\right)$ and the X_j's are mutually independent we conclude that

$$\mathcal{L}(W) = \frac{\sigma^2}{2}\chi_{2n}^2 .$$

■

The following theorem considers a complex random quadratic form involving a projection matrix. We find under certain assumptions that the distribution of this particular complex random quadratic form is a complex Wishart distribution.

Theorem 3.3
Let $\boldsymbol{X}$ be an $n \times p$ complex random matrix with $\mathcal{L}(\boldsymbol{X}) = \mathbb{C}\mathcal{N}_{n\times p}(\boldsymbol{\Theta}, \boldsymbol{I}_n \otimes \boldsymbol{H})$, where $\boldsymbol{\Theta} \in \mathbb{C}^{n\times p}$ and $\boldsymbol{H} \in \mathbb{C}_S^{p\times p}$. Further let $\boldsymbol{P}$ be an $n \times n$ complex matrix representing the orthogonal projection of $\mathbb{C}^n$ onto a k-dimensional subspace. If $\boldsymbol{P\Theta} = \boldsymbol{O}$, then

$$\mathcal{L}(\boldsymbol{X}^*\boldsymbol{P}\boldsymbol{X}) = \mathbb{C}\mathcal{W}_p(\boldsymbol{H}, k) .$$

Proof:
Let $\boldsymbol{X}$ be an $n \times p$ complex random matrix with $\mathcal{L}(\boldsymbol{X}) = \mathbb{C}\mathcal{N}_{n\times p}(\boldsymbol{\Theta}, \boldsymbol{I}_n \otimes \boldsymbol{H})$, where $\boldsymbol{\Theta} \in \mathbb{C}^{n\times p}$ and $\boldsymbol{H} \in \mathbb{C}_S^{p\times p}$. Further let $\boldsymbol{P}$ be an $n \times n$ complex matrix representing the orthogonal projection of $\mathbb{C}^n$ onto a k-dimensional subspace and assume that $\boldsymbol{P\Theta} = \boldsymbol{O}$.

Since $\boldsymbol{P}$ is Hermitian and idempotent there exists a unitary $n \times n$ matrix $\boldsymbol{U}$ such that

$$\boldsymbol{P} = \boldsymbol{U}\begin{pmatrix} \boldsymbol{I}_k & \boldsymbol{O} \\ \boldsymbol{O} & \boldsymbol{O} \end{pmatrix}\boldsymbol{U}^* . \tag{3.1}$$

Hereby it is seen that

$$\begin{aligned} \boldsymbol{X}^*\boldsymbol{P}\boldsymbol{X} &= \boldsymbol{X}^*\boldsymbol{U}\begin{pmatrix} \boldsymbol{I}_k & \boldsymbol{O} \\ \boldsymbol{O} & \boldsymbol{O} \end{pmatrix}\boldsymbol{U}^*\boldsymbol{X} \\ &= \boldsymbol{Y}^*\boldsymbol{Y} , \end{aligned}$$

where $\boldsymbol{Y} = (\boldsymbol{I}_k, \boldsymbol{O})\,\boldsymbol{U}^*\boldsymbol{X}$.

From Theorem 2.19 page 34 we find the distribution of $\boldsymbol{Y}$ as

$$\mathcal{L}(\boldsymbol{Y}) = \mathbb{C}\mathcal{N}_{k\times p}((\boldsymbol{I}_k, \boldsymbol{O})\,\boldsymbol{U}^*\boldsymbol{\Theta}, (\boldsymbol{I}_k, \boldsymbol{O})\,\boldsymbol{U}^*\boldsymbol{U}\,(\boldsymbol{I}_k, \boldsymbol{O})^* \otimes \boldsymbol{H}) \ .$$

Using that $\boldsymbol{P}\boldsymbol{\Theta} = \boldsymbol{O}$ and (3.1) page 41 we see

$$\begin{aligned}(\boldsymbol{I}_k, \boldsymbol{O})\,\boldsymbol{U}^*\boldsymbol{\Theta} &= (\boldsymbol{I}_k, \boldsymbol{O})\,\boldsymbol{U}^*\,(\boldsymbol{I}_n - \boldsymbol{P})\,\boldsymbol{\Theta} \\ &= (\boldsymbol{I}_k, \boldsymbol{O}) \begin{pmatrix} \boldsymbol{O} & \boldsymbol{O} \\ \boldsymbol{O} & \boldsymbol{I}_{n-k} \end{pmatrix} \boldsymbol{U}^*\boldsymbol{\Theta} \\ &= \boldsymbol{O}\,,\end{aligned}$$

whereby we obtain

$$\mathcal{L}(\boldsymbol{Y}) = \mathbb{C}\mathcal{N}_{k\times p}(\boldsymbol{O}, \boldsymbol{I}_k \otimes \boldsymbol{H}) \ .$$

Then we derive from Definition 3.1 page 40 that $\mathcal{L}(\boldsymbol{X}^*\boldsymbol{P}\boldsymbol{X}) = \mathbb{C}\mathcal{W}_p(\boldsymbol{H}, k)$. ■

In the next theorem we consider mutual independence of complex random matrices involving two projection matrices. By means of these we are able to derive mutual independence of some complex Wishart distributed random quadratic forms. These are used later when we establish the distribution of the likelihood ratio test statistics from the hypothesis test in the multivariate linear complex normal model.

Theorem 3.4
Let $\boldsymbol{X}$ be an $n \times p$ complex random matrix with $\mathcal{L}(\boldsymbol{X}) = \mathbb{C}\mathcal{N}_{n\times p}(\boldsymbol{\Theta}, \boldsymbol{I}_n \otimes \boldsymbol{H})$, where $\boldsymbol{\Theta} \in \mathbb{C}^{n\times p}$ and $\boldsymbol{H} \in \mathbb{C}_S^{p\times p}$. Further let $\boldsymbol{P}$ be an $n \times n$ complex matrix representing the orthogonal projection of $\mathbb{C}^n$ onto a k-dimensional subspace and let $\boldsymbol{P}_0$ be an $n \times n$ complex matrix representing the orthogonal projection of $\mathbb{C}^n$ onto a k_0-dimensional subspace of the k-dimensional subspace. It holds that

$$\boldsymbol{P}_0\boldsymbol{X}, (\boldsymbol{I}_n - \boldsymbol{P})\,\boldsymbol{X} \text{ and } (\boldsymbol{P} - \boldsymbol{P}_0)\,\boldsymbol{X}$$

are mutually independent.

Proof:
Let the assumptions in the theorem be satisfied. It holds that

$$\boldsymbol{P}_0\boldsymbol{I}_n\,(\boldsymbol{I}_n - \boldsymbol{P}_0)^* \otimes \boldsymbol{H} = \boldsymbol{O}\,,$$

thus from Theorem 2.20 page 34 we get that

$$\boldsymbol{P}_0\boldsymbol{X} \perp\!\!\!\perp (\boldsymbol{I}_n - \boldsymbol{P}_0)\,\boldsymbol{X}\,.$$

We know that there is a one-to-one correspondence between $(\boldsymbol{I}_n - \boldsymbol{P}_0)$ and $((\boldsymbol{I}_n - \boldsymbol{P}), (\boldsymbol{P} - \boldsymbol{P}_0))$, i.e.

$$\boldsymbol{P}_0\boldsymbol{X} \perp\!\!\!\perp ((\boldsymbol{I}_n - \boldsymbol{P})\,\boldsymbol{X}, (\boldsymbol{P} - \boldsymbol{P}_0)\,\boldsymbol{X})\ .$$

Since $(\boldsymbol{I}_n - \boldsymbol{P})(\boldsymbol{P} - \boldsymbol{P}_0)^* = \boldsymbol{O}$ we have by Theorem 2.20 that

$$(\boldsymbol{I}_n - \boldsymbol{P})\boldsymbol{X} \perp\!\!\!\perp (\boldsymbol{P} - \boldsymbol{P}_0)\boldsymbol{X} .$$

Altogether we have the mutual independence. ∎

Using Theorem 3.4 we get the corollary below.

Corollary 3.1
Let $\boldsymbol{X}$ be an $n \times p$ complex random matrix with $\mathcal{L}(\boldsymbol{X}) = \mathbb{C}\mathcal{N}_{n\times p}(\Theta, \boldsymbol{I}_n \otimes \boldsymbol{H})$, where $\Theta \in \mathbb{C}^{n\times p}$ and $\boldsymbol{H} \in \mathbb{C}_S^{p\times p}$. Further let $\boldsymbol{P}$ be an $n \times n$ complex matrix representing the orthogonal projection of $\mathbb{C}^n$ onto a k-dimensional subspace. It holds that

$$\boldsymbol{P}\boldsymbol{X} \perp\!\!\!\perp (\boldsymbol{I}_n - \boldsymbol{P})\boldsymbol{X} .$$

In the next theorem we see that the sum of two independent complex Wishart distributed random matrices with equal variance matrix is a complex Wishart distributed matrix with the same variance matrix.

Theorem 3.5 Sum of independent complex Wishart distributed matrices
Let $\boldsymbol{W}_1$ and $\boldsymbol{W}_2$ be independent $p \times p$ complex random matrices with $\mathcal{L}(\boldsymbol{W}_j) = \mathbb{C}\mathcal{W}_p(\boldsymbol{H}, n_j)$, where $\boldsymbol{H} \in \mathbb{C}_S^{p\times p}$, for $j = 1, 2$. It holds that

$$\mathcal{L}(\boldsymbol{W}_1 + \boldsymbol{W}_2) = \mathbb{C}\mathcal{W}_p(\boldsymbol{H}, n_1 + n_2) .$$

Proof:
Let $\boldsymbol{W}_1$ and $\boldsymbol{W}_2$ be independent $p \times p$ complex random matrices with $\mathcal{L}(\boldsymbol{W}_j) = \mathbb{C}\mathcal{W}_p(\boldsymbol{H}, n_j)$, where $\boldsymbol{H} \in \mathbb{C}_S^{p\times p}$, for $j = 1, 2$. The distribution of $\boldsymbol{W}_j$ is also the distribution of $\boldsymbol{X}_j^*\boldsymbol{X}_j$, where $\boldsymbol{X}_j$ is a complex random matrix of dimension $n_j \times p$ with $\mathcal{L}(\boldsymbol{X}_j) = \mathbb{C}\mathcal{N}_{n_j\times p}\big(\boldsymbol{O}, \boldsymbol{I}_{n_j} \otimes \boldsymbol{H}\big)$, $j = 1, 2$. Further $\boldsymbol{X}_1$ and $\boldsymbol{X}_2$ are independent.

Let $\boldsymbol{X}$ be the $(n_1 + n_2) \times p$ complex random matrix given by $\boldsymbol{X} = (\boldsymbol{X}_1^*, \boldsymbol{X}_2^*)^*$, then it holds that $\mathcal{L}(\boldsymbol{X}) = \mathbb{C}\mathcal{N}_{(n_1+n_2)\times p}(\boldsymbol{O}, \boldsymbol{I}_{n_1+n_2} \otimes \boldsymbol{H})$. Finally

$$\begin{aligned}\mathcal{L}(\boldsymbol{W}_1 + \boldsymbol{W}_2) &= \mathcal{L}(\boldsymbol{X}_1^*\boldsymbol{X}_1 + \boldsymbol{X}_2^*\boldsymbol{X}_2) \\ &= \mathcal{L}(\boldsymbol{X}^*\boldsymbol{X}) \\ &= \mathbb{C}\mathcal{W}_p(\boldsymbol{H}, n_1 + n_2) .\end{aligned}$$

∎

The next theorem contains some important properties of a $p \times p$ complex Wishart distributed random matrix. We determine the necessary conditions for a complex Wishart distributed random matrix to be positive definite with probability one. Further the distributions of some random matrices formed by a partition of a complex Wishart distributed random matrix are studied. Also independence of the matrices obtained is taken into consideration.

Theorem 3.6
Let W be a $p \times p$ complex random matrix with $\mathcal{L}(W) = \mathbb{C}\mathcal{W}_p(H, n)$, where $H \in \mathbb{C}_S^{p\times p}$. Let W and H be partitioned as

$$W = \begin{pmatrix} W_{11} & W_{12} \\ W_{21} & W_{22} \end{pmatrix} \text{ and } H = \begin{pmatrix} H_{11} & H_{12} \\ H_{21} & H_{22} \end{pmatrix},$$

where W_{jk} and H_{jk} are $p_j \times p_k$ for $j, k = 1, 2$ and $p = p_1 + p_2$.

1. *If $n \geq p$ and $H > O$, then $W > O$ with probability one.*
2. *If $n \geq p_2$ and $H_{22} > O$, then*

$$\mathcal{L}\left(W_{11} - W_{12}W_{22}^{-1}W_{21}\right) = \mathbb{C}\mathcal{W}_{p_1}\left(H_{11} - H_{12}H_{22}^{-1}H_{21}, n - p_2\right)$$

and

$$W_{11} - W_{12}W_{22}^{-1}W_{21} \perp\!\!\!\perp (W_{12}, W_{22}) \, .$$

3. *If in addition to part 2 $H_{12} = O$, then*

$$\begin{aligned} \mathcal{L}\left(W_{11} - W_{12}W_{22}^{-1}W_{21}\right) &= \mathbb{C}\mathcal{W}_{p_1}(H_{11}, n - p_2) \, , \\ \mathcal{L}\left(W_{12}W_{22}^{-1}W_{21}\right) &= \mathbb{C}\mathcal{W}_{p_1}(H_{11}, p_2) \end{aligned}$$

and

$$W_{11} - W_{12}W_{22}^{-1}W_{21}, \; W_{12}W_{22}^{-1}W_{21} \text{ and } W_{22}$$

are mutually independent.

Proof:
Let W be a complex random matrix with $\mathcal{L}(W) = \mathbb{C}\mathcal{W}_p(H, n)$, where $H \in \mathbb{C}_S^{p\times p}$. Further let W and H be partitioned as in the theorem.

The distribution of W is also the distribution of X^*X, where X is an $n \times p$ complex random matrix with $\mathcal{L}(X) = \mathbb{C}\mathcal{N}_{n\times p}(O, I_n \otimes H)$. Let X be partitioned according to the partition of W and H, i.e. $X = (X_1, X_2)$, where X_j is $n \times p_j$ for $j = 1, 2$ and $p = p_1 + p_2$. Then we have

$$X^*X = \begin{pmatrix} X_1^*X_1 & X_1^*X_2 \\ X_2^*X_1 & X_2^*X_2 \end{pmatrix}.$$

Re 1:
Assume $n \geq p$ and $H > O$. We show by induction on p that $X^*X > O$ with probability one, whereby we conclude that $W > O$ with probability one as $\mathcal{L}(W) = \mathcal{L}(X^*X)$.

Induction start:
For $p = 1$ we notice that $\mathcal{L}(\boldsymbol{X}^*\boldsymbol{X}) = \mathbb{C}\mathcal{W}_1(\boldsymbol{H}, n) = \frac{1}{2}\boldsymbol{H}\chi^2_{2n}$. Therefore, since $\boldsymbol{H} > 0$ and $n \geq 1$, we get that $\boldsymbol{X}^*\boldsymbol{X} > \boldsymbol{O}$ with probability one.

Induction step:
Assume that $\boldsymbol{X}^*\boldsymbol{X} > \boldsymbol{O}$ with probability one for $p = d - 1$. We seek to show that $\boldsymbol{X}^*\boldsymbol{X} > \boldsymbol{O}$ with probability one for $p = d$. Consider a partition as described for $p_1 = 1$ and $p_2 = p - 1$.

According to the induction assumption we see that $\boldsymbol{X}_2^*\boldsymbol{X}_2 > \boldsymbol{O}$ with probability one, thus $(\boldsymbol{X}_2^*\boldsymbol{X}_2)^{-1}$ exists with probability one. It holds that $\boldsymbol{X}^*\boldsymbol{X} > \boldsymbol{O}$ iff $\boldsymbol{X}_2^*\boldsymbol{X}_2 > \boldsymbol{O}$ and $\boldsymbol{X}_1^*\boldsymbol{X}_1 - \boldsymbol{X}_1^*\boldsymbol{X}_2(\boldsymbol{X}_2^*\boldsymbol{X}_2)^{-1}\boldsymbol{X}_2^*\boldsymbol{X}_1 > \boldsymbol{O}$. Therefore the only thing left to show is that $\boldsymbol{X}_1^*\boldsymbol{X}_1 - \boldsymbol{X}_1^*\boldsymbol{X}_2(\boldsymbol{X}_2^*\boldsymbol{X}_2)^{-1}\boldsymbol{X}_2^*\boldsymbol{X}_1 > \boldsymbol{O}$ with probability one.

Notice that

$$\boldsymbol{X}_1^*\boldsymbol{X}_1 - \boldsymbol{X}_1^*\boldsymbol{X}_2(\boldsymbol{X}_2^*\boldsymbol{X}_2)^{-1}\boldsymbol{X}_2^*\boldsymbol{X}_1 = \boldsymbol{X}_1^*(\boldsymbol{I}_n - \boldsymbol{P})\boldsymbol{X}_1,$$

where $\boldsymbol{P} = \boldsymbol{X}_2(\boldsymbol{X}_2^*\boldsymbol{X}_2)^{-1}\boldsymbol{X}_2^*$. The matrix $\boldsymbol{P}$ is Hermitian and idempotent with $\operatorname{tr}(\boldsymbol{P}) = p-1$. Therefore given $\boldsymbol{X}_2$ it represents the orthogonal projection of $\mathbb{C}^n$ onto a $(p-1)$-dimensional subspace.

From Theorem 2.25 page 37 we have, since $\boldsymbol{H}_{22} > \boldsymbol{O}$, that

$$\mathcal{L}(\boldsymbol{X}_1|\boldsymbol{X}_2) = \mathbb{C}\mathcal{N}_n\left(\boldsymbol{X}_2\boldsymbol{H}_{22}^{-1}\boldsymbol{H}_{21}, \boldsymbol{I}_n \otimes \left(\boldsymbol{H}_{11} - \boldsymbol{H}_{12}\boldsymbol{H}_{22}^{-1}\boldsymbol{H}_{21}\right)\right).$$

Since $(\boldsymbol{I}_n - \boldsymbol{P})\boldsymbol{X}_2\boldsymbol{H}_{22}^{-1}\boldsymbol{H}_{21} = \boldsymbol{O}$ we get according to Theorem 3.3 page 41 that

$$\begin{aligned}\mathcal{L}(\boldsymbol{X}_1^*(\boldsymbol{I}_n - \boldsymbol{P})\boldsymbol{X}_1|\boldsymbol{X}_2) &= \mathcal{L}\left(\boldsymbol{X}_1^*\boldsymbol{X}_1 - \boldsymbol{X}_1^*\boldsymbol{X}_2(\boldsymbol{X}_2^*\boldsymbol{X}_2)^{-1}\boldsymbol{X}_2^*\boldsymbol{X}_1\middle|\boldsymbol{X}_2\right)\\ &= \mathbb{C}\mathcal{W}_1\left(\boldsymbol{H}_{11} - \boldsymbol{H}_{12}\boldsymbol{H}_{22}^{-1}\boldsymbol{H}_{21}, n - (p-1)\right)\\ &= \frac{1}{2}\left(\boldsymbol{H}_{11} - \boldsymbol{H}_{12}\boldsymbol{H}_{22}^{-1}\boldsymbol{H}_{21}\right)\chi^2_{2(n-(p-1))}.\end{aligned}$$

This distribution does not depend on $\boldsymbol{X}_2$, thus

$$\mathcal{L}\left(\boldsymbol{X}_1^*\boldsymbol{X}_1 - \boldsymbol{X}_1^*\boldsymbol{X}_2(\boldsymbol{X}_2^*\boldsymbol{X}_2)^{-1}\boldsymbol{X}_2^*\boldsymbol{X}_1\right) = \frac{1}{2}\left(\boldsymbol{H}_{11} - \boldsymbol{H}_{12}\boldsymbol{H}_{22}^{-1}\boldsymbol{H}_{21}\right)\chi^2_{2(n-(p-1))}.$$

Since $\boldsymbol{H}_{11} - \boldsymbol{H}_{12}\boldsymbol{H}_{22}^{-1}\boldsymbol{H}_{21} > \boldsymbol{O}$ and $n \geq p$ we obtain from the distribution above that $\boldsymbol{X}_1^*\boldsymbol{X}_1 - \boldsymbol{X}_1^*\boldsymbol{X}_2(\boldsymbol{X}_2^*\boldsymbol{X}_2)^{-1}\boldsymbol{X}_2^*\boldsymbol{X}_1 > \boldsymbol{O}$ with probability one.

Re 2:
Assume that $n \geq p_2$ and $\boldsymbol{H}_{22} > \boldsymbol{O}$. Hence by part 1 we have that $\boldsymbol{X}_2^*\boldsymbol{X}_2 > \boldsymbol{O}$ with probability one and hereby $(\boldsymbol{X}_2^*\boldsymbol{X}_2)^{-1}$ exists with probability one.

By arguments similar as in the proof of part 1 we get

$$\mathcal{L}(\boldsymbol{X}_1|\boldsymbol{X}_2) = \mathbb{C}\mathcal{N}_{n\times p_1}\left(\boldsymbol{X}_2\boldsymbol{H}_{22}^{-1}\boldsymbol{H}_{21}, \boldsymbol{I}_n \otimes \left(\boldsymbol{H}_{11} - \boldsymbol{H}_{12}\boldsymbol{H}_{22}^{-1}\boldsymbol{H}_{21}\right)\right) \tag{3.2}$$

and

$$\mathcal{L}(\boldsymbol{X}_1^*(\boldsymbol{I}_n - \boldsymbol{P})\boldsymbol{X}_1 | \boldsymbol{X}_2) = \mathbb{C}\mathcal{W}_{p_1}\left(\boldsymbol{H}_{11} - \boldsymbol{H}_{12}\boldsymbol{H}_{22}^{-1}\boldsymbol{H}_{21}, n - p_2\right),$$

where $\boldsymbol{P}$ is defined as in part one.

Since this distribution does not depend on $\boldsymbol{X}_2$ we observe that

$$\mathcal{L}(\boldsymbol{X}_1^*(\boldsymbol{I}_n - \boldsymbol{P})\boldsymbol{X}_1) = \mathbb{C}\mathcal{W}_{p_1}\left(\boldsymbol{H}_{11} - \boldsymbol{H}_{12}\boldsymbol{H}_{22}^{-1}\boldsymbol{H}_{21}, n - p_2\right) \tag{3.3}$$

and

$$\boldsymbol{X}_1^*(\boldsymbol{I}_n - \boldsymbol{P})\boldsymbol{X}_1 \perp\!\!\!\perp \boldsymbol{X}_2 . \tag{3.4}$$

We get from Corollary 3.1 page 43 that

$$\boldsymbol{P}\boldsymbol{X}_1 \perp\!\!\!\perp (\boldsymbol{I}_n - \boldsymbol{P})\boldsymbol{X}_1 \mid \boldsymbol{X}_2 .$$

As it holds that

$$\boldsymbol{X}_1^*\boldsymbol{X}_2 = \boldsymbol{X}_1^*\boldsymbol{P}\boldsymbol{X}_2 = (\boldsymbol{P}\boldsymbol{X}_1)^*\boldsymbol{X}_2$$

we deduce from the fact that $\boldsymbol{P}$ is idempotent and Theorem 6.3 page 102 that

$$\boldsymbol{X}_1^*\boldsymbol{X}_2 \perp\!\!\!\perp \boldsymbol{X}_1^*(\boldsymbol{I}_n - \boldsymbol{P})\boldsymbol{X}_1 \mid \boldsymbol{X}_2 .$$

By further use of Theorem 6.3 together with (3.4) we deduce that

$$\boldsymbol{X}_1^*(\boldsymbol{I}_n - \boldsymbol{P})\boldsymbol{X}_1 \perp\!\!\!\perp (\boldsymbol{X}_1^*\boldsymbol{X}_2, \boldsymbol{X}_2^*\boldsymbol{X}_2) . \tag{3.5}$$

As it holds that $\mathcal{L}(\boldsymbol{W}) = \mathcal{L}(\boldsymbol{X}^*\boldsymbol{X})$ we finally conclude by (3.3) that

$$\mathcal{L}\left(\boldsymbol{W}_{11} - \boldsymbol{W}_{12}\boldsymbol{W}_{22}^{-1}\boldsymbol{W}_{21}\right) = \mathbb{C}\mathcal{W}_{p_1}\left(\boldsymbol{H}_{11} - \boldsymbol{H}_{12}\boldsymbol{H}_{22}^{-1}\boldsymbol{H}_{21}, n - p_2\right)$$

and from (3.5) that

$$\boldsymbol{W}_{11} - \boldsymbol{W}_{12}\boldsymbol{W}_{22}^{-1}\boldsymbol{W}_{21} \perp\!\!\!\perp (\boldsymbol{W}_{12}, \boldsymbol{W}_{22}) .$$

Re 3:
Assume that $n \geq p_2$, $\boldsymbol{H}_{22} > \boldsymbol{O}$ and $\boldsymbol{H}_{12} = \boldsymbol{O}$, then it follows immediately from part 2 that

$$\mathcal{L}\left(\boldsymbol{W}_{11} - \boldsymbol{W}_{12}\boldsymbol{W}_{22}^{-1}\boldsymbol{W}_{21}\right) = \mathbb{C}\mathcal{W}_{p_1}(\boldsymbol{H}_{11}, n - p_2) .$$

Because $\boldsymbol{H}_{12} = \boldsymbol{O}$ we obtain from (3.2) page 45 that

$$\mathcal{L}(\boldsymbol{X}_1 | \boldsymbol{X}_2) = \mathbb{C}\mathcal{N}_{n \times p_1}(\boldsymbol{O}, \boldsymbol{I}_n \otimes \boldsymbol{H}_{11}) .$$

By Theorem 3.3 page 41 we get

$$\mathcal{L}(\boldsymbol{X}_1^*\boldsymbol{P}\boldsymbol{X}_1 | \boldsymbol{X}_2) = \mathbb{C}\mathcal{W}_{p_1}(\boldsymbol{H}_{11}, p_2) .$$

Since this distribution does not depend of $\boldsymbol{X}_2$ we conclude that

$$\mathcal{L}(\boldsymbol{X}_1^*\boldsymbol{P}\boldsymbol{X}_1) = \mathbb{C}\mathcal{W}_{p_1}(\boldsymbol{H}_{11}, p_2)$$

and

$$\boldsymbol{X}_1^*\boldsymbol{P}\boldsymbol{X}_1 \perp\!\!\!\perp \boldsymbol{X}_2$$

implying that

$$\boldsymbol{X}_1^*\boldsymbol{P}\boldsymbol{X}_1 \perp\!\!\!\perp \boldsymbol{X}_2^*\boldsymbol{X}_2 .$$

As $\mathcal{L}(\boldsymbol{W}) = \mathcal{L}(\boldsymbol{X}^*\boldsymbol{X})$ we conclude that

$$\mathcal{L}\left(\boldsymbol{W}_{12}\boldsymbol{W}_{22}^{-1}\boldsymbol{W}_{21}\right) = \mathbb{C}\mathcal{W}_{p_1}(\boldsymbol{H}_{11}, p_2)$$

and

$$\boldsymbol{W}_{12}\boldsymbol{W}_{22}^{-1}\boldsymbol{W}_{21} \perp\!\!\!\perp \boldsymbol{W}_{22} .$$

Recall from part 2 that

$$\boldsymbol{W}_{11} - \boldsymbol{W}_{12}\boldsymbol{W}_{22}^{-1}\boldsymbol{W}_{21} \perp\!\!\!\perp (\boldsymbol{W}_{12}, \boldsymbol{W}_{22}) .$$

Hence we deduce that

$$\boldsymbol{W}_{11} - \boldsymbol{W}_{12}\boldsymbol{W}_{22}^{-1}\boldsymbol{W}_{21} \perp\!\!\!\perp \left(\boldsymbol{W}_{12}\boldsymbol{W}_{22}^{-1}\boldsymbol{W}_{21}, \boldsymbol{W}_{22}\right) ,$$

which completes the proof. ■

Using Theorem 3.6 we are able to deduce the density function of a $p \times p$ complex Wishart distributed random matrix w.r.t. Lebesgue measure on $\mathbb{C}_+^{p\times p}$. This density function is introduced in the following theorem, and we have chosen not to utilize this knowledge in the development of the remaining results.

Theorem 3.7 The density function of $\mathbb{C}\mathcal{W}_p(\boldsymbol{H}, n)$
Let $\boldsymbol{W}$ be a $p \times p$ complex random matrix with $\mathcal{L}(\boldsymbol{W}) = \mathbb{C}\mathcal{W}_p(\boldsymbol{H}, n)$, where $\boldsymbol{H} \in \mathbb{C}_+^{p\times p}$. If $n \geq p$, then the density function of $\boldsymbol{W}$ w.r.t. Lebesgue measure on $\mathbb{C}_+^{p\times p}$ is given as

$$f_{\boldsymbol{W}}(\boldsymbol{w}) = \frac{\det(\boldsymbol{w})^{n-p} \exp\left(-\operatorname{tr}\left(\boldsymbol{w}\boldsymbol{H}^{-1}\right)\right)}{\det(\boldsymbol{H})^n \pi^{\frac{p(p-1)}{2}} \prod_{j=1}^{p} \Gamma(n+1-j)} , \quad \boldsymbol{w} \in \mathbb{C}_+^{p\times p} . \tag{3.6}$$

Proof:
Let $\boldsymbol{W}$ be a $p \times p$ complex random matrix with $\mathcal{L}(\boldsymbol{W}) = \mathbb{C}\mathcal{W}_p(\boldsymbol{H}, n)$, where $\boldsymbol{H} \in \mathbb{C}_+^{p\times p}$, and assume that $n \geq p$.

We notice from Theorem 3.6 page 44 that $\boldsymbol{W} > \boldsymbol{O}$ with probability one, since $n \geq p$ and $\boldsymbol{H} > \boldsymbol{O}$. By induction on p we are able to find the density function of $\boldsymbol{W}$ w.r.t. Lebesgue measure on $\mathbb{C}_+^{p\times p}$.

Induction start:
For $p = 1$ we notice that

$$\mathcal{L}(\boldsymbol{W}) = \frac{1}{2}\boldsymbol{H}\chi^2_{2n} .$$

where $\boldsymbol{H} \in \mathbb{R}_+$. It holds that

$$\mathcal{L}(\boldsymbol{W}) = \mathcal{L}\left(\frac{1}{2}\boldsymbol{H}\boldsymbol{Y}\right) ,$$

where $\boldsymbol{Y}$ is a random variable with $\mathcal{L}(\boldsymbol{Y}) = \chi^2_{2n}$. Hereby the density function of $\boldsymbol{W}$ w.r.t. Lebesgue measure on $\mathbb{R}_+$ is

$$f_{\boldsymbol{W}}(\boldsymbol{w}) = \frac{1}{|\boldsymbol{J}|} f_{\boldsymbol{Y}}(\boldsymbol{y}) , \tag{3.7}$$

where $|\cdot|$ denotes the absolute value and $\boldsymbol{J}$ denotes the Jacobian given by

$$\boldsymbol{J} = \frac{1}{2}\boldsymbol{H} .$$

The density function of $\boldsymbol{Y}$ w.r.t. Lebesgue measure on $\mathbb{R}_+$ is known as

$$f_{\boldsymbol{Y}}(\boldsymbol{y}) = \frac{\boldsymbol{y}^{n-1}\exp\left(-\frac{1}{2}\boldsymbol{y}\right)}{2^n\Gamma(n)} , \boldsymbol{y} \in \mathbb{R}_+ .$$

Using this result in (3.7) we get

$$\begin{aligned} f_{\boldsymbol{W}}(\boldsymbol{w}) &= 2\boldsymbol{H}^{-1}\frac{\boldsymbol{y}^{n-1}\exp\left(-\frac{1}{2}\boldsymbol{y}\right)}{2^n\Gamma(n)} \\ &= 2\boldsymbol{H}^{-1}\frac{\left(2\boldsymbol{H}^{-1}\boldsymbol{w}\right)^{n-1}\exp\left(-\frac{1}{2}\left(2\boldsymbol{H}^{-1}\boldsymbol{w}\right)\right)}{2^n\Gamma(n)} \\ &= \frac{\boldsymbol{w}^{n-1}\exp\left(-\boldsymbol{w}\boldsymbol{H}^{-1}\right)}{\boldsymbol{H}^n\Gamma(n)} , \boldsymbol{w} \in \mathbb{R}_+ . \end{aligned}$$

This tells us that (3.6) page 47 is fulfilled for $p = 1$.

Induction step:
Assume that (3.6) is fulfilled for $p \leq d - 1$. We seek to show that (3.6) also holds for $p = d$. Let $\boldsymbol{W}$ and $\boldsymbol{H}$ be partitioned as in Theorem 3.6.

The distribution of $\boldsymbol{W}$ is also the distribution of $\boldsymbol{X}^*\boldsymbol{X}$, where $\boldsymbol{X}$ is an $n \times p$ complex random matrix with $\mathcal{L}(\boldsymbol{X}) = \mathbb{C}\mathcal{N}_{n\times p}(\boldsymbol{O}, \boldsymbol{I}_n \otimes \boldsymbol{H})$. Let $\boldsymbol{X}$ be partitioned according to the partition of $\boldsymbol{W}$ and $\boldsymbol{H}$, i.e. $\boldsymbol{X} = (\boldsymbol{X}_1, \boldsymbol{X}_2)$, where $\boldsymbol{X}_j$ is $n \times p_j$ for $j = 1, 2$ and $p = p_1 + p_2$. Then we have

$$\boldsymbol{X}^*\boldsymbol{X} = \begin{pmatrix} \boldsymbol{X}_1^*\boldsymbol{X}_1 & \boldsymbol{X}_1^*\boldsymbol{X}_2 \\ \boldsymbol{X}_2^*\boldsymbol{X}_1 & \boldsymbol{X}_2^*\boldsymbol{X}_2 \end{pmatrix} .$$

Since W is Hermitian, it is uniquely determined by (W_{11}, W_{21}, W_{22}). Hereby it follows that

$$f_W(w) = f_{W_{11},W_{21},W_{22}}(w_{11}, w_{21}, w_{22}) .$$

We introduce the notation

$$\begin{aligned} \widetilde{W}_{11} &= W_{11} - W_{12}W_{22}^{-1}W_{21} \\ \widetilde{H}_{11} &= H_{11} - H_{12}H_{22}^{-1}H_{21} . \end{aligned}$$

Since $n \geq p_2$ and $H_{22} > O$ we have from Theorem 3.6 that

(3.8) $$\widetilde{W}_{11} \perp\!\!\!\perp (W_{12}, W_{22})$$

and further

(3.9) $$\mathcal{L}\left(\widetilde{W}_{11}\right) = \mathbb{C}\mathcal{W}_{p_1}\left(\widetilde{H}_{11}, n - p_2\right) .$$

Moreover we know that

$$\mathcal{L}\left(\widetilde{W}_{22}\right) = \mathcal{L}\left(X_2^* X_2\right)$$

and from Theorem 2.23 page 36 it holds that

$$\mathcal{L}(X_2) = \mathbb{C}\mathcal{N}_{n \times p_2}(O, I_n \otimes H_{22}) .$$

Hence we deduce that

(3.10) $$\mathcal{L}(W_{22}) = \mathbb{C}\mathcal{W}_{p_2}(H_{22}, n) .$$

Theorem 2.25 page 37 states the conditional distribution of X_1 given X_2 as

$$\mathcal{L}(X_1 | X_2) = \mathbb{C}\mathcal{N}_{n \times p_1}\left(X_2 H_{22}^{-1} H_{21}, I_n \otimes \widetilde{H}_{11}\right) .$$

Using Theorem 2.19 page 34 we obtain

$$\mathcal{L}\left(\left(X_2^* X_2\right)^{-1} X_2^* X_1 \middle| X_2\right) = \mathbb{C}\mathcal{N}_{p_2 \times p_1}\left(H_{22}^{-1} H_{21}, \left(X_2^* X_2\right)^{-1} \otimes \widetilde{H}_{11}\right) ,$$

which leads to following distribution

$$\mathcal{L}\left(\left(X_2^* X_2\right)^{\frac{1}{2}} \left(\left(X_2^* X_2\right)^{-1} X_2^* X_1 - H_{22}^{-1} H_{21}\right) \middle| X_2\right) = \mathbb{C}\mathcal{N}_{p_2 \times p_1}\left(O, I_{p_2} \otimes \widetilde{H}_{11}\right) .$$

This distribution does not depend on X_2, whereby we get

(3.11) $$\mathcal{L}\left(W_{22}^{\frac{1}{2}}\left(W_{22}^{-1} W_{21} - H_{22}^{-1} H_{21}\right)\right) = \mathbb{C}\mathcal{N}_{p_2 \times p_1}\left(O, I_{p_2} \otimes \widetilde{H}_{11}\right) .$$

and

(3.12) $$W_{22}^{\frac{1}{2}}\left(W_{22}^{-1} W_{21} - H_{22}^{-1} H_{21}\right) \perp\!\!\!\perp W_{22} .$$

We let $\widetilde{W}_{21} = W_{22}^{\frac{1}{2}}\left(W_{22}^{-1}W_{21} - H_{22}^{-1}H_{21}\right)$ and $\widetilde{W}_{22} = W_{22}$. Combining (3.8) and (3.12) page 49 it appears that

$$\widetilde{W}_{11}, \widetilde{W}_{21} \text{ and } \widetilde{W}_{22}$$

are mutually independent. Therefore it holds that

$$f_{\widetilde{W}_{11},\widetilde{W}_{21},\widetilde{W}_{22}}(\widetilde{w}_{11},\widetilde{w}_{21},\widetilde{w}_{22}) = f_{\widetilde{W}_{11}}(\widetilde{w}_{11})\, f_{\widetilde{W}_{21}}(\widetilde{w}_{21})\, f_{\widetilde{W}_{22}}(\widetilde{w}_{22}) \; . \tag{3.13}$$

We have the following one-to-one correspondence

$$\begin{aligned} W_{21} &= \widetilde{W}_{22}^{\frac{1}{2}}\left(\widetilde{W}_{21} + \widetilde{W}_{22}^{\frac{1}{2}}H_{22}^{-1}H_{21}\right) \\ &= \left(\widetilde{W}_{22}^{\frac{1}{2}} \otimes I_{p_1}\right)\left(\widetilde{W}_{21}\right) + g_1\left(\widetilde{W}_{22}\right) \\ W_{11} &= \widetilde{W}_{11} + W_{12}\widetilde{W}_{22}^{-1}W_{21} \\ &= \widetilde{W}_{11} + g_2\left(\widetilde{W}_{21}, \widetilde{W}_{22}\right) \\ W_{22} &= \widetilde{W}_{22} \, , \end{aligned}$$

where g_1 and g_2 are suitable functions. Using this correspondence and (3.13) we can find the density function of W w.r.t. Lebesgue measure on $\mathbb{C}_+^{p\times p}$ as

$$f_{W_{11},W_{21},W_{22}}(w_{11},w_{21},w_{22}) = \frac{1}{|J|} f_{\widetilde{W}_{11}}(\widetilde{w}_{11})\, f_{\widetilde{W}_{21}}(\widetilde{w}_{21})\, f_{\widetilde{W}_{22}}(\widetilde{w}_{22}) \; , \tag{3.14}$$

where J denotes the Jacobian.

We consider the one-to-one correspondence between (W_{11}, W_{21}, W_{22}) and $\left(\widetilde{W}_{11}, \widetilde{W}_{21}, \widetilde{W}_{22}\right)$ as a transformation on the real vector space $\mathbb{R}^{p_1^2} \times \mathbb{R}^{2p_1p_2} \times \mathbb{R}^{p_2^2}$. Observing that

$$\begin{aligned} \{W_{21}\} &= \left\{\widetilde{W}_{22}^{\frac{1}{2}} \otimes I_{p_1}\right\}\left\{\widetilde{W}_{21}\right\} + \left\{g_1\left(\widetilde{W}_{22}\right)\right\} \\ \{W_{11}\} &= \left\{\widetilde{W}_{11}\right\} + \left\{g_2\left(\widetilde{W}_{21}, \widetilde{W}_{22}\right)\right\} \\ \{W_{22}\} &= \left\{\widetilde{W}_{22}\right\} \, , \end{aligned}$$

we obtain that J takes the form

$$J = \det \begin{pmatrix} I_{p_1^2} & * & * \\ O & \left\{\widetilde{W}_{22}^{\frac{1}{2}} \otimes I_{p_1}\right\} & * \\ O & O & I_{p_2^2} \end{pmatrix} .$$

The explicit evaluations of the $*$-entries are of no interest, since we are able to conclude that

$$\begin{aligned} J &= \det\left(\left\{\widetilde{W}_{22}^{\frac{1}{2}} \otimes I_{p_1}\right\}\right) \\ &= \det\left(\widetilde{W}_{22}^{\frac{1}{2}} \otimes I_{p_1}\right)^2 \\ &= \det\left(\widetilde{W}_{22}\right)^{p_1} . \end{aligned}$$

Since $\widetilde{W}_{22}$ is Hermitian we obtain that

$$|J| = \det\left(\widetilde{W}_{22}\right)^{p_1} .$$

From (3.9), (3.10) page 49 and the induction assumption we get

$$f_{\widetilde{W}_{11}}\left(\widetilde{w}_{11}\right) = \frac{\det\left(\widetilde{w}_{11}\right)^{n-p_2-p_1} \exp\left(-\operatorname{tr}\left(\widetilde{w}_{11}\widetilde{H}_{11}^{-1}\right)\right)}{\det\left(\widetilde{H}_{11}\right)^{n-p_2} \pi^{\frac{p_1(p_1-1)}{2}} \prod_{j=1}^{p_1} \Gamma\left(n-p_2+1-j\right)}$$

and

$$f_{\widetilde{W}_{22}}\left(\widetilde{w}_{22}\right) = \frac{\det\left(\widetilde{w}_{22}\right)^{n-p_2} \exp\left(-\operatorname{tr}\left(\widetilde{w}_{22}H_{22}^{-1}\right)\right)}{\det\left(H_{22}\right)^{n} \pi^{\frac{p_2(p_2-1)}{2}} \prod_{j=1}^{p_2} \Gamma\left(n+1-j\right)} .$$

Further we get from (3.11) page 49 and Theorem 2.17 page 33 that

$$f_{\widetilde{W}_{21}}\left(\widetilde{w}_{21}\right) = \pi^{-p_1p_2} \det\left(\widetilde{H}_{11}\right)^{-p_2} \exp\left(-\operatorname{tr}\left(\widetilde{w}_{21}\widetilde{H}_{11}^{-1}\widetilde{w}_{21}^{*}\right)\right) .$$

Inserting the above in (3.14) page 50 and performing some calculations we conclude that

$$f_{W_{11},W_{21},W_{22}}\left(w_{11}, w_{21}, w_{22}\right) = \frac{\det\left(w\right)^{n-p} \exp\left(-\operatorname{tr}\left(wH^{-1}\right)\right)}{\det\left(H\right)^{n} \pi^{\frac{p(p-1)}{2}} \prod_{j=1}^{p} \Gamma\left(n+1-j\right)} .$$

■

Considering a complex Wishart distributed random matrix it appears from the following theorem that the distribution of the determinant of it is proportional to the distribution of a product of mutually independent chi-square distributed random variables.

Theorem 3.8
Let W be a $p \times p$ complex random matrix with $\mathcal{L}(W) = \mathbb{C}\mathcal{W}_p(H, n)$, where $H \in \mathbb{C}_+^{p\times p}$. If $n \geq p$, then

$$\mathcal{L}\left(\frac{\det(W)}{\det(H)}\right) = \mathcal{L}\left(\prod_{j=1}^{p} V_j\right) ,$$

where $\mathcal{L}(V_j) = \frac{1}{2}\chi^2_{2(n-(p-j))}$ and the V_j's are mutually independent.

Proof:
Let W be a $p \times p$ complex random matrix with $\mathcal{L}(W) = \mathbb{C}\mathcal{W}_p(H, n)$, where $H \in \mathbb{C}_+^{p\times p}$, and assume that $n \geq p$. Note from Theorem 3.6 page 44 that $W > O$ with probability one, since $H > O$ and $n \geq p$.

Let $\boldsymbol{W}$ and $\boldsymbol{H}$ be partitioned as in Theorem 3.6 with $p_1 = 1$ and consider the fraction $\frac{\det(\boldsymbol{W})}{\det(\boldsymbol{H})}$, which is well defined as $\boldsymbol{H} > \boldsymbol{O}$. Using the partition of $\boldsymbol{W}$ and $\boldsymbol{H}$ it can be written as

$$\frac{\det(\boldsymbol{W})}{\det(\boldsymbol{H})} = \frac{\det(\boldsymbol{W}_{22})\left(\boldsymbol{W}_{11} - \boldsymbol{W}_{12}\boldsymbol{W}_{22}^{-1}\boldsymbol{W}_{21}\right)}{\det(\boldsymbol{H}_{22})\left(\boldsymbol{H}_{11} - \boldsymbol{H}_{12}\boldsymbol{H}_{22}^{-1}\boldsymbol{H}_{21}\right)} .$$

From Theorem 3.6 we know, since $n \geq p - 1$ and $\boldsymbol{H}_{22} > \boldsymbol{O}$, that

$$\mathcal{L}\left(\boldsymbol{W}_{11} - \boldsymbol{W}_{12}\boldsymbol{W}_{22}^{-1}\boldsymbol{W}_{21}\right) = \frac{1}{2}\left(\boldsymbol{H}_{11} - \boldsymbol{H}_{12}\boldsymbol{H}_{22}^{-1}\boldsymbol{H}_{21}\right)\chi^2_{2(n-(p-1))} .$$

Therefore letting

$$V_1 = \frac{\boldsymbol{W}_{11} - \boldsymbol{W}_{12}\boldsymbol{W}_{22}^{-1}\boldsymbol{W}_{21}}{\boldsymbol{H}_{11} - \boldsymbol{H}_{12}\boldsymbol{H}_{22}^{-1}\boldsymbol{H}_{21}}$$

we reach that

$$\mathcal{L}(V_1) = \frac{1}{2}\chi^2_{2(n-(p-1))} .$$

Consider the remaining part of the fraction, i.e. $\frac{\det(\boldsymbol{W}_{22})}{\det(\boldsymbol{H}_{22})}$. Using the arguments above successively until $\det(\boldsymbol{W})$ and $\det(\boldsymbol{H})$ are totally partitioned into products of real random variables and real numbers, respectively, it appears that

$$\mathcal{L}\left(\frac{\det(\boldsymbol{W})}{\det(\boldsymbol{H})}\right) = \mathcal{L}\left(\prod_{j=1}^{p} V_j\right) ,$$

where $\mathcal{L}(V_j) = \frac{1}{2}\chi^2_{2(n-(p-j))}$.

From Theorem 3.6 we also have that

$$\boldsymbol{W}_{11} - \boldsymbol{W}_{12}\boldsymbol{W}_{22}^{-1}\boldsymbol{W}_{21} \perp\!\!\!\perp (\boldsymbol{W}_{12}, \boldsymbol{W}_{22}) ,$$

whereby it holds that

$$\boldsymbol{W}_{11} - \boldsymbol{W}_{12}\boldsymbol{W}_{22}^{-1}\boldsymbol{W}_{21} \perp\!\!\!\perp \boldsymbol{W}_{22} .$$

This implies that

$$V_1 \perp\!\!\!\perp (V_2, V_3, \ldots, V_p) .$$

Using similar arguments successively on all the parts of $\frac{\det(\boldsymbol{W})}{\det(\boldsymbol{H})}$ we conclude that the V_j's for $j = 1, 2, \ldots, p$ are mutually independent. ∎

When regarding a complex random quadratic form involving a projection matrix a useful consequence of Theorem 3.6 page 44 is the following theorem. It states the necessary conditions for this complex random quadratic form to be positive definite.

Theorem 3.9
Let $\boldsymbol{X}$ be an $n \times p$ complex random matrix with $\mathcal{L}(\boldsymbol{X}) = \mathbb{C}\mathcal{N}_{n\times p}(\boldsymbol{\Theta}, \boldsymbol{I}_n \otimes \boldsymbol{H})$, where $\boldsymbol{\Theta} \in \mathbb{C}^{n\times p}$ and $\boldsymbol{H} \in \mathbb{C}_+^{p\times p}$. Further let $\boldsymbol{P}$ be an $n \times n$ complex matrix representing the orthogonal projection of $\mathbb{C}^n$ onto a k-dimensional subspace. If $\boldsymbol{P\Theta} = \boldsymbol{O}$ and $k \geq p$, then $\boldsymbol{X}^*\boldsymbol{PX} > \boldsymbol{O}$ with probability one.

Proof:
The theorem follows from Theorem 3.3 page 41 and Theorem 3.6 page 44. ■

The following example illustrates how multivariate complex distributions can be used within multiple time series.

Example 3.1
Let $\{\boldsymbol{X}_t\}_{t\in\mathbf{Z}}$ be a p-dimensional stationary Gaussian time series with auto-covariance function $\boldsymbol{R}(u) = (R_{jk}(u)) = \mathbb{C}(\boldsymbol{X}_{t+u}, \boldsymbol{X}_t)$, which fulfills

$$\sum_{u\in\mathbf{Z}} |R_{jk}(u)| < \infty, \quad j, k = 1, 2, \ldots, p. \tag{3.15}$$

This condition ensures no long range dependencies in the series, and it allows us to define

$$\boldsymbol{\Sigma}(\omega) = \frac{1}{2\pi} \sum_{u\in\mathbf{Z}} \exp(-i\omega u)\, \boldsymbol{R}(u), \quad \omega \in [0, \pi],$$

which is known as the spectral density matrix.

In order to estimate the spectral density we consider the finite Fourier transform of $\{\boldsymbol{X}_t\}_{t=0}^{T-1}$. This is given by

$$\widehat{\boldsymbol{X}}_T(\omega) = \sum_{t=0}^{T-1} \exp(-i\omega t)\, \boldsymbol{X}_t, \quad \omega \in [0, \pi],$$

and it holds that

$$\widehat{\boldsymbol{X}}_T(\omega) \in \mathbb{C}^p.$$

Let $j_m(T)$ for $m = 1, 2, \ldots, M$ be integers such that $0 < j_1(T) < j_2(T) < \cdots < j_M(T) < \frac{T}{2}$. Furthermore let $\omega_m(T) = \frac{2\pi j_m(T)}{T}$, where $\omega_m(T) \to \omega_m$ as $T \to \infty$ and $\omega_m \in]0, \pi[$ for $m = 1, 2, \ldots, M$. Then $\widehat{\boldsymbol{X}}_T(\omega_m(T))$ for $m = 1, 2, \ldots, M$ are asymptotically mutually independent p-dimensional complex random vectors asymptotically distributed as $\mathbb{C}\mathcal{N}_p(\boldsymbol{0}, 2\pi T\boldsymbol{\Sigma}(\omega_m))$, respectively.

Let

$$I_T(\omega) = \frac{1}{2\pi T} \widehat{\boldsymbol{X}}_T(\omega)\, \widehat{\boldsymbol{X}}_T^*(\omega), \quad \omega \in [0, \pi].$$

Then $I_T(\omega_m(T))$ for $m = 1, 2, \dots, M$ are asymptotically mutually independent $p \times p$ complex random matrices asymptotically distributed as $\mathbb{C}\mathcal{W}_p(\Sigma(\omega_m), 1)$, respectively. The result can be extracted from Brillinger (1975), which covers the more general case, where the series is not necessarily Gaussian, but fulfills cumulant conditions, which reduce to (3.15) page 53 in case of Gaussianity.

The result offers a way of estimating $\Sigma(\omega)$ for selected frequencies $0 < \omega_1 < \omega_2 < \cdots < \omega_K < \pi$. Define

$$\text{(3.16)} \qquad \widehat{\Sigma}_T(\omega_k) = \frac{1}{2N+1} \sum_{l=-N}^{N} I_T\left(\frac{2\pi\left(\left\lfloor \frac{\omega_k T}{2\pi} \right\rfloor + l\right)}{T}\right), \quad k = 1, 2, \dots, K,$$

where $\lfloor \cdot \rfloor$ denotes the integer part. Hereby we obtain asymptotically mutually independent $p \times p$ complex random matrices which are asymptotically distributed as $\mathbb{C}\mathcal{W}_p\left(\frac{1}{2N+1}\Sigma(\omega_k), 2N+1\right)$, respectively. ■

In the next example we consider a case where time-stationarity is replaced by "circular"-stationarity.

Example 3.2
Let $\boldsymbol{X} = (\boldsymbol{X}_0, \boldsymbol{X}_1, \dots, \boldsymbol{X}_{T-1})^*$ be a $T \times p$ random matrix which is multivariate normal distributed and invariant under cyclical permutation, i.e. for $j = 1, 2, \dots, T-1$

$$\mathcal{L}((\boldsymbol{X}_0, \boldsymbol{X}_1, \dots, \boldsymbol{X}_{T-1})^*) = \mathcal{L}((\boldsymbol{X}_j, \boldsymbol{X}_{j+1}, \dots, \boldsymbol{X}_{T-1}, \boldsymbol{X}_0, \boldsymbol{X}_1, \dots, \boldsymbol{X}_{j-1})^*) .$$

Let $\omega_j = \frac{2\pi j}{T}$, $j = 0, 1, \dots, \lfloor \frac{T}{2} \rfloor$ and define

$$\boldsymbol{Y}_j = \widehat{\boldsymbol{X}}_T(\omega_j), \quad j = 0, 1, \dots, \left\lfloor \frac{T}{2} \right\rfloor,$$

where

$$\widehat{\boldsymbol{X}}_T(\omega_j) = \sum_{t=0}^{T-1} \exp(-i\omega_j t) \boldsymbol{X}_t^* .$$

The $\boldsymbol{Y}_j$'s are then mutually independent p-dimensional complex random vectors which for $1 \leq j < \frac{T}{2}$ are complex normally distributed with mean zero. Furthermore $\boldsymbol{Y}_0$ and - in the case where T is even - $\boldsymbol{Y}_{\frac{T}{2}}$ are real normally distributed, the latter with mean zero.

The sitnation much resembles the time series set-up, where time-stationarity has been replaced by "circular"-stationarity. A lot of the distributional results obtained for the present set-up have an asymptotic analogue for time series as already illustrated for the finite Fourier transform in Example 3.1 page 53. ■

3.2 The Complex U-distribution

In this section we consider the complex U-distribution. We show that a complex U-distribution is equal to the distribution of a product of random variables, which are beta distributed and mutually independent. But we begin by defining the complex U-distribution in the same way as the real counterpart is defined.

Definition 3.2 The complex U-distribution
Let $\boldsymbol{V}$ and $\boldsymbol{W}$ be $p \times p$ complex random matrices with $\mathcal{L}(\boldsymbol{V}) = \mathbb{C}\mathcal{W}_p(\boldsymbol{H}, n)$ and $\mathcal{L}(\boldsymbol{W}) = \mathbb{C}\mathcal{W}_p(\boldsymbol{H}, m)$, where $\boldsymbol{H} \in \mathbb{C}_+^{p\times p}$. If $\boldsymbol{V}$ and $\boldsymbol{W}$ are independent and $n \geq p$, then the distribution of

$$U = \frac{\det(\boldsymbol{V})}{\det(\boldsymbol{V}+\boldsymbol{W})}$$

is called a complex U-distribution with parameters p, m and n. This is denoted by $\mathcal{L}(U) = \mathbb{C}\mathcal{U}(p, m, n)$.

The complex U-distribution does not depend on $\boldsymbol{H}$, which can be seen by the following considerations. The distributions of the independent complex matrices $\boldsymbol{V}$ and $\boldsymbol{W}$ are equal to the distributions of $\boldsymbol{X}^*\boldsymbol{X}$ and $\boldsymbol{Y}^*\boldsymbol{Y}$, respectively, where $\boldsymbol{X}$ and $\boldsymbol{Y}$ are independent complex random matrices of dimensions $n \times p$ and $m \times p$ and with $\mathcal{L}(\boldsymbol{X}) = \mathbb{C}\mathcal{N}_{n\times p}(\boldsymbol{O}, \boldsymbol{I}_n \otimes \boldsymbol{H})$ and $\mathcal{L}(\boldsymbol{Y}) = \mathbb{C}\mathcal{N}_{m\times p}(\boldsymbol{O}, \boldsymbol{I}_m \otimes \boldsymbol{H})$ for $\boldsymbol{H} \in \mathbb{C}_+^{p\times p}$. As $\boldsymbol{H}$ is positive definite we know that there exists a $p\times p$ complex matrix $\boldsymbol{H}^{\frac{1}{2}} > \boldsymbol{O}$ such that $\boldsymbol{H} = \left(\boldsymbol{H}^{\frac{1}{2}}\right)^2$. Therefore $\mathcal{L}(\boldsymbol{X}) = \mathcal{L}\left(\boldsymbol{Z}\boldsymbol{H}^{\frac{1}{2}}\right)$, where $\mathcal{L}(\boldsymbol{Z}) = \mathbb{C}\mathcal{N}_{n\times p}(\boldsymbol{O}, \boldsymbol{I}_n \otimes \boldsymbol{I}_p)$, which implies

$$\mathcal{L}(\boldsymbol{V}) = \mathcal{L}(\boldsymbol{X}^*\boldsymbol{X}) = \mathcal{L}\left(\boldsymbol{H}^{\frac{1}{2}}\boldsymbol{Z}^*\boldsymbol{Z}\boldsymbol{H}^{\frac{1}{2}}\right) = \mathcal{L}\left(\boldsymbol{H}^{\frac{1}{2}}\widetilde{\boldsymbol{V}}\boldsymbol{H}^{\frac{1}{2}}\right),$$

where $\mathcal{L}\left(\widetilde{\boldsymbol{V}}\right) = \mathbb{C}\mathcal{W}_p(\boldsymbol{I}_p, n)$. Similarly we see that

$$\mathcal{L}(\boldsymbol{W}) = \mathcal{L}\left(\boldsymbol{H}^{\frac{1}{2}}\widetilde{\boldsymbol{W}}\boldsymbol{H}^{\frac{1}{2}}\right),$$

where $\mathcal{L}\left(\widetilde{\boldsymbol{W}}\right) = \mathbb{C}\mathcal{W}_p(\boldsymbol{I}_p, m)$.

Hereby we are able to deduce that $\mathcal{L}(\boldsymbol{V}, \boldsymbol{W}) = \mathcal{L}\left(\boldsymbol{H}^{\frac{1}{2}}\widetilde{\boldsymbol{V}}\boldsymbol{H}^{\frac{1}{2}}, \boldsymbol{H}^{\frac{1}{2}}\widetilde{\boldsymbol{W}}\boldsymbol{H}^{\frac{1}{2}}\right)$, therefore

$$\mathcal{L}\left(\frac{\det(\boldsymbol{V})}{\det(\boldsymbol{V}+\boldsymbol{W})}\right) = \mathcal{L}\left(\frac{\det\left(\widetilde{\boldsymbol{V}}\right)}{\det\left(\widetilde{\boldsymbol{V}}+\widetilde{\boldsymbol{W}}\right)}\right)$$

and we deduce that the complex U-distribution does not depend on $\boldsymbol{H}$.

Further U is well defined since we from Theorem 3.6 page 44 know that $m+n \geq p$ and $\boldsymbol{H} > \boldsymbol{O}$ implies $\boldsymbol{V} + \boldsymbol{W} > \boldsymbol{O}$ with probability one. Hence $\det(\boldsymbol{V}+\boldsymbol{W}) > 0$ with probability one.

We seek to show that a complex U-distribution is equal to the distribution of a product of mutually independent beta distributed random variables. To show this result we begin by considering a complex random variable containing some univariate quadratic forms. This result is obtained by combining results on the complex Wishart distribution from the previous section.

Lemma 3.1

Let $\boldsymbol{X}$ be an n-dimensional complex random vector with $\mathcal{L}(\boldsymbol{X}) = \mathbb{C}\mathcal{N}_n(\boldsymbol{\theta}, \sigma^2 \boldsymbol{I}_n)$, where $\boldsymbol{\theta} \in \mathbb{C}^n$ and $\sigma^2 \in \overline{\mathbb{R}}_+$. Further let $\boldsymbol{P}$ be an $n \times n$ complex matrix representing the orthogonal projection of $\mathbb{C}^n$ onto a k-dimensional subspace and let $\boldsymbol{P}_0$ be an $n \times n$ complex matrix representing the orthogonal projection of $\mathbb{C}^n$ onto a k_0-dimensional subspace of the k-dimensional subspace. If $n - k \geq 1$ and $\boldsymbol{P}_0\boldsymbol{\theta} = \boldsymbol{\theta}$, then the distribution of $B = \frac{\boldsymbol{X}^*(\boldsymbol{I}_n - \boldsymbol{P})\boldsymbol{X}}{\boldsymbol{X}^*(\boldsymbol{I}_n - \boldsymbol{P}_0)\boldsymbol{X}}$ is given as

$$\mathcal{L}(B) = \mathcal{B}(n-k, k-k_0) \ ,$$

and it holds that

$$B, \ \boldsymbol{X}^*(\boldsymbol{I}_n - \boldsymbol{P}_0)\boldsymbol{X} \text{ and } \boldsymbol{P}_0\boldsymbol{X}$$

are mutually independent.

Proof:

Let the assumptions in the theorem be satisfied and assume that $n - k \geq 1$ and $\boldsymbol{P}_0\boldsymbol{\theta} = \boldsymbol{\theta}$.

Since $\boldsymbol{P}_0\boldsymbol{\theta} = \boldsymbol{\theta}$ implies $\boldsymbol{P}\boldsymbol{\theta} = \boldsymbol{\theta}$ and hereby $(\boldsymbol{I}_n - \boldsymbol{P})\boldsymbol{\theta} = 0$ it follows from Theorem 3.3 page 41 that

$$\mathcal{L}(\boldsymbol{X}^*(\boldsymbol{I}_n - \boldsymbol{P})\boldsymbol{X}) = \frac{\sigma^2}{2}\chi^2_{2(n-k)} \ .$$

Obviously $\boldsymbol{P} - \boldsymbol{P}_0$ is Hermitian and idempotent, so $\boldsymbol{P} - \boldsymbol{P}_0$ is a projection matrix. Furthermore, since

$$\operatorname{tr}(\boldsymbol{P} - \boldsymbol{P}_0) = \operatorname{tr}(\boldsymbol{P}) - \operatorname{tr}(\boldsymbol{P}_0) = k - k_0 \ ,$$

we have that $\boldsymbol{P} - \boldsymbol{P}_0$ represents the orthogonal projection of $\mathbb{C}^n$ onto a $(k - k_0)$-dimensional subspace. Using $(\boldsymbol{P} - \boldsymbol{P}_0)\boldsymbol{\theta} = 0$ we get according to Theorem 3.3 that

$$\begin{aligned} \mathcal{L}(\boldsymbol{X}^*(\boldsymbol{P} - \boldsymbol{P}_0)\boldsymbol{X}) &= \mathbb{C}\mathcal{W}_1\left(\sigma^2, k - k_0\right) \\ &= \frac{\sigma^2}{2}\chi^2_{2(k-k_0)} \ . \end{aligned}$$

Besides from Theorem 3.4 page 42 it follows that

$$(\boldsymbol{I}_n - \boldsymbol{P})\boldsymbol{X} \perp\!\!\!\perp (\boldsymbol{P} - \boldsymbol{P}_0)\boldsymbol{X} \ ,$$

which implies

$$\boldsymbol{X}^*(\boldsymbol{I}_n - \boldsymbol{P})\boldsymbol{X} \perp\!\!\!\perp \boldsymbol{X}^*(\boldsymbol{P} - \boldsymbol{P}_0)\boldsymbol{X} \ .$$

By letting

$$B = \frac{\boldsymbol{X}^* (\boldsymbol{I}_n - \boldsymbol{P}) \boldsymbol{X}}{\boldsymbol{X}^* (\boldsymbol{I}_n - \boldsymbol{P}_0) \boldsymbol{X}}$$

we obtain, since $\boldsymbol{X}^* (\boldsymbol{I}_n - \boldsymbol{P}_0) \boldsymbol{X} = \boldsymbol{X}^* (\boldsymbol{I}_n - \boldsymbol{P}) \boldsymbol{X} + \boldsymbol{X}^* (\boldsymbol{P} - \boldsymbol{P}_0) \boldsymbol{X}$, that

$$\mathcal{L}(B) = \mathcal{B}(n-k, k-k_0)$$

and

$$B \perp\!\!\!\perp \boldsymbol{X}^* (\boldsymbol{I}_n - \boldsymbol{P}_0) \boldsymbol{X} .$$

According to Theorem 3.4 we know that $\boldsymbol{P}_0\boldsymbol{X}$, $(\boldsymbol{I}_n - \boldsymbol{P}) \boldsymbol{X}$ and $(\boldsymbol{P} - \boldsymbol{P}_0) \boldsymbol{X}$ are mutually independent, whereby we deduce

$$\boldsymbol{P}_0\boldsymbol{X} \perp\!\!\!\perp (B, \boldsymbol{X}^* (\boldsymbol{I}_n - \boldsymbol{P}_0) \boldsymbol{X}) .$$

Hereby the proof is completed. ■

The following theorem establishes the correspondence between the complex U-distribution and the beta distribution.

Theorem 3.10 The relation to the beta distribution
Let $\boldsymbol{V}$ and $\boldsymbol{W}$ be $p \times p$ complex random matrices with $\mathcal{L}(\boldsymbol{V}) = \mathbb{C}\mathcal{W}_p(\boldsymbol{H}, n)$ and $\mathcal{L}(\boldsymbol{W}) = \mathbb{C}\mathcal{W}_p(\boldsymbol{H}, m)$, where $\boldsymbol{H} \in \mathbb{C}_+^{p\times p}$. If $\boldsymbol{V}$ and $\boldsymbol{W}$ are independent and $n \geq p$, then the distribution of $U = \frac{\det(\boldsymbol{V})}{\det(\boldsymbol{V}+\boldsymbol{W})}$ is given as

$$\mathcal{L}(U) = \mathcal{L}\left(\prod_{j=1}^{p} B_j\right),$$

where $\mathcal{L}(B_j) = \mathcal{B}(n-(p-j), m)$ and the B_j's are mutually independent for $j = 2, 3, \ldots, p$ and it holds that

$$U \perp\!\!\!\perp \boldsymbol{V} + \boldsymbol{W} .$$

Proof:
Let $\boldsymbol{V}$ and $\boldsymbol{W}$ be $p \times p$ complex random matrices with $\mathcal{L}(\boldsymbol{V}) = \mathbb{C}\mathcal{W}_p(\boldsymbol{H}, n)$ and $\mathcal{L}(\boldsymbol{W}) = \mathbb{C}\mathcal{W}_p(\boldsymbol{H}, m)$, where $\boldsymbol{H} \in \mathbb{C}_+^{p\times p}$. Further assume that $\boldsymbol{V}$ and $\boldsymbol{W}$ are independent and $n \geq p$. Let

$$U = \frac{\det(\boldsymbol{V})}{\det(\boldsymbol{V}+\boldsymbol{W})},$$

which is well defined because $n + m \geq p$ and $\boldsymbol{H} > \boldsymbol{O}$. We show the theorem by induction on p.

Induction start:
For $p = 1$ we know that

$$U = \frac{\boldsymbol{V}}{\boldsymbol{V} + \boldsymbol{W}} .$$

Furthermore it holds that

$$\mathcal{L}(\boldsymbol{V}) = \frac{1}{2}\boldsymbol{H}\chi^2_{2n} \text{ and } \mathcal{L}(\boldsymbol{W}) = \frac{1}{2}\boldsymbol{H}\chi^2_{2m} .$$

Since $\boldsymbol{V}$ and $\boldsymbol{W}$ are independent we deduce that

$$\mathcal{L}(U) = \mathcal{B}(n, m)$$

and

$$U \perp\!\!\!\perp \boldsymbol{V} + \boldsymbol{W} .$$

Induction step:
For $p = d - 1$ we assume that

$$\mathcal{L}(U) = \mathcal{L}\left(\prod_{j=2}^{d} B_j\right) ,$$

where $\mathcal{L}(B_j) = \mathcal{B}(n - (d - j), m)$ and the B_j's are mutually independent for $j = 2, 3, \ldots, d$. Furthermore in this case assume that

$$U \perp\!\!\!\perp \boldsymbol{V} + \boldsymbol{W} .$$

We seek to show for $p = d$ that

$$\mathcal{L}(U) = \mathcal{L}\left(\prod_{j=1}^{d} B_j\right) ,$$

where $\mathcal{L}(B_j) = \mathcal{B}(n - (d - j), m)$ and the B_j's are mutually independent for $j = 1, 2, \ldots, d$. Furthermore we seek to show that

$$U \perp\!\!\!\perp \boldsymbol{V} + \boldsymbol{W} .$$

Let $\boldsymbol{V}, \boldsymbol{W}$ and $\boldsymbol{H}$ be partitioned as

$$\boldsymbol{V} = \begin{pmatrix} \boldsymbol{V}_{11} & \boldsymbol{V}_{12} \\ \boldsymbol{V}_{21} & \boldsymbol{V}_{22} \end{pmatrix}, \ \boldsymbol{W} = \begin{pmatrix} \boldsymbol{W}_{11} & \boldsymbol{W}_{12} \\ \boldsymbol{W}_{21} & \boldsymbol{W}_{22} \end{pmatrix} \text{ and } \boldsymbol{H} = \begin{pmatrix} \boldsymbol{H}_{11} & \boldsymbol{H}_{12} \\ \boldsymbol{H}_{21} & \boldsymbol{H}_{22} \end{pmatrix} ,$$

where $\boldsymbol{V}_{jk}, \boldsymbol{W}_{jk}$ and $\boldsymbol{H}_{jk}$ are $p_j \times p_k$ for $j, k = 1, 2, p_1 = 1$ and $p_2 = p - 1$. Notice that $(\boldsymbol{V} + \boldsymbol{W})_{jk} = \boldsymbol{V}_{jk} + \boldsymbol{W}_{jk}$ for $j, k = 1, 2$.

Using the partition we get

$$\det(\boldsymbol{V}) = \det(\boldsymbol{V}_{22}) \left(\boldsymbol{V}_{11} - \boldsymbol{V}_{12}\boldsymbol{V}_{22}^{-1}\boldsymbol{V}_{21}\right)$$

and

$$\det(\boldsymbol{V}+\boldsymbol{W}) = \det((\boldsymbol{V}+\boldsymbol{W})_{22})\left((\boldsymbol{V}+\boldsymbol{W})_{11} - (\boldsymbol{V}+\boldsymbol{W})_{12}(\boldsymbol{V}+\boldsymbol{W})_{22}^{-1}(\boldsymbol{V}+\boldsymbol{W})_{21}\right),$$

where the inverses exist with probability one, since $\boldsymbol{V} > \boldsymbol{O}$ and $\boldsymbol{V}+\boldsymbol{W} > \boldsymbol{O}$ with probability one. Combining these results leads to

$$U = \frac{\det(\boldsymbol{V}_{22})\left(\boldsymbol{V}_{11} - \boldsymbol{V}_{12}\boldsymbol{V}_{22}^{-1}\boldsymbol{V}_{21}\right)}{\det((\boldsymbol{V}+\boldsymbol{W})_{22})\left((\boldsymbol{V}+\boldsymbol{W})_{11} - (\boldsymbol{V}+\boldsymbol{W})_{12}(\boldsymbol{V}+\boldsymbol{W})_{22}^{-1}(\boldsymbol{V}+\boldsymbol{W})_{21}\right)}.$$

For reasons of simplicity introduce

$$B_1 = \frac{\boldsymbol{V}_{11} - \boldsymbol{V}_{12}\boldsymbol{V}_{22}^{-1}\boldsymbol{V}_{21}}{(\boldsymbol{V}+\boldsymbol{W})_{11} - (\boldsymbol{V}+\boldsymbol{W})_{12}(\boldsymbol{V}+\boldsymbol{W})_{22}^{-1}(\boldsymbol{V}+\boldsymbol{W})_{21}}.$$

Now we show that

i. $\mathcal{L}(B_1) = \mathcal{B}(n-(p-1), m)$.

The distributions of the independent complex matrices $\boldsymbol{V}$ and $\boldsymbol{W}$ are equal to the distributions of $\boldsymbol{X}^*\boldsymbol{X}$ and $\boldsymbol{Y}^*\boldsymbol{Y}$, respectively, where $\boldsymbol{X}$ and $\boldsymbol{Y}$ are independent complex matrices of dimensions $n \times p$ and $m \times p$ and with $\mathcal{L}(\boldsymbol{X}) = \mathbb{C}\mathcal{N}_{n\times p}(\boldsymbol{O}, \boldsymbol{I}_n \otimes \boldsymbol{H})$ and $\mathcal{L}(\boldsymbol{Y}) = \mathbb{C}\mathcal{N}_{m\times p}(\boldsymbol{O}, \boldsymbol{I}_m \otimes \boldsymbol{H})$ for $\boldsymbol{H} \in \mathbb{C}_+^{p\times p}$. Observe that $\mathcal{L}(\boldsymbol{V}, \boldsymbol{W}) = \mathcal{L}(\boldsymbol{X}^*\boldsymbol{X}, \boldsymbol{Y}^*\boldsymbol{Y})$. Let $\boldsymbol{X}$ and $\boldsymbol{Y}$ be partitioned according to the partition of $\boldsymbol{V}$ and $\boldsymbol{W}$, i.e. $\boldsymbol{X} = (\boldsymbol{X}_1, \boldsymbol{X}_2)$ and $\boldsymbol{Y} = (\boldsymbol{Y}_1, \boldsymbol{Y}_2)$, where $\boldsymbol{X}_j$ is $n \times p_j$ and $\boldsymbol{Y}_j$ is $m \times p_j$ for $j = 1, 2$, $p_1 = 1$ and $p_2 = p-1$. Notice that

$$\boldsymbol{X}_1^*\boldsymbol{X}_1 - \boldsymbol{X}_1^*\boldsymbol{X}_2(\boldsymbol{X}_2^*\boldsymbol{X}_2)^{-1}\boldsymbol{X}_2^*\boldsymbol{X}_1 = \begin{pmatrix} \boldsymbol{X}_1 \\ \boldsymbol{Y}_1 \end{pmatrix}^* (\boldsymbol{I}_{n+m} - \boldsymbol{P}) \begin{pmatrix} \boldsymbol{X}_1 \\ \boldsymbol{Y}_1 \end{pmatrix},$$

where $\boldsymbol{P}$ is the $(n+m) \times (n+m)$ complex matrix given by

$$\boldsymbol{P} = \begin{pmatrix} \boldsymbol{X}_2(\boldsymbol{X}_2^*\boldsymbol{X}_2)^{-1}\boldsymbol{X}_2^* & \boldsymbol{O} \\ \boldsymbol{O} & \boldsymbol{I}_m \end{pmatrix}.$$

The matrix $\boldsymbol{P}$ is Hermitian and idempotent with $\operatorname{tr}(\boldsymbol{P}) = p-1+m$. Therefore given $(\boldsymbol{X}_2, \boldsymbol{Y}_2)$ it represents the orthogonal projection of $\mathbb{C}^{n+m}$ onto a $(p-1+m)$-dimensional subspace. Moreover observe that

$$\begin{aligned} &(\boldsymbol{X}_1^*\boldsymbol{X}_1 + \boldsymbol{Y}_1^*\boldsymbol{Y}_1) - (\boldsymbol{X}_1^*\boldsymbol{X}_2 + \boldsymbol{Y}_1^*\boldsymbol{Y}_2)(\boldsymbol{X}_2^*\boldsymbol{X}_2 + \boldsymbol{Y}_2^*\boldsymbol{Y}_2)^{-1}(\boldsymbol{X}_2^*\boldsymbol{X}_1 + \boldsymbol{Y}_2^*\boldsymbol{Y}_1) \\ &= \begin{pmatrix} \boldsymbol{X}_1 \\ \boldsymbol{Y}_1 \end{pmatrix}^* (\boldsymbol{I}_{n+m} - \boldsymbol{P}_0) \begin{pmatrix} \boldsymbol{X}_1 \\ \boldsymbol{Y}_1 \end{pmatrix}, \end{aligned}$$

where $\boldsymbol{P}_0$ is the $(n+m) \times (n+m)$ complex matrix given by

$$\boldsymbol{P}_0 = \begin{pmatrix} \boldsymbol{X}_2 \\ \boldsymbol{Y}_2 \end{pmatrix} (\boldsymbol{X}_2^*\boldsymbol{X}_2 + \boldsymbol{Y}_2^*\boldsymbol{Y}_2)^{-1} \begin{pmatrix} \boldsymbol{X}_2 \\ \boldsymbol{Y}_2 \end{pmatrix}^*.$$

Similarly as before we see given $(\boldsymbol{X}_2, \boldsymbol{Y}_2)$ that $\boldsymbol{P}_0$ represents the orthogonal projection of $\mathbb{C}^{n+m}$ onto a $(p-1)$-dimensional subspace. Since $\boldsymbol{P}\boldsymbol{P}_0 = \boldsymbol{P}_0\boldsymbol{P} = \boldsymbol{P}_0$ we know that the $(p-1)$-dimensional subspace, which $\boldsymbol{P}_0$ projects onto, is a subspace of the $(p-1+m)$-dimensional subspace, which $\boldsymbol{P}$ projects onto.

Theorem 2.25 page 37 tells us that

$$\mathcal{L}(\boldsymbol{X}_1 | \boldsymbol{X}_2) = \mathbb{C}\mathcal{N}_n\left(\boldsymbol{X}_2\boldsymbol{H}_{22}^{-1}\boldsymbol{H}_{21}, \boldsymbol{I}_n \otimes \left(\boldsymbol{H}_{11} - \boldsymbol{H}_{12}\boldsymbol{H}_{22}^{-1}\boldsymbol{H}_{21}\right)\right)$$

and

$$\mathcal{L}(\boldsymbol{Y}_1 | \boldsymbol{Y}_2) = \mathbb{C}\mathcal{N}_m\left(\boldsymbol{Y}_2\boldsymbol{H}_{22}^{-1}\boldsymbol{H}_{21}, \boldsymbol{I}_m \otimes \left(\boldsymbol{H}_{11} - \boldsymbol{H}_{12}\boldsymbol{H}_{22}^{-1}\boldsymbol{H}_{21}\right)\right) .$$

Since $\boldsymbol{X}$ and $\boldsymbol{Y}$ are independent we hereby deduce

$$\mathcal{L}\left(\begin{pmatrix} \boldsymbol{X}_1 \\ \boldsymbol{Y}_1 \end{pmatrix} \middle| \begin{pmatrix} \boldsymbol{X}_2 \\ \boldsymbol{Y}_2 \end{pmatrix}\right) = \mathbb{C}\mathcal{N}_{n+m}\left(\begin{pmatrix} \boldsymbol{X}_2 \\ \boldsymbol{Y}_2 \end{pmatrix} \boldsymbol{H}_{22}^{-1}\boldsymbol{H}_{21}, \boldsymbol{I}_{n+m} \otimes \left(\boldsymbol{H}_{11} - \boldsymbol{H}_{12}\boldsymbol{H}_{22}^{-1}\boldsymbol{H}_{21}\right)\right) .$$

Now we introduce

$$\tilde{B}_1 = \frac{\begin{pmatrix} \boldsymbol{X}_1 \\ \boldsymbol{Y}_1 \end{pmatrix}^* (\boldsymbol{I}_{n+m} - \boldsymbol{P}) \begin{pmatrix} \boldsymbol{X}_1 \\ \boldsymbol{Y}_1 \end{pmatrix}}{\begin{pmatrix} \boldsymbol{X}_1 \\ \boldsymbol{Y}_1 \end{pmatrix}^* (\boldsymbol{I}_{n+m} - \boldsymbol{P}_0) \begin{pmatrix} \boldsymbol{X}_1 \\ \boldsymbol{Y}_1 \end{pmatrix}} .$$

Since $n+m-(p-1+m) = n-p+1 \geq 1$ and $\boldsymbol{P}_0 (\boldsymbol{X}_2^*, \boldsymbol{Y}_2^*)^* \boldsymbol{H}_{22}^{-1}\boldsymbol{H}_{21} = (\boldsymbol{X}_2^*, \boldsymbol{Y}_2^*)^* \boldsymbol{H}_{22}^{-1}\boldsymbol{H}_{21}$ we get from Lemma 3.1 page 56 that

$$\mathcal{L}\left(\tilde{B}_1 \middle| \begin{pmatrix} \boldsymbol{X}_2 \\ \boldsymbol{Y}_2 \end{pmatrix}\right) = \mathcal{B}(n-(p-1), m) ,$$

which does not depend on $(\boldsymbol{X}_2, \boldsymbol{Y}_2)$, thus

$$\mathcal{L}\left(\tilde{B}_1\right) = \mathcal{B}(n-(p-1), m)$$

and

$$\tilde{B}_1 \perp\!\!\!\perp (\boldsymbol{X}_2, \boldsymbol{Y}_2) .$$

It holds that

$$\mathcal{L}\left(\tilde{B}_1\right) = \mathcal{L}(B_1) ,$$

therefore it follows that

$$\mathcal{L}(B_1) = \mathcal{B}(n-(p-1), m) .$$

Now we are able to deduce

ii. $\mathcal{L}(U) = \mathcal{L}\left(\prod_{j=1}^{p} B_j\right)$, where $\mathcal{L}(B_j) = \mathcal{B}(n-(p-j), m)$ and the B_j's are mutually independent for $j = 1, 2, \ldots, p$.

From the induction assumption we have that

$$\mathcal{L}\left(\frac{\det(\boldsymbol{V}_{22})}{\det((\boldsymbol{V}+\boldsymbol{W})_{22})}\right) = \mathcal{L}\left(\prod_{j=2}^{p} B_j\right),$$

where $\mathcal{L}(B_j) = \mathcal{B}(n-(p-j), m)$ and the B_j's are mutually independent for $j = 2, 3, \ldots, p$. Hereby we observe that the corresponding $\widetilde{B}_j$'s are mutually independent for $j = 2, 3, \ldots, p$.

Because

$$\widetilde{B}_1 \perp\!\!\!\perp (\boldsymbol{X}_2, \boldsymbol{Y}_2)$$

we obtain that

$$\widetilde{B}_1 \perp\!\!\!\perp \left(\widetilde{B}_2, \widetilde{B}_3, \ldots, \widetilde{B}_p\right),$$

whereby the $\widetilde{B}_j$'s are mutually independent for $j = 1, 2, \ldots, p$. So we deduce that the corresponding B_j's are also mutually independent for $j = 1, 2, \ldots, p$. Thus we conclude

$$\mathcal{L}(U) = \mathcal{L}\left(\prod_{j=1}^{p} B_j\right),$$

where $\mathcal{L}(B_j) = \mathcal{B}(n-(p-j), m)$ for $j = 1, 2, \ldots, p$.

Next step is to show

iii. $\widetilde{B}_1$, $(\boldsymbol{X}_1^*\boldsymbol{X}_1 + \boldsymbol{Y}_1^*\boldsymbol{Y}_1) - (\boldsymbol{X}_1^*\boldsymbol{X}_2 + \boldsymbol{Y}_1^*\boldsymbol{Y}_2)(\boldsymbol{X}_2^*\boldsymbol{X}_2 + \boldsymbol{Y}_2^*\boldsymbol{Y}_2)^{-1}(\boldsymbol{X}_2^*\boldsymbol{X}_1 + \boldsymbol{Y}_2^*\boldsymbol{Y}_1)$, $(\boldsymbol{X}_2^*\boldsymbol{X}_2 + \boldsymbol{Y}_2^*\boldsymbol{Y}_2)^{\frac{1}{2}}\left((\boldsymbol{X}_2^*\boldsymbol{X}_2 + \boldsymbol{Y}_2^*\boldsymbol{Y}_2)^{-1}(\boldsymbol{X}_2^*\boldsymbol{X}_1 + \boldsymbol{Y}_2^*\boldsymbol{Y}_1) - \boldsymbol{H}_{22}^{-1}\boldsymbol{H}_{21}\right)$ and $(\boldsymbol{X}_2, \boldsymbol{Y}_2)$ are mutually independent.

First we observe that $(\boldsymbol{X}_2^*\boldsymbol{X}_2 + \boldsymbol{Y}_2^*\boldsymbol{Y}_2) > \boldsymbol{O}$ with probability one implying that there exists with probability one a $(p-1) \times (p-1)$ complex matrix $(\boldsymbol{X}_2^*\boldsymbol{X}_2 + \boldsymbol{Y}_2^*\boldsymbol{Y}_2)^{\frac{1}{2}} > \boldsymbol{O}$, such that $(\boldsymbol{X}_2^*\boldsymbol{X}_2 + \boldsymbol{Y}_2^*\boldsymbol{Y}_2) = \left((\boldsymbol{X}_2^*\boldsymbol{X}_2 + \boldsymbol{Y}_2^*\boldsymbol{Y}_2)^{\frac{1}{2}}\right)^2$.

Using the conditional distribution of $(\boldsymbol{X}_1^*, \boldsymbol{Y}_1^*)^*$ given $(\boldsymbol{X}_2^*, \boldsymbol{Y}_2^*)^*$ stated in the proof of i. and Theorem 2.19 page 34 we obtain

$$\begin{aligned}
&\mathcal{L}\left(\left(\begin{pmatrix}\boldsymbol{X}_2\\ \boldsymbol{Y}_2\end{pmatrix}^* \begin{pmatrix}\boldsymbol{X}_2\\ \boldsymbol{Y}_2\end{pmatrix}\right)^{-1} \begin{pmatrix}\boldsymbol{X}_2\\ \boldsymbol{Y}_2\end{pmatrix}^* \begin{pmatrix}\boldsymbol{X}_1\\ \boldsymbol{Y}_1\end{pmatrix} \middle| \begin{pmatrix}\boldsymbol{X}_2\\ \boldsymbol{Y}_2\end{pmatrix}\right) \\
=\ & \mathcal{L}\left((\boldsymbol{X}_2^*\boldsymbol{X}_2 + \boldsymbol{Y}_2^*\boldsymbol{Y}_2)^{-1}(\boldsymbol{X}_2^*\boldsymbol{X}_1 + \boldsymbol{Y}_2^*\boldsymbol{Y}_1) \middle| \begin{pmatrix}\boldsymbol{X}_2\\ \boldsymbol{Y}_2\end{pmatrix}\right) \\
=\ & \mathbb{C}\mathcal{N}_{p-1}\left(\boldsymbol{H}_{22}^{-1}\boldsymbol{H}_{21}, (\boldsymbol{X}_2^*\boldsymbol{X}_2 + \boldsymbol{Y}_2^*\boldsymbol{Y}_2)^{-1} \otimes \left(\boldsymbol{H}_{11} - \boldsymbol{H}_{12}\boldsymbol{H}_{22}^{-1}\boldsymbol{H}_{21}\right)\right),
\end{aligned}$$

whereby it holds that

(3.17)

$$\mathcal{L}\left((X_2^*X_2+Y_2^*Y_2)^{\frac{1}{2}}\left((X_2^*X_2+Y_2^*Y_2)^{-1}(X_2^*X_1+Y_2^*Y_1)-H_{22}^{-1}H_{21}\right)\middle|\begin{pmatrix}X_2\\Y_2\end{pmatrix}\right)$$
$$=\mathbb{C}\mathcal{N}_{p-1}\left(O,I_{p-1}\otimes\left(H_{11}-H_{12}H_{22}^{-1}H_{21}\right)\right).$$

Since $(I_{n+m}-P_0)(X_2^*,Y_2^*)^*H_{22}^{-1}H_{21}=O$ it also follows from the conditional distribution of $(X_1^*,Y_1^*)^*$ given $(X_2^*,Y_2^*)^*$ and Theorem 3.3 page 41 that

$$\mathcal{L}\left(\begin{pmatrix}X_1\\Y_1\end{pmatrix}^*(I_{n+m}-P_0)\begin{pmatrix}X_1\\Y_1\end{pmatrix}\middle|\begin{pmatrix}X_2\\Y_2\end{pmatrix}\right)=\frac{1}{2}\left(H_{11}-H_{12}H_{22}^{-1}H_{21}\right)\chi^2_{2(n+m-(p-1))}. \tag{3.18}$$

Observing that

$$\begin{pmatrix}X_1\\Y_1\end{pmatrix}^*(I_{n+m}-P_0)\begin{pmatrix}X_1\\Y_1\end{pmatrix}$$
$$=(X_1^*X_1+Y_1^*Y_1)-(X_1^*X_2+Y_1^*Y_2)(X_2^*X_2+Y_2^*Y_2)^{-1}(X_2^*X_1+Y_2^*Y_1)$$

and that

$$\left(\begin{pmatrix}X_2\\Y_2\end{pmatrix}^*\begin{pmatrix}X_2\\Y_2\end{pmatrix}\right)^{-1}\begin{pmatrix}X_2\\Y_2\end{pmatrix}^*\begin{pmatrix}X_1\\Y_1\end{pmatrix}=\left(\begin{pmatrix}X_2\\Y_2\end{pmatrix}^*\begin{pmatrix}X_2\\Y_2\end{pmatrix}\right)^{-1}\begin{pmatrix}X_2\\Y_2\end{pmatrix}^*P_0\begin{pmatrix}X_1\\Y_1\end{pmatrix}$$

we deduce from Lemma 3.1 page 56 that

$$\tilde{B}_1,\ (X_1^*X_1+Y_1^*Y_1)-(X_1^*X_2+Y_1^*Y_2)(X_2^*X_2+Y_2^*Y_2)^{-1}(X_2^*X_1+Y_2^*Y_1),$$
$$\text{and } (X_2^*X_2+Y_2^*Y_2)^{\frac{1}{2}}\left((X_2^*X_2+Y_2^*Y_2)^{-1}(X_2^*X_1+Y_2^*Y_1)-H_{22}^{-1}H_{21}\right)$$

are mutually independent given (X_2,Y_2). The conditional distributions in (3.17) and (3.18) do not depend on (X_2,Y_2), whereby we see that

$$(X_2^*X_2+Y_2^*Y_2)^{\frac{1}{2}}\left((X_2^*X_2+Y_2^*Y_2)^{-1}(X_2^*X_1+Y_2^*Y_1)-H_{22}^{-1}H_{21}\right)\perp\!\!\!\perp(X_2,Y_2)$$

and

$$(X_1^*X_1+Y_1^*Y_1)-(X_1^*X_2+Y_1^*Y_2)(X_2^*X_2+Y_2^*Y_2)^{-1}(X_2^*X_1+Y_2^*Y_1)\perp\!\!\!\perp(X_2,Y_2).$$

Recall from the proof of i. that

$$\tilde{B}_1\perp\!\!\!\perp(X_2,Y_2).$$

Combining all the independence according to Theorem 6.3 page 102 we obtain iii.

Finally we conclude

iv. U and $V+W$ are independent.

From the induction assumption we know that

$$\frac{\det(\boldsymbol{V}_{22})}{\det((\boldsymbol{V}+\boldsymbol{W})_{22})} \perp\!\!\!\perp (\boldsymbol{V}+\boldsymbol{W})_{22}\,,$$

therefore we get

$$\frac{\det(\boldsymbol{X}_2^*\boldsymbol{X}_2)}{\det((\boldsymbol{X}_2^*\boldsymbol{X}_2+\boldsymbol{Y}_2^*\boldsymbol{Y}_2))} \perp\!\!\!\perp \boldsymbol{X}_2^*\boldsymbol{X}_2+\boldsymbol{Y}_2^*\boldsymbol{Y}_2\,.$$

Since the expressions above both are functions of $(\boldsymbol{X}_2, \boldsymbol{Y}_2)$ we deduce from iii. that

$$\begin{gathered}\widetilde{B}_1,\ (\boldsymbol{X}_1^*\boldsymbol{X}_1+\boldsymbol{Y}_1^*\boldsymbol{Y}_1)-(\boldsymbol{X}_1^*\boldsymbol{X}_2+\boldsymbol{Y}_1^*\boldsymbol{Y}_2)(\boldsymbol{X}_2^*\boldsymbol{X}_2+\boldsymbol{Y}_2^*\boldsymbol{Y}_2)^{-1}(\boldsymbol{X}_2^*\boldsymbol{X}_1+\boldsymbol{Y}_2^*\boldsymbol{Y}_1)\\ (\boldsymbol{X}_2^*\boldsymbol{X}_2+\boldsymbol{Y}_2^*\boldsymbol{Y}_2)^{\frac{1}{2}}\left((\boldsymbol{X}_2^*\boldsymbol{X}_2+\boldsymbol{Y}_2^*\boldsymbol{Y}_2)^{-1}(\boldsymbol{X}_2^*\boldsymbol{X}_1+\boldsymbol{Y}_2^*\boldsymbol{Y}_1)-\boldsymbol{H}_{22}^{-1}\boldsymbol{H}_{21}\right),\\ \frac{\det(\boldsymbol{X}_2^*\boldsymbol{X}_2)}{\det((\boldsymbol{X}_2^*\boldsymbol{X}_2+\boldsymbol{Y}_2^*\boldsymbol{Y}_2))} \text{ and } \boldsymbol{X}_2^*\boldsymbol{X}_2+\boldsymbol{Y}_2^*\boldsymbol{Y}_2\end{gathered}$$

are mutually independent. For $(\boldsymbol{X}_2^*\boldsymbol{X}_2+\boldsymbol{Y}_2^*\boldsymbol{Y}_2)^{-\frac{1}{2}}=\left((\boldsymbol{X}_2^*\boldsymbol{X}_2+\boldsymbol{Y}_2^*\boldsymbol{Y}_2)^{\frac{1}{2}}\right)^{-1}$ observe that

$$\begin{aligned}(\boldsymbol{X}_2^*\boldsymbol{X}_1+\boldsymbol{Y}_2^*\boldsymbol{Y}_1) &= (\boldsymbol{X}_2^*\boldsymbol{X}_2+\boldsymbol{Y}_2^*\boldsymbol{Y}_2)^{\frac{1}{2}}\left((\boldsymbol{X}_2^*\boldsymbol{X}_2+\boldsymbol{Y}_2^*\boldsymbol{Y}_2)^{-\frac{1}{2}}(\boldsymbol{X}_2^*\boldsymbol{X}_1+\boldsymbol{Y}_2^*\boldsymbol{Y}_1)\right.\\ &\quad \left.-(\boldsymbol{X}_2^*\boldsymbol{X}_2+\boldsymbol{Y}_2^*\boldsymbol{Y}_2)^{\frac{1}{2}}\boldsymbol{H}_{22}^{-1}\boldsymbol{H}_{21}+(\boldsymbol{X}_2^*\boldsymbol{X}_2+\boldsymbol{Y}_2^*\boldsymbol{Y}_2)^{\frac{1}{2}}\boldsymbol{H}_{22}^{-1}\boldsymbol{H}_{21}\right)\end{aligned}$$

and further that

$$\begin{aligned}(\boldsymbol{X}_1^*\boldsymbol{X}_1+\boldsymbol{Y}_1^*\boldsymbol{Y}_1) &= (\boldsymbol{X}_1^*\boldsymbol{X}_1+\boldsymbol{Y}_1^*\boldsymbol{Y}_1)-(\boldsymbol{X}_1^*\boldsymbol{X}_2+\boldsymbol{Y}_1^*\boldsymbol{Y}_2)(\boldsymbol{X}_2^*\boldsymbol{X}_2+\boldsymbol{Y}_2^*\boldsymbol{Y}_2)^{-1}(\boldsymbol{X}_2^*\boldsymbol{X}_1+\boldsymbol{Y}_2^*\boldsymbol{Y}_1)\\ &\quad +(\boldsymbol{X}_1^*\boldsymbol{X}_2+\boldsymbol{Y}_1^*\boldsymbol{Y}_2)(\boldsymbol{X}_2^*\boldsymbol{X}_2+\boldsymbol{Y}_2^*\boldsymbol{Y}_2)^{-1}(\boldsymbol{X}_2^*\boldsymbol{X}_1+\boldsymbol{Y}_2^*\boldsymbol{Y}_1)\,.\end{aligned}$$

Hence we deduce that $\boldsymbol{X}^*\boldsymbol{X}+\boldsymbol{Y}^*\boldsymbol{Y}$ is a function of

$$\begin{gathered}(\boldsymbol{X}_1^*\boldsymbol{X}_1+\boldsymbol{Y}_1^*\boldsymbol{Y}_1)-(\boldsymbol{X}_1^*\boldsymbol{X}_2+\boldsymbol{Y}_1^*\boldsymbol{Y}_2)(\boldsymbol{X}_2^*\boldsymbol{X}_2+\boldsymbol{Y}_2^*\boldsymbol{Y}_2)^{-1}(\boldsymbol{X}_2^*\boldsymbol{X}_1+\boldsymbol{Y}_2^*\boldsymbol{Y}_1),\\ (\boldsymbol{X}_2^*\boldsymbol{X}_2+\boldsymbol{Y}_2^*\boldsymbol{Y}_2)^{\frac{1}{2}}\left((\boldsymbol{X}_2^*\boldsymbol{X}_2+\boldsymbol{Y}_2^*\boldsymbol{Y}_2)^{-1}(\boldsymbol{X}_2^*\boldsymbol{X}_1+\boldsymbol{Y}_2^*\boldsymbol{Y}_1)-\boldsymbol{H}_{22}^{-1}\boldsymbol{H}_{21}\right)\\ \text{and } \boldsymbol{X}_2^*\boldsymbol{X}_2+\boldsymbol{Y}_2^*\boldsymbol{Y}_2\,.\end{gathered}$$

This tells us that

$$\widetilde{B}_1,\ \frac{\det(\boldsymbol{X}_2^*\boldsymbol{X}_2)}{\det(\boldsymbol{X}_2^*\boldsymbol{X}_2+\boldsymbol{Y}_2^*\boldsymbol{Y}_2)} \text{ and } \boldsymbol{X}^*\boldsymbol{X}+\boldsymbol{Y}^*\boldsymbol{Y}$$

are mutually independent. Hence we obtain that

$$B_1,\ \frac{\det(\boldsymbol{V}_{22})}{\det((\boldsymbol{V}+\boldsymbol{W})_{22})} \text{ and } \boldsymbol{V}+\boldsymbol{W}$$

are mutually independent, whereby we finally deduce that

$$U \perp\!\!\!\perp \boldsymbol{V}+\boldsymbol{W}\,.$$

■

We are now able to prove the following theorem. It states the distribution of a complex random variable containing some quadratic forms. This complex random variable appears to be the likelihood ratio test statistic of a later considered test in multivariate linear complex normal models.

Theorem 3.11
Let X be an $n \times p$ complex random matrix with $\mathcal{L}(X) = \mathbb{C}\mathcal{N}_{n\times p}(\Theta, I_n \otimes H)$, where $\Theta \in \mathbb{C}^{n\times p}$ and $H \in \mathbb{C}_{+}^{p\times p}$. Further let P be an $n \times n$ complex matrix representing the orthogonal projection of $\mathbb{C}^n$ onto a k-dimensional subspace and let P_0 be an $n \times n$ complex matrix representing the orthogonal projection of $\mathbb{C}^n$ onto a k_0-dimensional subspace of the k-dimensional subspace. If $n-k \geq p$ and $P_0\Theta = \Theta$, then the distribution of $U = \frac{\det(X^*(I_n-P)X)}{\det(X^*(I_n-P_0)X)}$ is given as

$$\mathcal{L}(U) = \mathbb{C}\mathcal{U}(p, k-k_0, n-k) \ ,$$

and it holds that

$$U,\ X^*(I_n - P_0)X \text{ and } P_0 X$$

are mutually independent.

Proof:
Let the assumptions in the theorem be satisfied and assume that $n-k \geq p$ and $P_0\Theta = \Theta$.

First we define

$$\begin{aligned} U &= \frac{\det(X^*(I_n - P)X)}{\det(X^*(I_n - P_0)X)} \\ &= \frac{\det(X^*(I_n - P)X)}{\det(X^*(I_n - P)X + X^*(P - P_0)X)} \ . \end{aligned}$$

As in the proof of Lemma 3.1 page 56 we deduce that

$$\mathcal{L}(X^*(I_n - P)X) = \mathbb{C}\mathcal{W}_p(H, n-k)$$

and

$$\mathcal{L}(X^*(P - P_0)X) = \mathbb{C}\mathcal{W}_p(H, k-k_0) \ .$$

According to Theorem 3.4 page 42 we know that

$$P_0X, (I_n - P)X \text{ and } (P - P_0)X \tag{3.19}$$

are mutually independent, hence

$$X^*(I_n - P)X \perp\!\!\!\perp X^*(P - P_0)X$$

and we deduce from Definition 3.2 page 55 that

$$\mathcal{L}(U) = \mathbb{C}\mathcal{U}(p, k-k_0, n-k) \ .$$

From the independence in (3.19) it appears that

$$P_0X \perp\!\!\!\perp (U, X^*(I_n - P_0)X) \ ,$$

and from Theorem 3.10 page 57 it holds that

$$U \perp\!\!\!\perp X^* (I_n - P_0) X ,$$

whereby the proof is completed. ∎

Using Theorem 3.10 page 57 we are able to state that a complex U-distribution with $m = 1$ is a beta distribution.

Theorem 3.12 The relation to the beta distribution
Let V and W be $p \times p$ complex random matrices with $\mathcal{L}(V) = \mathbb{C}\mathcal{W}_p(H, n)$ and $\mathcal{L}(W) = \mathbb{C}\mathcal{W}_p(H, 1)$, where $H \in \mathbb{C}_+^{p \times p}$. If V and W are independent and $n \geq p$, then the distribution of $U = \frac{\det(V)}{\det(V+W)}$ is given as

$$\mathcal{L}(U) = \mathcal{B}(n - (p-1), p) .$$

Proof:
Let V and W be $p \times p$ complex random matrices with $\mathcal{L}(V) = \mathbb{C}\mathcal{W}_p(H, n)$ and $\mathcal{L}(W) = \mathbb{C}\mathcal{W}_p(H, 1)$, where $H \in \mathbb{C}_+^{p \times p}$. Further assume that V and W are independent and that $n \geq p$.

Moreover let $S, T_1, T_2, \ldots, T_{p-1}$ and T_p be mutually independent random variables distributed as $\mathcal{L}(S) = \chi^2_{2(n-(p-1))}$ and $\mathcal{L}(T_k) = \chi^2_2$ for $k = 1, 2, \ldots, p$. We define for $j = 1, 2, \ldots, p$ the random variables

$$\widetilde{B}_j = \frac{S + \sum_{k=1}^{j-1} T_k}{S + \sum_{k=1}^{j} T_k} ,$$

then it holds that

$$\mathcal{L}\left(\widetilde{B}_j\right) = \mathcal{B}(n - (p-j), 1) .$$

Furthermore for $j = 1, 2, \ldots, p$ we have

$$\widetilde{B}_j \perp\!\!\!\perp S + \sum_{k=1}^{j} T_k ,$$

whereby it follows that

$$\widetilde{B}_1, S + T_1, T_2, T_3, \ldots, T_{p-1} \text{ and } T_p$$

are mutually independent. This implies that $\widetilde{B}_1$ and $\left(\widetilde{B}_2, \widetilde{B}_3, \ldots, \widetilde{B}_p\right)$ are independent. Using similar arguments on $\widetilde{B}_2, \widetilde{B}_3, \ldots, \widetilde{B}_{p-1}$ and $\widetilde{B}_p$ we obtain that

$$\widetilde{B}_1, \widetilde{B}_2, \ldots, \widetilde{B}_{p-1} \text{ and } \widetilde{B}_p$$

are mutually independent. Observe that

$$\prod_{j=1}^{p} \widetilde{B}_j = \prod_{j=1}^{p} \frac{S + \sum_{k=1}^{j-1} T_k}{S + \sum_{k=1}^{j} T_k} = \frac{S}{S + \sum_{k=1}^{p} T_k} ,$$

where $\mathcal{L}\left(\sum_{k=1}^{p} T_k\right) = \chi^2_{2p}$. Since S and $\sum_{k=1}^{j} T_k$ are independent it hereby follows that

$$\mathcal{L}\left(\prod_{j=1}^{p} \widetilde{B}_j\right) = \mathcal{B}(n-(p-1),p) \ .$$

According to Theorem 3.10 page 57 we have for $U = \frac{\det(\boldsymbol{V})}{\det(\boldsymbol{V}+\boldsymbol{W})}$ that

$$\mathcal{L}(U) = \mathcal{L}\left(\prod_{j=1}^{p} B_j\right) ,$$

where $\mathcal{L}(B_j) = \mathcal{B}(n-(p-j),1)$ and the B_j's are mutually independent for $j = 1, 2, \ldots, p$. Hence the B_j's fulfill the conditions which hold for the $\widetilde{B}_j$'s, whereby we deduce

$$\mathcal{L}\left(\prod_{j=1}^{p} B_j\right) = \mathcal{B}(n-(p-1),p) \ .$$

■

4

Multivariate Linear Complex Normal Models

In this chapter we consider linear models for the multivariate complex normal distribution. The results of linear models are widely known from the literature. In our presentation we make extensive use of vector space considerations and matrix algebra. First we define complex MANOVA models and then maximum likelihood estimation of the parameters in the complex MANOVA model is considered. We determine the maximum likelihood estimators and their distributions. We find that these estimators are expressed by means of a projection matrix representing the orthogonal projection onto the vector space involved in definition of the complex MANOVA model. Besides independence of the estimators are stated and we also derive the normal equations. Finally likelihood ratio test concerning the mean and test of independence in complex MANOVA models are presented. In both tests we find the likelihood ratio test statistic and its distribution. It turns out that the test statistics both are complex *U*-distributed.

4.1 Complex MANOVA Models

Consider n independent p-dimensional complex random vectors $\boldsymbol{X}_1, \boldsymbol{X}_2, \ldots, \boldsymbol{X}_n$ distributed as $\mathcal{L}(\boldsymbol{X}_j) = \mathbb{C}\mathcal{N}_p(\boldsymbol{\theta}_j, \boldsymbol{H})$, where $\boldsymbol{\theta}_j \in \mathbb{C}^p$ and $\boldsymbol{H} \in \mathbb{C}_+^{p \times p}$ for $j = 1, 2, \ldots, n$. The distribution of the $n \times p$ complex random matrix $\boldsymbol{X}$ given by $\boldsymbol{X} = (\boldsymbol{X}_1, \boldsymbol{X}_2, \ldots, \boldsymbol{X}_n)^*$ is

$$\mathcal{L}(\boldsymbol{X}) = \mathbb{C}\mathcal{N}_{n \times p}(\Theta, \boldsymbol{I}_n \otimes \boldsymbol{H}) ,$$

where $\Theta = (\boldsymbol{\theta}_1, \boldsymbol{\theta}_2, \ldots, \boldsymbol{\theta}_n)^* \in \mathbb{C}^{n \times p}$ and $\boldsymbol{H} \in \mathbb{C}_+^{p \times p}$.

We define the complex MANOVA models in the same way as the real MANOVA models are defined. This means that we let Θ belong to a linear subspace of $\mathbb{C}^{n \times p}$ given by

$$M = \{\Theta \in \mathbb{C}^{n \times p} \mid \exists \boldsymbol{C} \in \mathbb{C}^{k \times p} : \Theta = (\boldsymbol{Z} \otimes \boldsymbol{I}_p)(\boldsymbol{C})\} ,$$

where $\boldsymbol{Z}$ is an $n \times k$ known complex matrix with $k \leq n$ satisfying that the range of the linear transformation $\boldsymbol{Z} \otimes \boldsymbol{I}_p$ is M. We denote the range of $\boldsymbol{Z} \otimes \boldsymbol{I}_p$ by $\mathcal{R}[\boldsymbol{Z} \otimes \boldsymbol{I}_p]$, i.e. $M = \mathcal{R}[\boldsymbol{Z} \otimes \boldsymbol{I}_p]$. This model of $\boldsymbol{X}$ is called a *multivariate linear complex normal model* or a *complex multivariate analysis of variance model* (complex MANOVA model).

The above leads to the following definition of a complex MANOVA model.

Definition 4.1 The complex MANOVA model
Let $\boldsymbol{X} = (\boldsymbol{X}_1, \boldsymbol{X}_2, \ldots, \boldsymbol{X}_n)^*$ be an $n \times p$ complex random matrix. If it holds that

1. *$X_1, X_2, \ldots, X_n$ are mutually independent,*
2. *$\mathcal{L}(X_j) = \mathbb{C}\mathcal{N}_p(\theta_j, H)$, where $\theta_j \in \mathbb{C}^p, H \in \mathbb{C}_+^{p \times p}$, $j = 1, 2, \ldots, n$,*
3. *$\Theta = (\theta_1, \theta_2, \ldots, \theta_n)^*$ belongs to a known linear subspace of $\mathbb{C}^{n \times p}$ given by*
$$M = \{\Theta \in \mathbb{C}^{n \times p} \mid \exists C \in \mathbb{C}^{k \times p} : \Theta = (Z \otimes I_p)(C)\},$$

then we say that X is described by a complex MANOVA model.

The linear subspace M together with $\mathbb{C}_+^{p \times p}$ is called the *parameter set* for the complex MANOVA model.

Observe that Θ also can be expressed as

$$\begin{aligned} \Theta &= (Z \otimes I_p)(C) \\ &= ZC. \end{aligned}$$

Thus all the columns of Θ belong to a common subspace of $\mathbb{C}^n$, let us call it N, where $N = \mathcal{R}[Z]$. Note that N is spanned by the columns of Z and that $M = N \otimes \mathbb{C}^p$. Often Z is referred to as the *design matrix.*

Furthermore M is spanned by the kp $n \times p$ complex matrices represented by the kp np-dimensional columns of $Z \otimes I_p$. If moreover these columns are independent, they form a basis for M and the elements of C are the coefficients of Θ in this particular basis. In this case $Z \otimes I_p$ has full rank and the dimension of M is kp.

From the definition of $Z \otimes I_p$ we notice that this $np \times kp$ complex matrix has full rank, namely kp, iff the $n \times k$ complex matrix Z has full rank, namely k. In the following theory we will *assume that Z has full rank,* which implies that the columns of Z form a basis of N, and that the elements in a column of C are the coefficients of the corresponding column of Θ in this particular basis. Moreover in this case the dimension of the subspace N is k.

It should be noticed that the matrix Z is not uniquely determined.

Example 4.1
Let $X = (X_1, X_2, \ldots, X_n)^*$ be an $n \times p$ complex random matrix with $\mathcal{L}(X) = \mathbb{C}\mathcal{N}_{n \times p}(\Theta, I_n \otimes H)$, where $\Theta \in M$ and $H \in \mathbb{C}_+^{p \times p}$. Assume that M is given by $M = \mathcal{R}[Z \otimes I_p]$, where

$$Z = (1_n, T) = \begin{pmatrix} 1 & T_1 \\ 1 & T_2 \\ \vdots & \vdots \\ 1 & T_n \end{pmatrix}$$

and $T_1, T_2, \ldots, T_{n-1}$ and T_n represent n not all equal values of an explanatory variable of X. Hence Θ can be expressed as

$$\Theta = ZC,$$

where $C = (\alpha, \beta)^*$ and $\alpha, \beta \in \mathbb{C}^p$. In other words

$$\mathbb{E}(X) = \begin{pmatrix} 1 & T_1 \\ 1 & T_2 \\ \vdots & \vdots \\ 1 & T_n \end{pmatrix} \begin{pmatrix} \alpha^* \\ \beta^* \end{pmatrix} = \begin{pmatrix} \alpha^* + \beta^* T_1 \\ \alpha^* + \beta^* T_2 \\ \vdots \\ \alpha^* + \beta^* T_n \end{pmatrix},$$

i.e. $\mathbb{E}(X_j) = \alpha + \beta \overline{T}_j$ for $j = 1, 2, \ldots, n$. For $p = 1$ the complex matrix Z is the design matrix of a linear regression model. ■

Later we will consider maximum likelihood estimation and hypothesis testing in the complex MANOVA model with parameter set $M \times \mathbb{C}_+^{p \times p}$. This involves orthogonal projections onto M and N.

Let P_M be an $np \times np$ complex matrix representing the orthogonal projection of $\mathbb{C}^{n \times p}$ onto M. Because Z has full rank, the projection matrix P_M can explicitly be expressed in terms of Z as

$$P_M = Z\,(Z^* Z)^{-1}\,Z^* \otimes I_p\,.$$

It holds that

$$\mathcal{R}\,[P_M] = \mathcal{R}\,[Z \otimes I_p] = M\,.$$

Analogously an $n \times n$ complex matrix representing the orthogonal projection of $\mathbb{C}^n$ onto N is

$$P_N = Z\,(Z^* Z)^{-1}\,Z^*$$

and we know that

$$\mathcal{R}\,[P_N] = \mathcal{R}\,[Z] = N\,.$$

Hereby P_M can be expressed as

$$P_M = P_N \otimes I_p\,.$$

The $np \times np$ complex matrix $P_M^{\perp} = I_{np} - P_M$ represents the orthogonal projection of $\mathbb{C}^{n \times p}$ onto the $(n-k)\,p$-dimensional subspace $M^{\perp}$, which is the orthogonal complement of M w.r.t. the inner product on $\mathbb{C}^{n \times p}$. We observe that

$$\begin{aligned} P_M^{\perp} &= I_{np} - P_M \\ &= (I_n \otimes I_p) - (P_N \otimes I_p) \\ &= (I_n - P_N) \otimes I_p \\ &= P_N^{\perp} \otimes I_p\,, \end{aligned}$$

where $P_N^{\perp}$ is an $n \times n$ complex matrix representing the orthogonal projection of $\mathbb{C}^n$ onto the $(n-k)$-dimensional subspace $N^{\perp}$, which is the orthogonal complement of N w.r.t. the inner product on $\mathbb{C}^n$.

4.2 Maximum Likelihood Estimation in Complex MANOVA Models

In this section we find the maximum likelihood estimators of the parameters in the complex MANOVA model with parameter set $M \times \mathbb{C}_+^{p \times p}$. This model is stated in Section 4.1 page 67. Additionally we consider the distributions of the estimators and independence of them. Goodman (1963) derives the maximum likelihood estimator of the variance matrix in the complex MANOVA model with parameter set $\{O\} \times \mathbb{C}_+^{p \times p}$ and he also derives the distribution of the estimator. Extensions to more general maximum likelihood estimation are made by Giri (1965) and Khatri (1965*a*). Note that we use the word estimator in the meaning of a function of a complex random matrix, where the function is determined by the fact that each observation matrix corresponds to a unique value of the function. Further the term estimate is used to the specify the value of the estimator for an observation of the complex random matrix.

Theorem 4.1 The maximum likelihood estimators
Let $\boldsymbol{X}$ be an $n \times p$ complex random matrix described by the complex MANOVA model with parameter set $M \times \mathbb{C}_+^{p \times p}$, where $M = N \otimes \mathbb{C}^p$ and N is a subspace of $\mathbb{C}^n$. Let $\boldsymbol{P}_N$ and $\boldsymbol{P}_N^\perp$ be $n \times n$ complex matrices representing the orthogonal projections of $\mathbb{C}^n$ onto N and $N^\perp$, respectively. The following properties hold.

1. $\left(\boldsymbol{P}_N \boldsymbol{X}, \boldsymbol{X}^* \boldsymbol{P}_N^\perp \boldsymbol{X}\right)$ *is a sufficient statistic for* $(\Theta, \boldsymbol{H})$.

2. *If* $n - k \geq p$, *then* $\left(\widehat{\Theta}, \widehat{\boldsymbol{H}}\right) = \left(\boldsymbol{P}_N \boldsymbol{X}, \frac{1}{n} \boldsymbol{X}^* \boldsymbol{P}_N^\perp \boldsymbol{X}\right)$ *is the maximum likelihood estimator of* $(\Theta, \boldsymbol{H})$.

Proof:
Let $\boldsymbol{X}$ be described by the complex MANOVA model with parameter set $M \times \mathbb{C}_+^{p \times p}$, where $M = N \otimes \mathbb{C}^p$ and N is a subspace of $\mathbb{C}^n$. Furthermore let $\boldsymbol{P}_N$ and $\boldsymbol{P}_N^\perp$ be $n \times n$ complex matrices representing the orthogonal projections of $\mathbb{C}^n$ onto N and $N^\perp$, respectively. Finally let $\boldsymbol{x} = (\boldsymbol{x}_1, \boldsymbol{x}_2, \ldots, \boldsymbol{x}_n)^*$ represent an observation matrix of $\boldsymbol{X}$

Re 1:
According to the factorization criterion for sufficiency (Silvey 1975, p. 27) a necessary and sufficient condition for $\left(\boldsymbol{P}_N \boldsymbol{X}, \boldsymbol{X}^* \boldsymbol{P}_N^\perp \boldsymbol{X}\right)$ to be sufficient for $(\Theta, \boldsymbol{H})$ is that there exist functions g and h such that

$$f_{\boldsymbol{X}}(\boldsymbol{x} | \Theta, \boldsymbol{H}) = g(\boldsymbol{x})\, h\left(\left(\boldsymbol{P}_N \boldsymbol{x}, \boldsymbol{x}^* \boldsymbol{P}_N^\perp \boldsymbol{x}\right), (\Theta, \boldsymbol{H})\right) .$$

From Theorem 2.17 page 33 we have that the density function of $\boldsymbol{X}$ w.r.t. Lebesgue measure on $\mathbb{C}^{n \times p}$ can be written as

$$f_{\boldsymbol{X}}(\boldsymbol{x} | \Theta, \boldsymbol{H}) = \pi^{-np} \det(\boldsymbol{H})^{-n} \exp\left(-\operatorname{tr}\left((\boldsymbol{x} - \Theta)^* (\boldsymbol{x} - \Theta) \boldsymbol{H}^{-1}\right)\right) .$$

Since $\Theta \in M$ it holds that $\boldsymbol{P}_M\Theta = \Theta$. Hence $(\boldsymbol{P}_N \otimes \boldsymbol{I}_p)\,\Theta = \boldsymbol{P}_N\Theta = \Theta$ and we get

$$\begin{aligned}(\boldsymbol{x}-\Theta)^*(\boldsymbol{x}-\Theta) &= \left(\boldsymbol{P}_N\boldsymbol{x}+\boldsymbol{P}_N^{\perp}\boldsymbol{x}-\Theta\right)^*\left(\boldsymbol{P}_N\boldsymbol{x}+\boldsymbol{P}_N^{\perp}\boldsymbol{x}-\Theta\right)\\ &= \left(\boldsymbol{P}_N(\boldsymbol{x}-\Theta)+\boldsymbol{P}_N^{\perp}\boldsymbol{x}\right)^*\left(\boldsymbol{P}_N(\boldsymbol{x}-\Theta)+\boldsymbol{P}_N^{\perp}\boldsymbol{x}\right)\\ &= \left(\boldsymbol{P}_N(\boldsymbol{x}-\Theta)\right)^*\boldsymbol{P}_N(\boldsymbol{x}-\Theta)+\boldsymbol{x}^*\boldsymbol{P}_N^{\perp}\boldsymbol{x}\,,\end{aligned}$$

where the last conclusion is obtained by using that $\boldsymbol{P}_N\boldsymbol{P}_N^{\perp} = \boldsymbol{P}_N^{\perp}\boldsymbol{P}_N = \boldsymbol{O}$. Then the density function of $\boldsymbol{X}$ w.r.t. Lebesgue measure on $\mathbb{C}^{n\times p}$ is

$$f_{\boldsymbol{X}}(\boldsymbol{x}|\,\Theta,\boldsymbol{H}) = \pi^{-np}\det(\boldsymbol{H})^{-n}\exp\!\left(-\mathrm{tr}\!\left(\left(\boldsymbol{P}_N(\boldsymbol{x}-\Theta)\right)^*\boldsymbol{P}_N(\boldsymbol{x}-\Theta)\,\boldsymbol{H}^{-1}+\boldsymbol{x}^*\boldsymbol{P}_N^{\perp}\boldsymbol{x}\boldsymbol{H}^{-1}\right)\right).$$

Thus we conclude, by letting

$$g(\boldsymbol{x}) = \pi^{-np}$$

and

$$h\!\left(\left(\boldsymbol{P}_N\boldsymbol{x},\boldsymbol{x}^*\boldsymbol{P}_N^{\perp}\boldsymbol{x}\right),(\Theta,\boldsymbol{H})\right) = \det(\boldsymbol{H})^{-n}\exp\!\left(-\mathrm{tr}\!\left(\left(\boldsymbol{P}_N(\boldsymbol{x}-\Theta)\right)^*\boldsymbol{P}_N(\boldsymbol{x}-\Theta)\boldsymbol{H}^{-1}+\boldsymbol{x}^*\boldsymbol{P}_N^{\perp}\boldsymbol{x}\boldsymbol{H}^{-1}\right)\right),$$

that $\left(\boldsymbol{P}_N\boldsymbol{X},\boldsymbol{X}^*\boldsymbol{P}_N^{\perp}\boldsymbol{X}\right)$ is a sufficient statistic for $(\Theta,\boldsymbol{H})$.

Re 2:
Assume that $n-k \geq p$. The likelihood function for $(\Theta,\boldsymbol{H})$ is from the proof of part 1 given as

$$L(\Theta,\boldsymbol{H}|\,\boldsymbol{x}) = \pi^{-np}\det(\boldsymbol{H})^{-n}\exp\left(-\mathrm{tr}\!\left(\boldsymbol{P}_N(\boldsymbol{x}-\Theta)\,\boldsymbol{H}^{-1}(\boldsymbol{P}_N(\boldsymbol{x}-\Theta))^*+\boldsymbol{x}^*\boldsymbol{P}_N^{\perp}\boldsymbol{x}\boldsymbol{H}^{-1}\right)\right).$$

Since $\boldsymbol{H}^{-1} > \boldsymbol{O}$ we have that $\boldsymbol{P}_N(\boldsymbol{x}-\Theta)\,\boldsymbol{H}^{-1}(\boldsymbol{P}_N(\boldsymbol{x}-\Theta))^* \geq \boldsymbol{O}$, which yields that $\mathrm{tr}\left(\boldsymbol{P}_N(\boldsymbol{x}-\Theta)\,\boldsymbol{H}^{-1}(\boldsymbol{P}_N(\boldsymbol{x}-\Theta))^*\right) \geq 0$. Thus

$$L(\Theta,\boldsymbol{H}|\,\boldsymbol{x}) \;\leq\; \pi^{-np}\det(\boldsymbol{H})^{-n}\exp\left(-\,\mathrm{tr}\left(\boldsymbol{x}^*\boldsymbol{P}_N^{\perp}\boldsymbol{x}\boldsymbol{H}^{-1}\right)\right)$$

with equality iff $\Theta = \boldsymbol{P}_N\boldsymbol{x}$. Hence we conclude that $L(\Theta,\boldsymbol{H}|\,\boldsymbol{x})$ is maximized uniquely for any $\boldsymbol{H} > \boldsymbol{O}$ when $\Theta = \boldsymbol{P}_N\boldsymbol{x}$. Let $\widehat{\Theta} = \boldsymbol{P}_N\boldsymbol{X}$. Then we have observed for a given observation matrix $\boldsymbol{x}$ that

$$L\left(\widehat{\Theta},\boldsymbol{H}\middle|\,\boldsymbol{x}\right) \geq L(\Theta,\boldsymbol{H}|\,\boldsymbol{x})$$

for all $\boldsymbol{H} \in \mathbb{C}_+^{p\times p}$ with equality iff $\Theta = \widehat{\Theta}$.

The next step is to maximize $L\left(\widehat{\Theta},\boldsymbol{H}\middle|\,\boldsymbol{x}\right)$ subject to $\boldsymbol{H} > \boldsymbol{O}$. First we notice that the log-likelihood function of $\left(\widehat{\Theta},\boldsymbol{H}\right)$ is given by

$$l\left(\widehat{\Theta},\boldsymbol{H}\middle|\,\boldsymbol{x}\right) = -np\log\pi - n\left(\log(\det(\boldsymbol{H}))+\mathrm{tr}\left(\frac{1}{n}\boldsymbol{x}^*\boldsymbol{P}_N^{\perp}\boldsymbol{x}\boldsymbol{H}^{-1}\right)\right).$$

Note that $\boldsymbol{H}$ has rank p. Since $\boldsymbol{H}^{-1}$ is positive definite there exists a positive definite matrix $\boldsymbol{H}^{-\frac{1}{2}}$ such that $\boldsymbol{H}^{-1} = \left(\boldsymbol{H}^{-\frac{1}{2}}\right)^2$. Let $\lambda_1,\lambda_2,\ldots,\lambda_p$ be the eigenvalues of $\frac{1}{n}\boldsymbol{x}^*\boldsymbol{P}_N^{\perp}\boldsymbol{x}\boldsymbol{H}^{-1}$ that is of $\frac{1}{n}\boldsymbol{H}^{-\frac{1}{2}}\boldsymbol{x}^*\boldsymbol{P}_N^{\perp}\boldsymbol{x}\boldsymbol{H}^{-\frac{1}{2}}$. Since $\frac{1}{n}\boldsymbol{H}^{-\frac{1}{2}}\boldsymbol{x}^*\boldsymbol{P}_N^{\perp}\boldsymbol{x}\boldsymbol{H}^{-\frac{1}{2}}$ is Hermitian the eigenvalues are real. As $\boldsymbol{P}_N^{\perp}\boldsymbol{\Theta} = \boldsymbol{O}$, $n - k \geq p$ and $\boldsymbol{H} > \boldsymbol{O}$ we have from Theorem 3.9 page 53 that $\boldsymbol{x}^*\boldsymbol{P}_N^{\perp}\boldsymbol{x} > \boldsymbol{O}$ with probability one. Since $\operatorname{rank}\left(\boldsymbol{H}^{-\frac{1}{2}}\right) = p$ it holds that $\frac{1}{n}\boldsymbol{H}^{-\frac{1}{2}}\boldsymbol{x}^*\boldsymbol{P}_N^{\perp}\boldsymbol{x}\boldsymbol{H}^{-\frac{1}{2}} > \boldsymbol{O}$ and the real eigenvalues $\lambda_1, \lambda_2, \ldots, \lambda_p$ are positive. By using the well-known result for positive real numbers that $\log u - u \leq -1$, with equality iff $u = 1$, we get

$$\begin{aligned}
&\log(\det(\boldsymbol{H})) + \operatorname{tr}\left(\frac{1}{n}\boldsymbol{x}^*\boldsymbol{P}_N^{\perp}\boldsymbol{x}\boldsymbol{H}^{-1}\right) - \log\left(\det\left(\frac{1}{n}\boldsymbol{x}^*\boldsymbol{P}_N^{\perp}\boldsymbol{x}\right)\right) - \operatorname{tr}\left(\frac{1}{n}\boldsymbol{x}^*\boldsymbol{P}_N^{\perp}\boldsymbol{x}\left(\frac{1}{n}\boldsymbol{x}^*\boldsymbol{P}_N^{\perp}\boldsymbol{x}\right)^{-1}\right) \\
&= -\log\left(\det\left(\frac{1}{n}\boldsymbol{x}^*\boldsymbol{P}_N^{\perp}\boldsymbol{x}\boldsymbol{H}^{-1}\right)\right) + \operatorname{tr}\left(\frac{1}{n}\boldsymbol{x}^*\boldsymbol{P}_N^{\perp}\boldsymbol{x}\boldsymbol{H}^{-1}\right) - p \\
&= -\log\left(\prod_{j=1}^{p}\lambda_j\right) + \sum_{j=1}^{p}\lambda_j - p \\
&= \sum_{j=1}^{p}(-\log\lambda_j + \lambda_j - 1) \\
&\geq 0,
\end{aligned}$$

with equality iff $\lambda_j = 1$ for $j = 1, 2, \ldots, p$. Therefore

$$\log(\det(\boldsymbol{H})) + \operatorname{tr}\left(\frac{1}{n}\boldsymbol{x}^*\boldsymbol{P}_N^{\perp}\boldsymbol{x}\boldsymbol{H}^{-1}\right) \geq \log\left(\det\left(\frac{1}{n}\boldsymbol{x}^*\boldsymbol{P}_N^{\perp}\boldsymbol{x}\right)\right) + \operatorname{tr}\left(\frac{1}{n}\boldsymbol{x}^*\boldsymbol{P}_N^{\perp}\boldsymbol{x}\left(\frac{1}{n}\boldsymbol{x}^*\boldsymbol{P}_N^{\perp}\boldsymbol{x}\right)^{-1}\right),$$

and equality occurs iff $\boldsymbol{H} = \frac{1}{n}\boldsymbol{x}^*\boldsymbol{P}_N^{\perp}\boldsymbol{x}$. Thus $l\left(\widehat{\boldsymbol{\Theta}}, \boldsymbol{H}\middle|\boldsymbol{x}\right)$ is maximized uniquely when $\boldsymbol{H} = \frac{1}{n}\boldsymbol{x}^*\boldsymbol{P}_N^{\perp}\boldsymbol{x}$. Let $\widehat{\boldsymbol{H}} = \frac{1}{n}\boldsymbol{X}^*\boldsymbol{P}_N^{\perp}\boldsymbol{X}$. Altogether we have seen for a given observation matrix $\boldsymbol{x}$ that

$$L\left(\widehat{\boldsymbol{\Theta}}, \widehat{\boldsymbol{H}}\middle|\boldsymbol{x}\right) \geq L\left(\widehat{\boldsymbol{\Theta}}, \boldsymbol{H}\middle|\boldsymbol{x}\right) \geq L\left(\boldsymbol{\Theta}, \boldsymbol{H}\middle|\boldsymbol{x}\right)$$

with equalities iff $\boldsymbol{\Theta} = \widehat{\boldsymbol{\Theta}}$ and $\boldsymbol{H} = \widehat{\boldsymbol{H}}$. Hereby $\widehat{\boldsymbol{\Theta}} = \boldsymbol{P}_N\boldsymbol{X}$ and $\widehat{\boldsymbol{H}} = \frac{1}{n}\boldsymbol{X}^*\boldsymbol{P}_N^{\perp}\boldsymbol{X}$ are the maximum likelihood estimators of $\boldsymbol{\Theta}$ and $\boldsymbol{H}$, respectively. ■

Notice that the maximum value of the likelihood function is

$$L\left(\widehat{\boldsymbol{\Theta}}, \widehat{\boldsymbol{H}}\middle|\boldsymbol{x}\right) = \pi^{-np}\det\left(\frac{1}{n}\boldsymbol{x}^*\boldsymbol{P}_N^{\perp}\boldsymbol{x}\right)^{-n}\exp(-np), \tag{4.1}$$

where for a given observation matrix of $\boldsymbol{X}$ it holds that $\frac{1}{n}\boldsymbol{x}^*\boldsymbol{P}_N^{\perp}\boldsymbol{x}$ is the unique maximum likelihood estimate of $\boldsymbol{H}$.

Next we find some equations from which we are able to determine the maximum likelihood estimator $\widehat{\boldsymbol{\Theta}}$. Observe that the columns of $\boldsymbol{Z}$ form a basis for N and the columns of $(\boldsymbol{I}_n - \boldsymbol{P}_N)\boldsymbol{X}$ belong to $N^{\perp}$, whereby we have that

$$\begin{aligned}
\boldsymbol{Z}^*(\boldsymbol{I}_n - \boldsymbol{P}_N)\boldsymbol{X} &= \boldsymbol{O} \\
&\Updownarrow \\
\boldsymbol{Z}^*\boldsymbol{X} - \boldsymbol{Z}^*\boldsymbol{P}_N\boldsymbol{X} &= \boldsymbol{O}.
\end{aligned}$$

Using that $\widehat{\Theta} = \boldsymbol{P}_N\boldsymbol{X}$ and that $\widehat{\Theta}$ also can be expressed as $\widehat{\Theta} = \boldsymbol{Z}\widehat{\boldsymbol{C}}$, where $\widehat{\boldsymbol{C}}$ for a given $\boldsymbol{Z}$ is the maximum likelihood estimator of $\boldsymbol{C}$, it follows that

$$\begin{aligned} \boldsymbol{Z}^*\boldsymbol{X} - \boldsymbol{Z}^*\boldsymbol{Z}\widehat{\boldsymbol{C}} &= \boldsymbol{O} \\ \Updownarrow \qquad & \\ \boldsymbol{Z}^*\boldsymbol{Z}\widehat{\boldsymbol{C}} &= \boldsymbol{Z}^*\boldsymbol{X} . \end{aligned}$$

The last relation is called the *normal equations* and from these we are able to determine $\widehat{\boldsymbol{C}}$ and hereby $\widehat{\Theta}$. When $\boldsymbol{Z}$ has full rank, as assumed, $\widehat{\boldsymbol{C}}$ is for a given $\boldsymbol{Z}$ uniquely determined as

$$\widehat{\boldsymbol{C}} = (\boldsymbol{Z}^*\boldsymbol{Z})^{-1}\boldsymbol{Z}^*\boldsymbol{X} .$$

The above leads to the following theorem.

Theorem 4.2
Let $\boldsymbol{X}$ be an $n \times p$ complex random matrix described by the complex MANOVA model with parameter set $M \times \mathbb{C}_+^{p\times p}$, where $M = N \otimes \mathbb{C}^p$ and N is a subspace of $\mathbb{C}^n$. If $\boldsymbol{Z}$ is an $n \times k$ complex matrix such that $k \leq n$, $\mathrm{rank}(\boldsymbol{Z}) = k$ and $\mathcal{R}[\boldsymbol{Z}] = N$, then the maximum likelihood estimator of $\boldsymbol{C}$ is given by

$$\widehat{\boldsymbol{C}} = (\boldsymbol{Z}^*\boldsymbol{Z})^{-1}\boldsymbol{Z}^*\boldsymbol{X} .$$

4.2.1 Distributions of the Maximum Likelihood Estimators

In this section we derive the distribution of the maximum likelihood estimators $\widehat{\Theta}, \widehat{\boldsymbol{C}}$ and $\widehat{\boldsymbol{H}}$. Besides independence of $\widehat{\Theta}$ and $\widehat{\boldsymbol{H}}$ is proved.

Theorem 4.3 Distributional results of the maximum likelihood estimators
Let $\boldsymbol{X}$ be an $n \times p$ complex random matrix described by the complex MANOVA model with parameter set $M \times \mathbb{C}_+^{p\times p}$, where $M = N \otimes \mathbb{C}^p$ and N is a subspace of $\mathbb{C}^n$. Furthermore let $\boldsymbol{P}_N$ and $\boldsymbol{P}_N^\perp$ be $n \times n$ complex matrices representing the orthogonal projections of $\mathbb{C}^n$ onto N and $N^\perp$, respectively, and let $\boldsymbol{Z}$ be an $n \times k$ complex matrix such that $k \leq n$, $\mathrm{rank}(\boldsymbol{Z}) = k$ and $\mathcal{R}[\boldsymbol{Z}] = N$. The respective distributions of the maximum likelihood estimators $\widehat{\Theta} = \boldsymbol{P}_N\boldsymbol{X}$, $\widehat{\boldsymbol{C}} = (\boldsymbol{Z}^*\boldsymbol{Z})^{-1}\boldsymbol{Z}^*\boldsymbol{X}$ and $n\widehat{\boldsymbol{H}} = \boldsymbol{X}^*\boldsymbol{P}_N^\perp\boldsymbol{X}$ are

$$\begin{aligned} \mathcal{L}\left(\widehat{\Theta}\right) &= \mathbb{C}\mathcal{N}_{n\times p}(\Theta, \boldsymbol{P}_N \otimes \boldsymbol{H}) , \\ \mathcal{L}\left(\widehat{\boldsymbol{C}}\right) &= \mathbb{C}\mathcal{N}_{n\times p}\left(\boldsymbol{C}, (\boldsymbol{Z}^*\boldsymbol{Z})^{-1} \otimes \boldsymbol{H}\right) \end{aligned}$$

and

$$\mathcal{L}\left(n\widehat{\boldsymbol{H}}\right) = \mathbb{C}\mathcal{W}_p(\boldsymbol{H}, n-k) .$$

Furthermore it holds that

$$\widehat{\Theta} \perp\!\!\!\perp \widehat{\boldsymbol{H}} .$$

Proof:
Let $\boldsymbol{X}$ be an $n \times p$ complex random matrix described by the complex MANOVA model with parameter set $M \times \mathbb{C}_{+}^{p \times p}$, where $M = N \otimes \mathbb{C}^{p}$ and N is a subspace of $\mathbb{C}^{n}$. Furthermore let $\boldsymbol{P}_N$ and $\boldsymbol{P}_N^{\perp}$ be $n \times n$ complex matrices representing the orthogonal projections of $\mathbb{C}^n$ onto N and $N^{\perp}$, respectively, and let $\boldsymbol{Z}$ be an $n \times k$ complex matrix such that $k \leq n$, rank $(\boldsymbol{Z}) = k$ and $\mathcal{R}[\boldsymbol{Z}] = N$.

From Theorem 2.19 page 34 we have

$$\begin{aligned}
\mathcal{L}\left(\widehat{\boldsymbol{\Theta}}\right) &= \mathcal{L}(\boldsymbol{P}_N \boldsymbol{X}) \\
&= \mathbb{C}\mathcal{N}_{n \times p}(\boldsymbol{P}_N \boldsymbol{\Theta}, \boldsymbol{P}_N \boldsymbol{P}_N^{*} \otimes \boldsymbol{H}) \\
&= \mathbb{C}\mathcal{N}_{n \times p}(\boldsymbol{\Theta}, \boldsymbol{P}_N \otimes \boldsymbol{H})
\end{aligned}$$

and further

$$\begin{aligned}
\mathcal{L}\left(\widehat{\boldsymbol{C}}\right) &= \mathcal{L}\left((\boldsymbol{Z}^{*}\boldsymbol{Z})^{-1} \boldsymbol{Z}^{*}\boldsymbol{X}\right) \\
&= \mathbb{C}\mathcal{N}_{n \times p}\left((\boldsymbol{Z}^{*}\boldsymbol{Z})^{-1} \boldsymbol{Z}^{*}\boldsymbol{Z}\boldsymbol{C}, (\boldsymbol{Z}^{*}\boldsymbol{Z})^{-1} \boldsymbol{Z}^{*}\boldsymbol{Z}\,(\boldsymbol{Z}^{*}\boldsymbol{Z})^{-1} \otimes \boldsymbol{H}\right) \\
&= \mathbb{C}\mathcal{N}_{n \times p}\left(\boldsymbol{C}, (\boldsymbol{Z}^{*}\boldsymbol{Z})^{-1} \otimes \boldsymbol{H}\right) .
\end{aligned}$$

We have $n\widehat{\boldsymbol{H}} = \boldsymbol{X}^{*}\boldsymbol{P}_N^{\perp}\boldsymbol{X}$ and $\boldsymbol{P}_N^{\perp}\boldsymbol{\Theta} = \boldsymbol{O}$. Then according to Theorem 3.3 page 41 the distribution of $n\widehat{\boldsymbol{H}}$ is given as

$$\mathcal{L}\left(n\widehat{\boldsymbol{H}}\right) = \mathbb{C}\mathcal{W}_p(\boldsymbol{H}, n-k) .$$

Finally we observe from Corollary 3.1 page 43 that $\boldsymbol{P}_N\boldsymbol{X}$ and $\boldsymbol{P}_N^{\perp}\boldsymbol{X}$ are independent. Further we have that $\boldsymbol{X}^{*}\boldsymbol{P}_N^{\perp}\boldsymbol{X} = \left(\boldsymbol{P}_N^{\perp}\boldsymbol{X}\right)^{*}\left(\boldsymbol{P}_N^{\perp}\boldsymbol{X}\right)$, thus $\widehat{\boldsymbol{\Theta}} = \boldsymbol{P}_N\boldsymbol{X}$ and $\widehat{\boldsymbol{H}} = \frac{1}{n}\boldsymbol{X}^{*}\boldsymbol{P}_N^{\perp}\boldsymbol{X}$ are independent. ∎

Remark from the theorem above that

$$\mathbb{E}\left(\widehat{\boldsymbol{\Theta}}\right) = \boldsymbol{\Theta} \text{ and } \mathbb{E}\left(\widehat{\boldsymbol{C}}\right) = \boldsymbol{C} ,$$

i.e. $\widehat{\boldsymbol{\Theta}}$ and $\widehat{\boldsymbol{C}}$ are unbiased maximum likelihood estimators of $\boldsymbol{\Theta}$ and $\boldsymbol{C}$, respectively. Whereas the maximum likelihood estimator $\widehat{\boldsymbol{H}}$ is biased, since from Theorem 3.1 page 40 it holds that

$$\mathbb{E}\left(\widehat{\boldsymbol{H}}\right) = \frac{n-k}{n}\boldsymbol{H} .$$

Therefore the statistic given by

$$\boldsymbol{S} = \frac{1}{n-k}\boldsymbol{X}^{*}\boldsymbol{P}_N^{\perp}\boldsymbol{X} \tag{4.2}$$

is often used as an unbiased estimator for $\boldsymbol{H}$.

4.3 Hypothesis Testing in Complex MANOVA Models

In this section we consider hypothesis testing in the complex MANOVA model with parameter set $M \times \mathbb{C}_+^{p\times p}$. This model is defined in Section 4.1 page 67. We consider a hypothesis concerning the structure of the subspace of $\mathbb{C}^{n\times p}$, which Θ belongs to. Further the hypothesis of independence of parts of a complex random matrix is considered. Various tests based on the multivariate complex normal distribution are treated by Giri (1965), Khatri (1965*a*) and Khatri (1965*b*).

Let $\boldsymbol{X} = (\boldsymbol{X}_1, \boldsymbol{X}_2, \ldots, \boldsymbol{X}_n)^*$ be an $n \times p$ complex random matrix described by the complex MANOVA model with parameter set $M \times \mathbb{C}_+^{p\times p}$, where $M = N \otimes \mathbb{C}^p$ and N is a subspace of $\mathbb{C}^n$. Assume that $\boldsymbol{H} \in \mathbb{C}_+^{p\times p}$ is unknown and let $\boldsymbol{x} = (\boldsymbol{x}_1, \boldsymbol{x}_2, \ldots, \boldsymbol{x}_n)^*$ represent an observation matrix of $\boldsymbol{X}$.

4.3.1 Likelihood Ratio Test Concerning the Mean

Let the $n \times k$ complex design matrix $\boldsymbol{Z}$ be partitioned as

$$\boldsymbol{Z} = (\boldsymbol{Z}_0, \boldsymbol{Z}_1) ,$$

where $\boldsymbol{Z}_0$ and $\boldsymbol{Z}_1$ are complex matrices of dimensions $n \times k_0$ and $n \times (k - k_0)$, respectively. From the assumption that $\boldsymbol{Z}$ has full rank k we see both $\boldsymbol{Z}_0$ and $\boldsymbol{Z}_1$ have full rank k_0 and $k - k_0$, respectively. Assume that $n - k \geq p$ to ensure the existence of the maximum likelihood estimator of $\boldsymbol{H}$. The matrices represented by the columns of $\boldsymbol{Z}_0 \otimes \boldsymbol{I}_p$ form a basis for the subspace M_0 of M given by

$$M_0 = \{\Theta \in \mathbb{C}^{n\times p} \mid \exists \boldsymbol{C}_0 \in \mathbb{C}^{k_0\times p} : \Theta = (\boldsymbol{Z}_0 \otimes \boldsymbol{I}_p)(\boldsymbol{C}_0)\} ,$$

i.e. $M_0 = \mathcal{R}[\boldsymbol{Z}_0 \otimes \boldsymbol{I}_p]$. The dimension of M_0 is $k_0 p$. Furthermore we denote the range of $\boldsymbol{Z}_0$ by N_0, i.e. $N_0 = \mathcal{R}[\boldsymbol{Z}_0]$. Thus N_0 has dimension k_0 and is a subspace of N. Moreover $M_0 = N_0 \otimes \mathbb{C}^p$.

Since the columns of $\boldsymbol{Z}$ form a basis for M, it will always be possible to rearrange the columns of $\boldsymbol{Z}$ or to introduce some linear combinations of the columns, as long as the matrix is $n \times k$ with rank k, to obtain a wanted partition of $\boldsymbol{Z}$ in the above form.

We wish to test the null hypothesis

$$\mathrm{H}_0 : \Theta \in M_0$$

under the hypothesis

$$\mathrm{H} : \Theta \in M .$$

To perform this test we use the likelihood ratio test which consists in rejecting H_0 if the likelihood ratio, defined by

$$U(\boldsymbol{x}) = \frac{\sup_{\boldsymbol{\Theta}\in M_0, \boldsymbol{H}\in\mathbb{C}_+^{p\times p}} L(\boldsymbol{\Theta},\boldsymbol{H}|\,\boldsymbol{x})}{\sup_{\boldsymbol{\Theta}\in M, \boldsymbol{H}\in\mathbb{C}_+^{p\times p}} L(\boldsymbol{\Theta},\boldsymbol{H}|\,\boldsymbol{x})},$$

is smaller than some chosen constant providing the size of the test.

As stated in Theorem 4.1 page 70 the maximum likelihood estimators of $\boldsymbol{\Theta}$ and $\boldsymbol{H}$ under H are given by

$$\widehat{\boldsymbol{\Theta}} = \boldsymbol{P}_N\boldsymbol{X} \text{ and } n\widehat{\boldsymbol{H}} = \boldsymbol{X}^*\boldsymbol{P}_N^{\perp}\boldsymbol{X},$$

where $\boldsymbol{P}_N$ and $\boldsymbol{P}_N^{\perp}$ are $n \times n$ complex matrices representing the orthogonal projections of $\mathbb{C}^n$ onto N and $N^{\perp}$, respectively.

The maximum of the likelihood function under H is by (4.1) page 72 given as

$$L\left(\widehat{\boldsymbol{\Theta}},\widehat{\boldsymbol{H}}\middle|\,\boldsymbol{x}\right) = \pi^{-np}\det\left(\frac{1}{n}\boldsymbol{x}^*\boldsymbol{P}_N^{\perp}\boldsymbol{x}\right)^{-n}\exp(-np),$$

where for a given observation matrix of $\boldsymbol{X}$ it holds that $\frac{1}{n}\boldsymbol{x}^*\boldsymbol{P}_N^{\perp}\boldsymbol{x}$ is the unique maximum likelihood estimate of $\boldsymbol{H}$ under H.

Considering the maximum likelihood estimators of $\boldsymbol{\Theta}$ and $\boldsymbol{H}$ under H_0, which we denote by $\widehat{\boldsymbol{\Theta}}_0$ and $\widehat{\boldsymbol{H}}_0$, respectively, we get by an argumentation similar to the one used to find the maximum likelihood estimators under H that

$$\widehat{\boldsymbol{\Theta}}_0 = \boldsymbol{P}_{N_0}\boldsymbol{X} \text{ and } n\widehat{\boldsymbol{H}}_0 = \boldsymbol{X}^*\boldsymbol{P}_{N_0}^{\perp}\boldsymbol{X},$$

where $\boldsymbol{P}_{N_0}$ and $\boldsymbol{P}_{N_0}^{\perp}$ are $n \times n$ complex matrices representing the orthogonal projections of $\mathbb{C}^n$ onto N_0 and $N_0^{\perp}$, respectively.

Hereby the maximum of the likelihood function under H_0 is

$$L\left(\widehat{\boldsymbol{\Theta}}_0,\widehat{\boldsymbol{H}}_0\middle|\,\boldsymbol{x}\right) = \pi^{-np}\det\left(\frac{1}{n}\boldsymbol{x}^*\boldsymbol{P}_{N_0}^{\perp}\boldsymbol{x}\right)^{-n}\exp(-np),$$

where for a given observation matrix of $\boldsymbol{X}$ it holds that $\frac{1}{n}\boldsymbol{x}^*\boldsymbol{P}_{N_0}^{\perp}\boldsymbol{x}$ is the unique maximum likelihood estimate of $\boldsymbol{H}$ under H_0.

Combining these results we get the likelihood ratio as

$$U(\boldsymbol{x}) = \frac{\det\left(\frac{1}{n}\boldsymbol{x}^*\boldsymbol{P}_{N_0}^{\perp}\boldsymbol{x}\right)^{-n}}{\det\left(\frac{1}{n}\boldsymbol{x}^*\boldsymbol{P}_N^{\perp}\boldsymbol{x}\right)^{-n}}$$

implying an equivalent likelihood ratio given as

$$U^{\frac{1}{n}}(\boldsymbol{x}) = \frac{\det\left(\boldsymbol{x}^*\boldsymbol{P}_N^{\perp}\boldsymbol{x}\right)}{\det\left(\boldsymbol{x}^*\boldsymbol{P}_{N_0}^{\perp}\boldsymbol{x}\right)}.$$

We reject H_0 for small values of $U^{\frac{1}{n}}(\boldsymbol{x})$.

The likelihood ratio test statistic is from the above given by

$$U^{\frac{1}{n}}(\boldsymbol{X}) = \frac{\det\left(\boldsymbol{X}^*\boldsymbol{P}_N^{\perp}\boldsymbol{X}\right)}{\det\left(\boldsymbol{X}^*\boldsymbol{P}_{N_0}^{\perp}\boldsymbol{X}\right)} = \frac{\det\left(\widehat{\boldsymbol{H}}\right)}{\det\left(\widehat{\boldsymbol{H}}_0\right)} .$$

Then the likelihood ratio test of size α of H_0 under H is determined by the critical region $U^{\frac{1}{n}}(\boldsymbol{x}) \leq q$, where q fulfills $P\left(U^{\frac{1}{n}}(\boldsymbol{X}) \leq q \middle| \mathrm{H}_0\right) = \alpha$.

To determine q the distribution of $U^{\frac{1}{n}}(\boldsymbol{X})$ under H_0 is necessary. Using that $n-k \geq p$ and that $\boldsymbol{P}_{N_0}\Theta = \Theta$ under H_0 we deduce by Theorem 3.11 page 64 that

$$\mathcal{L}\left(U^{\frac{1}{n}}(\boldsymbol{X})\right) = \mathbb{C}\mathcal{U}(p, k-k_0, n-k) .$$

Besides we get that

$$U^{\frac{1}{n}}(\boldsymbol{X}), \ \boldsymbol{X}^*\boldsymbol{P}_{N_0}^{\perp}\boldsymbol{X} \text{ and } \boldsymbol{P}_{N_0}\boldsymbol{X}$$

are mutually independent. This tells us that the likelihood ratio test statistic and the maximum likelihood estimators of Θ and $\boldsymbol{H}$ are mutually independent under H_0.

Even though the distribution of $U^{\frac{1}{n}}(\boldsymbol{X})$ under H_0 is known we are not able to determine q, since there exist no tables on the complex U-distribution. According to Theorem 3.10 page 57 it follows that the distribution of the likelihood ratio test statistic also is the distribution of a product of p mutually independent beta distributed random variables, i.e.

$$\mathcal{L}\left(U^{\frac{1}{n}}(\boldsymbol{X})\right) = \mathcal{L}\left(\prod_{j=1}^{p} B_j\right) ,$$

where $\mathcal{L}(B_j) = \mathcal{B}(n-k-(p-j), k-k_0)$ and the B_j's for $j = 1, 2, \ldots, p$ are mutually independent. Knowing this, an approximation of the distribution of $-2\log U(\boldsymbol{X})$ can be found e.g. by methods described in Jensen (1991). Besides we know that if $k - k_0 = 1$, then according to Theorem 3.12 page 65

$$\mathcal{L}\left(U^{\frac{1}{n}}(\boldsymbol{X})\right) = \mathcal{B}(n-k-(p-1), p)$$

and the exact quantiles can be found. Altogether we have obtained the following theorem.

Theorem 4.4 Likelihood ratio test concerning the mean
The likelihood ratio test in the complex MANOVA model with parameter set $M \times \mathbb{C}_+^{p\times p}$ *of the null hypothesis*

$$\mathrm{H}_0 : \Theta \in M_0$$

under the hypothesis

$$\mathrm{H}: \Theta \in M ,$$

where $M_0 \subset M \subseteq \mathbb{C}^{n\times p}$*, consists in rejecting* H_0 *if*

$$U^{\frac{1}{n}}(\boldsymbol{x}) \leq q .$$

The likelihood ratio test statistic is given by

$$U^{\frac{1}{n}}(\boldsymbol{X}) = \frac{\det\left(\widehat{\boldsymbol{H}}\right)}{\det\left(\widehat{\boldsymbol{H}}_0\right)} ,$$

where $\widehat{\boldsymbol{H}}$ *and* $\widehat{\boldsymbol{H}}_0$ *are the maximum likelihood estimators of* $\boldsymbol{H}$ *under* H *and* H_0*, respectively. Under* H_0 *the distribution of* $U^{\frac{1}{n}}(\boldsymbol{X})$ *is*

$$\mathcal{L}\left(U^{\frac{1}{n}}(\boldsymbol{X})\right) = \mathbb{C}\mathcal{U}(p, k-k_0, n-k) ,$$

and it holds that

$$U^{\frac{1}{n}}(\boldsymbol{X}) , \widehat{\Theta}_0 \text{ and } \widehat{\boldsymbol{H}}_0$$

are mutually independent, where $\widehat{\Theta}_0$ *is the maximum likelihood estimator of* Θ *under* H_0*. The constant* q *is chosen to provide a test of size* α*, i.e.* q *must fulfill* $P\left(U^{\frac{1}{n}}(\boldsymbol{X}) \leq q \middle| \mathrm{H}_0\right) = \alpha$*.*

If the null hypothesis examined previously is accepted it may be relevant to examine the hypothesis $\Theta = \Theta_0$, where $\Theta_0 \in M_0$ is a particular complex matrix. This can be done by letting

$$\begin{aligned} \boldsymbol{X} &:= \boldsymbol{X} - \Theta_0 , \\ M &:= M_0 , \\ M_0 &:= \{\boldsymbol{O}\} \end{aligned}$$

in Theorem 4.4. Here the subspace N_0 is spanned by the null vector, which does not have full rank. Therefore we are not able to express the projection matrix $\boldsymbol{P}_{N_0}$ by the usual formula. However a projection matrix onto the null space is the null matrix, i.e. $\boldsymbol{P}_{N_0} = \boldsymbol{O}$. In the following example we consider such a test.

Example 4.2
Let $\boldsymbol{X}$ be an $n \times p$ complex random matrix described by the complex MANOVA model with parameter set $M \times \mathbb{C}_+^{p\times p}$, where $M = N \otimes \mathbb{C}^p$ and N is a subspace of $\mathbb{C}^n$. Assume that we accept the null hypothesis $\Theta \in M_0$, where M_0 is given by

$$M_0 = \left\{\Theta \in \mathbb{C}^{n\times p} \,|\, \exists \boldsymbol{\theta} \in \mathbb{C}^p : \Theta = \mathbf{1}_n \boldsymbol{\theta}^*\right\} , \tag{4.3}$$

i.e. we have a sample of size n from $\mathbb{C}\mathcal{N}_p(\boldsymbol{\theta}, \boldsymbol{H})$. Further more assume that $n > p$.

We wish to examine the hypothesis $\Theta = \Theta_0 = \mathbf{1}_n \theta_0^*$, where $\Theta_0 \in M_0$ is a known complex matrix. According to the remark above this can be done by letting $\boldsymbol{X} := \boldsymbol{X} - \Theta_0$, M as in (4.3) page 78 and $M_0 = \{\boldsymbol{O}\}$ in Theorem 4.4 page 77. Using these transformations we have that $M = N \otimes \mathbb{C}^p$ and $M_0 = N_0 \otimes \mathbb{C}^p$, where $N = \mathcal{R}[\mathbf{1}_n]$ and $N_0 = \mathcal{R}[0]$. Hereby $\boldsymbol{P}_N = \frac{1}{n}\mathbf{1}_n\mathbf{1}_n^*$ and $\boldsymbol{P}_{N_0} = \boldsymbol{O}$. Since $\operatorname{tr}(\boldsymbol{P}_N) = 1$ and $\operatorname{tr}(\boldsymbol{P}_{N_0}) = 0$ the dimensions of N and N_0 are one and zero, respectively. Further the problem becomes to test the null hypothesis

$$\mathrm{H}_0 \ : \ \Theta \in \{\boldsymbol{O}\}$$

under the hypothesis

$$\mathrm{H} \ : \ \Theta \in M \, .$$

We find the likelihood ratio test statistic as

$$\begin{aligned} U^{\frac{1}{n}}(\boldsymbol{X}) &= \frac{\det\left(\widehat{\boldsymbol{H}}\right)}{\det\left(\widehat{\boldsymbol{H}}_0\right)} \\ &= \frac{\det\left(\boldsymbol{X}^*\boldsymbol{P}_N^{\perp}\boldsymbol{X}\right)}{\det\left(\boldsymbol{X}^*\boldsymbol{P}_{N_0}^{\perp}\boldsymbol{X}\right)} \\ &= \frac{\det\left(\boldsymbol{X}^*\boldsymbol{X} - \frac{1}{n}\boldsymbol{X}^*\mathbf{1}_n\mathbf{1}_n^*\boldsymbol{X}\right)}{\det\left(\boldsymbol{X}^*\boldsymbol{X}\right)} \, . \end{aligned} \tag{4.4}$$

Under H_0 the distribution of this likelihood ratio statistic is

$$\mathcal{L}\left(U^{\frac{1}{n}}(\boldsymbol{X})\right) = \mathbb{C}\mathcal{U}(p, 1, n-1) \ ,$$

which according to Theorem 3.12 page 65 is equivalent to

$$\mathcal{L}\left(U^{\frac{1}{n}}(\boldsymbol{X})\right) = \mathcal{B}(n-p, p) \ .$$

Let $n\widetilde{\boldsymbol{X}} = \boldsymbol{X}^*\mathbf{1}_n$ and $\boldsymbol{W} = \boldsymbol{X}^*\boldsymbol{X} - n\widetilde{\boldsymbol{X}}\widetilde{\boldsymbol{X}}^*$. Rewriting (4.4) we hereby get

$$U^{\frac{1}{n}}(\boldsymbol{X}) = \frac{\det(\boldsymbol{W})}{\det\left(\boldsymbol{W} + n\widetilde{\boldsymbol{X}}\widetilde{\boldsymbol{X}}^*\right)} \, .$$

Since $\boldsymbol{W}$ also can be expressed as $\boldsymbol{W} = \boldsymbol{X}^*\boldsymbol{P}_N^{\perp}\boldsymbol{X}$ and since $\boldsymbol{P}_N^{\perp}\Theta = \boldsymbol{O}$ and $n - 1 \geq p$ we know from Theorem 3.9 page 53 that $\boldsymbol{W} > \boldsymbol{O}$ with probability one. Then also $\boldsymbol{W}^{-1} > \boldsymbol{O}$ with probability one and there exists with probability one a complex random matrix $\boldsymbol{W}^{-\frac{1}{2}} > \boldsymbol{O}$, such that $\boldsymbol{W}^{-1} = \left(\boldsymbol{W}^{-\frac{1}{2}}\right)^2$. Using this we obtain

$$U^{\frac{1}{n}}(\boldsymbol{X}) = \det\left(\boldsymbol{I}_p + n\boldsymbol{W}^{-\frac{1}{2}}\widetilde{\boldsymbol{X}}\widetilde{\boldsymbol{X}}^*\boldsymbol{W}^{-\frac{1}{2}}\right)^{-1} \, .$$

Clearly $\boldsymbol{W}^{-\frac{1}{2}}\widetilde{\boldsymbol{X}}\widetilde{\boldsymbol{X}}^*\boldsymbol{W}^{-\frac{1}{2}}$ is Hermitian and

$$\operatorname{rank}\left(\boldsymbol{W}^{-\frac{1}{2}}\widetilde{\boldsymbol{X}}\widetilde{\boldsymbol{X}}^*\boldsymbol{W}^{-\frac{1}{2}}\right) = \operatorname{rank}\left(\boldsymbol{W}^{-\frac{1}{2}}\widetilde{\boldsymbol{X}}\right) = 1 \, ,$$

whereby

$$\begin{aligned} U^{\frac{1}{n}}(\boldsymbol{X}) &= \left(1 + n\,\mathrm{tr}\left(\boldsymbol{W}^{-\frac{1}{2}}\widetilde{\boldsymbol{X}}\widetilde{\boldsymbol{X}}^{*}\boldsymbol{W}^{-\frac{1}{2}}\right)\right)^{-1} \\ &= \left(1 + n\widetilde{\boldsymbol{X}}^{*}\boldsymbol{W}^{-1}\widetilde{\boldsymbol{X}}\right)^{-1}. \end{aligned}$$

Using the relationship between the beta distribution and the F-distribution (Seber 1984, p. 33) we deduce that

$$\mathcal{L}\left(\frac{n-p}{p} n\widetilde{\boldsymbol{X}}^{*}\boldsymbol{W}^{-1}\widetilde{\boldsymbol{X}}\right) = \mathcal{F}_{2p,2(n-p)} .$$

This testing problem has previously been treated by Giri (1965). In the real case the test is known as Hotelling's T^2 test. ■

4.3.2 Likelihood Ratio Test for Independence

We will now consider the test of independence among two parts of $\boldsymbol{X}$ described by the complex MANOVA model with parameter set $M \times \mathbb{C}_{+}^{p\times p}$, where $M = N \otimes \mathbb{C}^{p}$ and N is a subspace of $\mathbb{C}^{n}$.

Consider the partition of $\boldsymbol{X}$ given by

$$\boldsymbol{X} = (\boldsymbol{X}_1, \boldsymbol{X}_2) ,$$

where $\boldsymbol{X}_j$ is an $n \times p_j$ complex random matrix for $j = 1, 2$ and $p = p_1 + p_2$. Further let $\boldsymbol{\Theta}$ and $\boldsymbol{H}$ be partitioned likewise, i.e.

$$\boldsymbol{\Theta} = (\boldsymbol{\Theta}_1, \boldsymbol{\Theta}_2) \text{ and } \boldsymbol{H} = \begin{pmatrix} \boldsymbol{H}_{11} & \boldsymbol{H}_{12} \\ \boldsymbol{H}_{21} & \boldsymbol{H}_{22} \end{pmatrix},$$

where $\boldsymbol{\Theta}_j$ is $n \times p_j$ and $\boldsymbol{H}_{jk}$ is $p_j \times p_k$ for $j, k = 1, 2$. By Theorem 2.23 page 36 we get $\mathcal{L}(\boldsymbol{X}_j) = \mathbb{C}\mathcal{N}_{n\times p_j}(\boldsymbol{\Theta}_j, \boldsymbol{I}_n \otimes \boldsymbol{H}_{jj})$. Since $\boldsymbol{X}$ is described by the complex MANOVA model with parameter set $M \times \mathbb{C}_{+}^{p\times p}$ the matrices $\boldsymbol{\Theta}_j$ belong to the subspaces M_j of M given by

$$M_j = \{\boldsymbol{\Theta}_j \in \mathbb{C}^{n\times p_j} \mid \exists \boldsymbol{C}_j \in \mathbb{C}^{k\times p_j} : \boldsymbol{\Theta}_j = \left(\boldsymbol{Z} \otimes \boldsymbol{I}_{p_j}\right)(\boldsymbol{C}_j)\} , \; j = 1, 2 ,$$

where $\boldsymbol{Z}$ is the known $n \times k$ complex matrix with $n \geq k$ from the complex MANOVA model. It satisfies that the range of the linear transformation $\boldsymbol{Z} \otimes \boldsymbol{I}_{p_j}$ is M_j. Moreover $M_j = N \otimes \mathbb{C}^{p_j}$. We also have that $\boldsymbol{H}_{jj} \in \mathbb{C}_{+}^{p_j\times p_j}$. Assume that $n - k \geq p$ to ensure existence of the maximum likelihood estimator of $\boldsymbol{H}$.

We wish to test if $\boldsymbol{X}_1$ and $\boldsymbol{X}_2$ are independent. According to Theorem 2.22 page 35 this is equivalent to test the null hypothesis

$$\mathrm{H}_0 : \boldsymbol{H}_{12} = \boldsymbol{O}$$

against the alternative hypothesis

$$\mathrm{H} : \boldsymbol{H}_{12} \neq \boldsymbol{O} .$$

The likelihood ratio for this test is defined by

$$U(\boldsymbol{x}) = \frac{\sup_{\Theta \in M, \boldsymbol{H} \in \mathbb{C}_{+}^{p \times p}, \boldsymbol{H}_{12} = \boldsymbol{O}} L(\Theta, \boldsymbol{H} | \boldsymbol{x})}{\sup_{\Theta \in M, \boldsymbol{H} \in \mathbb{C}_{+}^{p \times p}} L(\Theta, \boldsymbol{H} | \boldsymbol{x})} .$$

According to (4.1) page 72 the maximum of the likelihood function for $(\Theta, \boldsymbol{H}) \in M \times \mathbb{C}_{+}^{p \times p}$ is

$$L\left(\widehat{\Theta}, \widehat{\boldsymbol{H}} \middle| \boldsymbol{x}\right) = \pi^{-np} \det\left(\frac{1}{n}\boldsymbol{x}^* \boldsymbol{P}_N^{\perp} \boldsymbol{x}\right)^{-n} \exp(-np) .$$

Under H_0 the matrices $\boldsymbol{X}_1$ and $\boldsymbol{X}_2$ are independent, therefore the likelihood function factorizes. Furthermore the parameter $(\Theta_1, \boldsymbol{H}_{11})$ associated with $\boldsymbol{X}_1$ varies independently of the parameter $(\Theta_2, \boldsymbol{H}_{22})$ associated with $\boldsymbol{X}_2$, so the likelihood function can be maximized by separately maximizing each factor. The maximum of the likelihood function under H_0 is therefore

$$\begin{aligned} L\left(\widehat{\Theta}, \widehat{\boldsymbol{H}} \middle| \boldsymbol{x}\right) &= L\left(\widehat{\Theta}_1, \widehat{\boldsymbol{H}}_{11} \middle| \boldsymbol{x}_1\right) L\left(\widehat{\Theta}_2, \widehat{\boldsymbol{H}}_{22} \middle| \boldsymbol{x}_2\right) \\ &= \prod_{j=1}^{2} \pi^{-np_j} \det\left(\frac{1}{n}\boldsymbol{x}_j^* \boldsymbol{P}_N^{\perp} \boldsymbol{x}_j\right)^{-n} \exp(-np_j) \\ &= \pi^{-np} \left(\prod_{j=1}^{2} \det\left(\frac{1}{n}\boldsymbol{x}_j^* \boldsymbol{P}_N^{\perp} \boldsymbol{x}_j\right)^{-n}\right) \exp(-np) , \end{aligned}$$

where for a given observation matrix of $\boldsymbol{X}_j$ it holds that $\frac{1}{n}\boldsymbol{x}_j^* \boldsymbol{P}_N^{\perp} \boldsymbol{x}_j$ for $j = 1, 2$ is the unique maximum likelihood estimate of $\boldsymbol{H}_{jj}$ found by considering the complex MANOVA model describing $\boldsymbol{X}_j$. The likelihood ratio therefore becomes

$$U(\boldsymbol{x}) = \frac{\prod_{j=1}^{2} \det\left(\frac{1}{n}\boldsymbol{x}_j^* \boldsymbol{P}_N^{\perp} \boldsymbol{x}_j\right)^{-n}}{\det\left(\frac{1}{n}\boldsymbol{x}^* \boldsymbol{P}_N^{\perp} \boldsymbol{x}\right)^{-n}} ,$$

which implies an equivalent likelihood ratio given as

$$U^{\frac{1}{n}}(\boldsymbol{x}) = \frac{\det\left(\boldsymbol{x}^* \boldsymbol{P}_N^{\perp} \boldsymbol{x}\right)}{\prod_{j=1}^{2} \det\left(\boldsymbol{x}_j^* \boldsymbol{P}_N^{\perp} \boldsymbol{x}_j\right)} .$$

We reject H_0 for small values of $U^{\frac{1}{n}}(\boldsymbol{x})$. The likelihood ratio test statistic is given by

$$\begin{aligned} U^{\frac{1}{n}}(\boldsymbol{X}) &= \frac{\det\left(\boldsymbol{X}^* \boldsymbol{P}_N^{\perp} \boldsymbol{X}\right)}{\prod_{j=1}^{2} \det\left(\boldsymbol{X}_j^* \boldsymbol{P}_N^{\perp} \boldsymbol{X}_j\right)} \\ &= \frac{\det\left(\widehat{\boldsymbol{H}}\right)}{\prod_{j=1}^{2} \det\left(\widehat{\boldsymbol{H}}_{jj}\right)} . \end{aligned}$$

The likelihood ratio test of size α of H_0 against H is determined by the critical region $U^{\frac{1}{n}}(\boldsymbol{x}) \leq q$, where q is a constant fulfilling $P\left(U^{\frac{1}{n}}(\boldsymbol{X}) \leq q \middle| \mathrm{H}_0\right) = \alpha$. To determine the critical region we need to know the distribution of the test statistic under H_0.

Let $\boldsymbol{W}$ be the complex random matrix given by

$$\boldsymbol{W} = \boldsymbol{X}^* \boldsymbol{P}_N^{\perp} \boldsymbol{X} = \begin{pmatrix} \boldsymbol{X}_1^* \boldsymbol{P}_N^{\perp} \boldsymbol{X}_1 & \boldsymbol{X}_1^* \boldsymbol{P}_N^{\perp} \boldsymbol{X}_2 \\ \boldsymbol{X}_2^* \boldsymbol{P}_N^{\perp} \boldsymbol{X}_1 & \boldsymbol{X}_2^* \boldsymbol{P}_N^{\perp} \boldsymbol{X}_2 \end{pmatrix} = \begin{pmatrix} \boldsymbol{W}_{11} & \boldsymbol{W}_{12} \\ \boldsymbol{W}_{21} & \boldsymbol{W}_{22} \end{pmatrix} .$$

Since $\boldsymbol{P}_N^{\perp} \Theta_2 = \boldsymbol{O}$ and $n - k \geq p_2$ we obtain from Theorem 3.9 page 53 that $\boldsymbol{W}_{22} > \boldsymbol{O}$ with probability one. This means that $\boldsymbol{W}_{22}^{-1}$ exists with probability one and thereby we have $\det(\boldsymbol{W}) = \det(\boldsymbol{W}_{22}) \det\left(\boldsymbol{W}_{11} - \boldsymbol{W}_{12}\boldsymbol{W}_{22}^{-1}\boldsymbol{W}_{21}\right)$. Then $U^{\frac{1}{n}}(\boldsymbol{X})$ can be rewritten as

$$\begin{aligned} U^{\frac{1}{n}}(\boldsymbol{X}) &= \frac{\det\left(\boldsymbol{X}^* \boldsymbol{P}_N^{\perp} \boldsymbol{X}\right)}{\det\left(\boldsymbol{X}_1^* \boldsymbol{P}_N^{\perp} \boldsymbol{X}_1\right) \det\left(\boldsymbol{X}_2^* \boldsymbol{P}_N^{\perp} \boldsymbol{X}_2\right)} \\ &= \frac{\det(\boldsymbol{W})}{\det(\boldsymbol{W}_{11}) \det(\boldsymbol{W}_{22})} \\ &= \frac{\det\left(\boldsymbol{W}_{11} - \boldsymbol{W}_{12}\boldsymbol{W}_{22}^{-1}\boldsymbol{W}_{21}\right)}{\det\left(\boldsymbol{W}_{11} - \boldsymbol{W}_{12}\boldsymbol{W}_{22}^{-1}\boldsymbol{W}_{21} + \boldsymbol{W}_{12}\boldsymbol{W}_{22}^{-1}\boldsymbol{W}_{21}\right)} . \end{aligned}$$

In Theorem 4.3 page 73 we have shown that

$$\mathcal{L}\left(n\widehat{\boldsymbol{H}}\right) = \mathcal{L}(\boldsymbol{W}) = \mathbb{C}\mathcal{W}_p(\boldsymbol{H}, n-k) .$$

Moreover we have $n - k \geq p_2$ and $\boldsymbol{H}_{22} > \boldsymbol{O}$. Further $\boldsymbol{H}_{12} = \boldsymbol{O}$ under H_0, thus by using Theorem 3.6 page 44 we get

$$\mathcal{L}\left(\boldsymbol{W}_{11} - \boldsymbol{W}_{12}\boldsymbol{W}_{22}^{-1}\boldsymbol{W}_{21}\right) = \mathbb{C}\mathcal{W}_{p_1}(\boldsymbol{H}_{11}, n-k-p_2)$$

and

$$\mathcal{L}\left(\boldsymbol{W}_{12}\boldsymbol{W}_{22}^{-1}\boldsymbol{W}_{21}\right) = \mathbb{C}\mathcal{W}_{p_1}(\boldsymbol{H}_{11}, p_2) .$$

Besides it holds that

$$\boldsymbol{W}_{11} - \boldsymbol{W}_{12}\boldsymbol{W}_{22}^{-1}\boldsymbol{W}_{21} \perp\!\!\!\perp \boldsymbol{W}_{12}\boldsymbol{W}_{22}^{-1}\boldsymbol{W}_{21} .$$

Hence it follows from Definition 3.2 page 55, since $n - k - p_2 \geq p_1$ and $\boldsymbol{H}_{11} > \boldsymbol{O}$, that

$$\mathcal{L}\left(U^{\frac{1}{n}}(\boldsymbol{X})\right) = \mathbb{C}\mathcal{U}(p_1, p_2, n-k-p_2) .$$

Again the distribution of $U^{\frac{1}{n}}(\boldsymbol{X})$ under H_0 is a complex $\mathcal{U}$-distribution and as on page 77 an approximation of the distribution of $-2\log U(\boldsymbol{X})$ can be found.

The test of independence is summarized in the following theorem.

Theorem 4.5 Likelihood ratio test of independence
The likelihood ratio test in the complex MANOVA model with parameter set $M \times \mathbb{C}_+^{p\times p}$ *of the null hypothesis*

$$\mathrm{H}_0 : \boldsymbol{H}_{12} = \boldsymbol{O}$$

against the alternative hypothesis

$$\mathrm{H} : \boldsymbol{H}_{12} \neq \boldsymbol{O} \,,$$

consists in rejecting H_0 *if*

$$U^{\frac{1}{n}}(\boldsymbol{x}) \leq q \,.$$

The likelihood ratio test statistic is given by

$$U^{\frac{1}{n}}(\boldsymbol{X}) = \frac{\det\left(\widehat{\boldsymbol{H}}\right)}{\prod_{j=1}^{2}\det\left(\widehat{\boldsymbol{H}}_{jj}\right)} \,,$$

where $\widehat{\boldsymbol{H}}$ *is the maximum likelihood estimator of* $\boldsymbol{H}$ *under* H *and* $\widehat{\boldsymbol{H}}_{jj}$ *is the maximum likelihood estimator of* $\boldsymbol{H}_{jj}$ *for* $j = 1, 2$. *Under* H_0 *the distribution of* $U^{\frac{1}{n}}(\boldsymbol{X})$ *is*

$$\mathcal{L}\left(U^{\frac{1}{n}}(\boldsymbol{X})\right) = \mathbb{C}\mathcal{U}(p_1, p_2, n-k-p_2) \,.$$

The constant q *is chosen to provide a test of size* α*, i.e.* q *must fulfill* $P\left(U^{\frac{1}{n}}(\boldsymbol{X}) \leq q \middle| \mathrm{H}_0\right) = \alpha$.

In the following example we illustrate how test of interchangeability can be made by performing tests of independence.

Example 4.3
Let $\left\{(X_t, Y_t)^T\right\}_{t\in\mathbb{Z}}$ be a bivariate Gaussian time series with spectral density matrix.

Consider the hypothesis that X_t and Y_t are interchangeable, i.e. the distributional properties of $\left\{(X_t, Y_t)^T\right\}_{t\in\mathbb{Z}}$ and $\left\{(Y_t, X_t)^T\right\}_{t\in\mathbb{Z}}$ are the same. By transforming to the series

$$\left\{(X_t', Y_t')^T\right\}_{t\in\mathbb{Z}} = \left\{\left((X_t + Y_t), (X_t - Y_t)\right)^T\right\}_{t\in\mathbb{Z}}$$

this is equivalent to independence between $\{X_t'\}_{t\in\mathbb{Z}}$ and $\{Y_t'\}_{t\in\mathbb{Z}}$, which in terms of the spectral density matrix takes the form

$$\Sigma'(\omega) = \begin{pmatrix} \sigma_{X'X'}(\omega) & 0 \\ 0 & \sigma_{Y'Y'}(\omega) \end{pmatrix} .$$

For given estimates $\widehat{\Sigma}_T'(\omega_k)$, $k = 1, 2, \ldots, K$, of the form (3.16) page 54, the hypothesis of interchangeability can be tested by performing K tests of independence as described in Theorem 4.5. ■

The hypothesis test concerning complex covariance structure of a real random vector is to test the null hypothesis

$$\mathrm{H}_0 : \boldsymbol{\Sigma} = \begin{pmatrix} \boldsymbol{B} & -\boldsymbol{A} \\ \boldsymbol{A} & \boldsymbol{B} \end{pmatrix}$$

under the hypothesis

$$\mathrm{H} : \boldsymbol{\Sigma} = \begin{pmatrix} \boldsymbol{\Sigma}_{11} & \boldsymbol{\Sigma}_{21}^{\top} \\ \boldsymbol{\Sigma}_{21} & \boldsymbol{\Sigma}_{22} \end{pmatrix} .$$

This test is described by Andersson, Brøns & Jensen (1983). A further step is to test the null hypothesis

$$\mathrm{H}_0 : \boldsymbol{A} = \boldsymbol{O}$$

against the alternative hypothesis

$$\mathrm{H} : \boldsymbol{A} \neq \boldsymbol{O} .$$

This is the test for reality of a variance matrix with complex covariance structure. It is treated by Khatri (1965*b*) and Andersson et al. (1983).

5

Simple Undirected Graphs

When we consider graphical models for the multivariate complex normal distribution we formulate the models in terms of simple undirected graphs. This chapter presents the concept of simple undirected graphs. As the main purpose is to define and introduce the later needed results, the presentation is at times short and compact. The well known results are stated without proof, but references containing further information are given. Results which are not quite well known are treated in more detail. First of all we define a simple undirected graph and associated basic definitions. Afterwards we consider the concepts separation, decomposition and decomposability of simple undirected graphs. This involves investigation of chordless 4-cycles, running intersection property orderings, the maximum cardinality search algorithm and an algorithm to determine decomposability of a simple undirected graph. Then we move on to definition of collapsibility and we treat the concept a regular edge. We observe that a simple undirected decomposable graph and a decomposable subgraph of it with one edge less differ by a regular edge. Finally we state some decompositions of subgraphs in a simple undirected graph containing a regular edge. We begin by defining a simple undirected graph.

Definition 5.1 Simple undirected graph
A simple undirected graph $\mathcal{G}$ is a pair (V, E), where V is a finite set of elements called vertices and E is a subset of the set of unordered pairs of distinct vertices in V. The elements of E are called edges.

Normally the vertices are labeled by integers. Often it can be helpful to illustrate a simple undirected graph by a picture, where a circle represents a vertex and a line joining a pair of distinct circles represents an edge between the corresponding vertices. This visualization of a simple undirected graph is illustrated in Example 5.1. Remark that two graphs can be identical even if the illustrations do not look alike. This is due to the fact that different placing of the vertices can result in different pictures.

Example 5.1
Let $\mathcal{G} = (V, E)$ be the simple undirected graph with $V = \{1, 2, 3, 4, 5\}$ and $E = \{\{1, 2\}, \{1, 4\}, \{1, 5\}, \{2, 3\}, \{3, 4\}, \{4, 5\}\}$. Then $\mathcal{G}$ can be illustrated as in Figure 5.1 page 86. ■

Definition 5.2 Adjacency of two vertices
Let $\mathcal{G} = (V, E)$ be a simple undirected graph and let $\alpha \neq \beta \in V$. If $\{\alpha, \beta\} \in E$, then α and β are adjacent or neighbours. This is also written as $\alpha \sim \beta$.

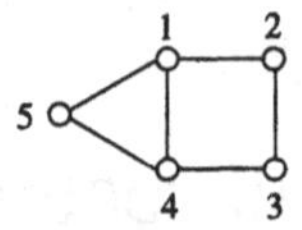

Figure 5.1: *Illustration of the simple undirected graph from Example 5.1.*

If $\{\alpha, \beta\} \notin E$, i.e. α and β are nonadjacent, we write $\alpha \not\sim \beta$.

Definition 5.3 Boundary
Let $\mathcal{G} = (V, E)$ be a simple undirected graph. The boundary of $A \subseteq V$ is denoted by $\text{bd}(A)$ *and consists of the vertices which are not in A, but adjacent to some vertex in A, i.e.*

$$\text{bd}(A) = \{\beta \in V \setminus A \mid \exists \alpha \in A : \{\alpha, \beta\} \in E\} .$$

Definition 5.4 Closure
Let $\mathcal{G} = (V, E)$ be a simple undirected graph. The closure of $A \subseteq V$ is denoted by $\text{cl}(A)$ *and defined as*

$$\text{cl}(A) = A \cup \text{bd}(A) .$$

Example 5.2
Let $\mathcal{G} = (V, E)$ be the simple undirected graph defined in Example 5.1 page 85. To illustrate the concepts boundary and closure let A be the subset of V given by $A = \{1, 5\}$. Then $\text{bd}(A) = \{2, 4\}$ and $\text{cl}(A) = \{1, 2, 4, 5\}$. ■

Definition 5.5 Induced subgraph
Let $\mathcal{G} = (V, E)$ be a simple undirected graph. A subset $A \subseteq V$ induces the subgraph $\mathcal{G}_A = (A, E_A)$, where E_A contains exactly those edges in E which connects vertices from A, i.e.

$$E_A = E \cap (A \times A) .$$

Remark that $A \times A$ untraditionally denotes the set of all unordered pairs of elements from A.

Example 5.3
Let $\mathcal{G} = (V, E)$ be the simple undirected graph with $V = \{1, 2, 3, 4, 5\}$ and $E = \{\{1, 2\}, \{2, 3\}, \{2, 4\}, \{2, 5\}, \{3, 4\}, \{4, 5\}\}$. The subset $A = \{1, 2, 4\}$ induces the subgraph $\mathcal{G}_A = (A, E_A)$, where $E_A = \{\{1, 2\}, \{2, 4\}\}$. The simple undirected graphs $\mathcal{G}$ and $\mathcal{G}_A$ are visualized in Figure 5.2 page 87.

■

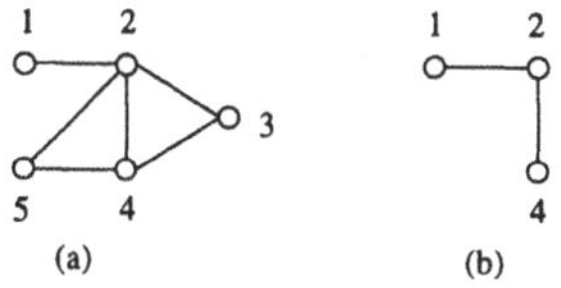

Figure 5.2: *In (a) an illustration of the simple undirected graph $\mathcal{G}$ from Example 5.3 is given. In (b) the induced subgraph $\mathcal{G}_A$ of $\mathcal{G}$ is illustrated.*

Definition 5.6 Complete graph
A simple undirected graph $\mathcal{G} = (V, E)$ is complete if all vertices are mutually adjacent.

Definition 5.7 Complete subset
Let $\mathcal{G} = (V, E)$ be a simple undirected graph. A subset $A \subseteq V$ is complete if it induces a complete subgraph of $\mathcal{G}$.

Example 5.4
Let $\mathcal{G} = (V, E)$ be the simple undirected graph defined in Example 5.3 page 86. Complete subsets of V is e.g. $\{2, 4, 5\}$ and $\{2, 3\}$ as they induce complete subgraphs of $\mathcal{G}$. ■

Definition 5.8 Union and intersection
The union of two simple undirected graphs (V, E) and (W, F) is defined by

$$(V, E) \cup (W, F) = (V \cup W, E \cup F) \ ,$$

and the intersection of two simple undirected graphs (V, E) and (W, F) is defined by

$$(V, E) \cap (W, F) = (V \cap W, E \cap F) \ .$$

Besides we define a special kind of union as it is advantageous in subsequent consideration.

Definition 5.9 Direct union
The union of two simple undirected graphs (V, E) and (W, F) is direct if $(V, E) \cap (W, F)$ is a complete graph. The direct union of (V, E) and (W, F) is denoted by

$$(V, E) \ \dot{\cup} \ (W, F) \ .$$

Example 5.5
This example illustrates the concepts union, intersection and direct union of two simple undirected graphs (V, E) and (W, F). Let $V = \{1, 2, 3, 4\}$ and $E = \{\{1, 2\}, \{1, 3\}, \{2, 3\}, \{2, 4\}, \{3, 4\}\}$ and further let $W = \{2, 3, 4, 5, 6\}$ and $F = \{\{2, 3\}, \{2, 4\}, \{3, 4\}, \{4, 5\}, \{4, 6\}, \{5, 6\}\}$. These graphs are visualized in Figure 5.3 page 88 together with their union and intersection. We see that $(V, E) \cap (W, F)$ is complete, hence the union is direct. ■

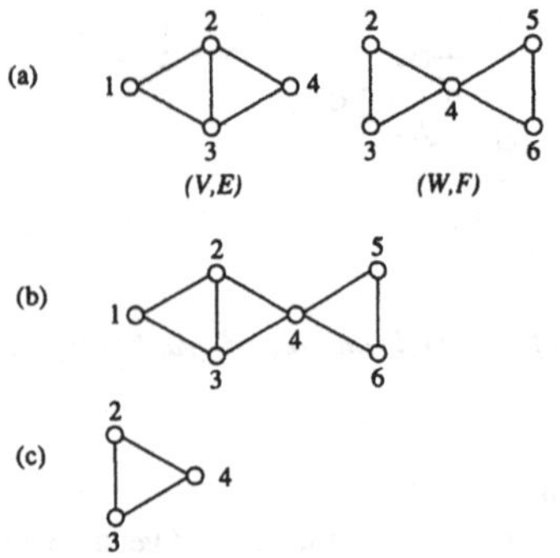

Figure 5.3: *In (a) illustrations of the two simple undirected graphs (V, E) and (W, F) from Example 5.5 are given. In (b) the union $(V, E) \cup (W, F)$ is illustrated and in (c) we see the intersection $(V, E) \cap (W, F)$.*

Definition 5.10 Path
Let $\mathcal{G} = (V, E)$ be a simple undirected graph. A path of length n in $\mathcal{G}$ is a sequence of distinct vertices $\alpha_0, \alpha_1, \dots, \alpha_n$, where $\alpha_j \in V$, such that $\{\alpha_{j-1}, \alpha_j\} \in E$ for all $j = 1, 2, \dots, n$.

Note that a path can have length zero.

Definition 5.11 Cycle
Let $\mathcal{G} = (V, E)$ be a simple undirected graph. An n-cycle in $\mathcal{G}$ is a path of length n with the modification that $\alpha_0 = \alpha_n$.

Definition 5.12 Chord
Let $\mathcal{G} = (V, E)$ be a simple undirected graph. A chord in an n-cycle in $\mathcal{G}$ is an edge between $\alpha_j \neq \alpha_k \in V$ in the cycle such that $\{\alpha_j, \alpha_k\} \in E$, but $k \neq j-1, j+1$ (modulo n).

Example 5.6
Let $\mathcal{G} = (V, E)$ be the simple undirected graph with $V = \{1, 2, 3, 4, 5\}$ and $E = \{\{1, 2\}, \{1, 5\}, \{2, 3\}, \{2, 5\}, \{3, 4\}, \{3, 5\}\}$. Then $\mathcal{G}$ can be illustrated as in Figure 5.4.

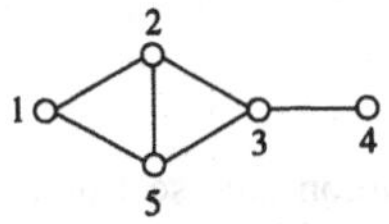

Figure 5.4: *Illustration of the simple undirected graph from Example 5.6.*

By looking at the illustration of $\mathcal{G}$ we observe that

- e.g. the sequence of vertices $1, 2, 3, 4$ is a path of length 3 in $\mathcal{G}$.
- e.g. the sequence $1, 2, 3, 5, 1$ is a 4-cycle in $\mathcal{G}$ and the edge $\{2, 5\}$ is a chord in this cycle. ■

We define a relation which indicates the existence of a path between two vertices. Obviously this relation is an equivalence relation.

Theorem 5.1 Equivalence relation
Let $\mathcal{G} = (V, E)$ be a simple undirected graph. The relation $\sim_p$ given by

$$\forall \alpha, \beta \in V : \alpha \sim_p \beta \Leftrightarrow \exists \text{ a path } \alpha_0, \alpha_1, \ldots, \alpha_n \text{ in } \mathcal{G} \text{ with } \alpha_0 = \alpha \text{ and } \alpha_n = \beta ,$$

is an equivalence relation.

We write $\alpha \not\sim_p \beta$ whenever there is no path from α to β in $\mathcal{G}$. The equivalence relation above leads to the following definition of the connectivity components in a simple undirected graph.

Definition 5.13 Connectivity components
Let $\mathcal{G} = (V, E)$ be a simple undirected graph. The subgraphs induced by the equivalence classes of $\sim_p$ are called the connectivity components of $\mathcal{G}$.

Note that if there is only one equivalence class we say that $\mathcal{G}$ is *connected.*

Example 5.7
To illustrate the concept of connectivity components of a simple undirected graph let $\mathcal{G} = (V, E)$ be given by $V = \{1, 2, 3, 4, 5, 6, 7\}$ and $E = \{\{1, 2\}, \{1, 4\}, \{2, 3\}, \{2, 4\}, \{3, 4\}, \{5, 6\}, \{6, 7\}\}$. Then $\mathcal{G}$ can be illustrated as in Figure 5.5.

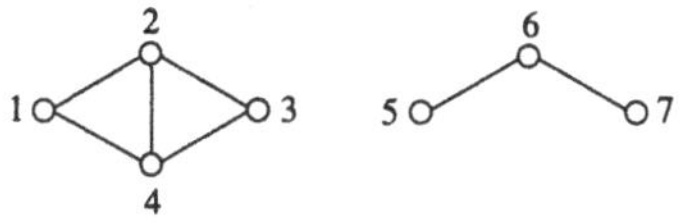

Figure 5.5: *Illustration of the simple undirected graph with two connectivity components from Example 5.7.*

We see that $\mathcal{G}$ has two connectivity components. ■

Definition 5.14 Separation
Let $\mathcal{G} = (V, E)$ be a simple undirected graph. Two subsets $A, B \subseteq V$ are separated by $S \subseteq V$ if all paths from A to B go via S, i.e. the paths intersect S at some vertex.

Remark that if A, B and S are disjoint then the statement A and B are separated by S equivalently can be given as A and B are in different connectivity components of $\mathcal{G}_{V \setminus S}$.

Example 5.8
This example investigates the concept separation of a simple undirected graph. Let $\mathcal{G} = (V, E)$ be a simple undirected graph with $V = \{1, 2, 3, 4, 5, 6, 7, 8, 9\}$ and $E = \{\{1, 2\}, \{2, 3\}, \{2, 4\}, \{4, 5\}, \{4, 6\}, \{6, 7\}, \{6, 8\}, \{7, 9\}, \{8, 9\}\}$. A representation of $\mathcal{G}$ is found in Figure 5.6.

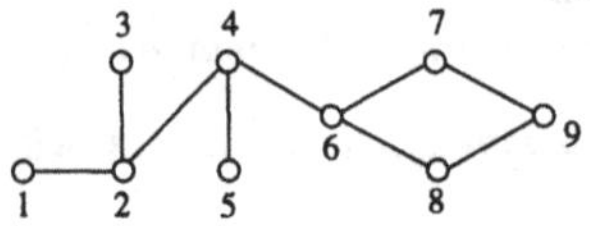

Figure 5.6: *Illustration of the concept of separation. The graph is defined in Example 5.8.*

Let A, B and S be subsets of V defined by $A = \{2, 3, 4\}$, $B = \{6, 7, 8\}$ and $S = \{4, 5\}$. We observe that A and B are separated by S. ■

After the concept separation is established we can define a decomposition of a simple undirected graph.

Definition 5.15 Decomposition
Let $\mathcal{G} = (V, E)$ be a simple undirected graph and let A and B be subsets of V. If it holds that

1. $V = A \cup B$,
2. $A \cap B$ *is a complete subset of* V,
3. $A \setminus B$ *and* $B \setminus A$ *are separated by* $A \cap B$,

then A and B form a decomposition of $\mathcal{G}$.

In Definition 5.15 we observe that the union of $\mathcal{G}_A = (A, E_A)$ and $\mathcal{G}_B = (B, E_B)$ is direct and that $V = A \cup B$ and $E = E_A \cup E_B$, i.e. $\mathcal{G} = \mathcal{G}_A \dot{\cup} \mathcal{G}_B$. In this case we say that A and B decompose $\mathcal{G}$ into the components $\mathcal{G}_A$ and $\mathcal{G}_B$. Conversely if we have two simple undirected graphs (V, E) and (W, F) with a direct union we observe that V and W form a decomposition of $(V, E) \dot{\cup} (W, F)$.

Example 5.9
To illustrate the concept decomposition of a simple undirected graph let $\mathcal{G} = (V, E)$ be given by $V = \{1, 2, 3, 4, 5, 6, 7, 8\}$ and $E = \{\{1, 2\}, \{2, 3\}, \{3, 4\}, \{3, 5\}, \{4, 5\}, \{4, 6\}, \{5, 7\}, \{6, 7\},$

$\{6,8\}, \{7,8\}\}$. The graph $\mathcal{G}$ is visualized in Figure 5.7. Let A and B be the two subsets of V given by $A = \{1,2,3,4,5\}$ and $B = \{3,4,5,6,7,8\}$. We see that A and B form a decomposition of $\mathcal{G}$.

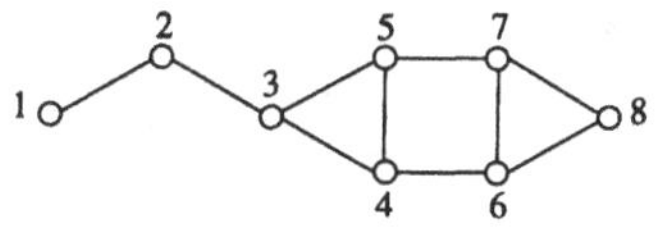

Figure 5.7: *Illustration of the simple undirected graph given in Example 5.9.*

■

To define decomposability of a simple undirected graph we need the definition of a clique.

Definition 5.16 Clique
Let $\mathcal{G} = (V, E)$ be a simple undirected graph. A complete subset of V which is maximal w.r.t. inclusion is called a clique, i.e.

$$(C \text{ is complete and } C \subset C' \Rightarrow C' \text{ is not complete}) \Leftrightarrow C \text{ is a clique},$$

where $C, C' \subseteq V$.

We denote the set of cliques in a simple undirected graph by $\mathcal{C}$. Note that a complete graph has only one clique. A simple undirected graph that can be successively decomposed into its cliques is called *decomposable*. The following definition states this recursively.

Definition 5.17 Decomposability
A simple undirected graph $\mathcal{G} = (V, E)$ is said to be decomposable if it is complete or if there exists a decomposition formed by proper subsets A and B of V into decomposable subgraphs $\mathcal{G}_A$ and $\mathcal{G}_B$.

We can decide whether or not a simple undirected graph is decomposable by examining the cycles of it. This result is stated in the theorem below and can be found in e.g. Leimer (1989).

Theorem 5.2
A simple undirected graph $\mathcal{G} = (V, E)$ is decomposable iff it contains no cycle of length greater than 3 without a chord.

An induced subgraph of a simple undirected decomposable graph is also decomposable. This result follows immediately from Theorem 5.2.

Corollary 5.1
Let $\mathcal{G} = (V, E)$ be a simple undirected decomposable graph. For $A \subseteq V$ the induced subgraph $\mathcal{G}_A = (A, E_A)$ is decomposable.

Using the method, where we consider the cycles, to decide decomposability of a simple undirected graph with many vertices and edges is quite often a complicated matter. In this situation another method, which in addition is constructive, is to determine whether the set of cliques in a simple undirected graph can be ordered by a running intersection property (RIP) ordering. Such an ordering of the cliques is defined below.

Definition 5.18 RIP-ordering
Let $\mathcal{G} = (V, E)$ be a simple undirected graph. An ordering $C_1, C_2, \ldots, C_m$ of the cliques in $\mathcal{G}$ is said to be a RIP-ordering if

$$C_j \cap D_j \subseteq C_q \textit{ for some } q < j, \; j = 2, 3, \ldots, m ,$$

where $D_j = \bigcup_{k<j} C_k$.

Theorem 5.3
A simple undirected graph $\mathcal{G} = (V, E)$ is decomposable iff its cliques $\mathcal{C}$ can be ordered as a RIP-ordering.

The result of Theorem 5.3 can be found in Leimer (1989). The following two algorithms state together a method to test decomposability of a simple undirected graph. If the graph is decomposable we get in addition a RIP-ordering of the cliques. The algorithms were introduced by Tarjan & Yannakakis (1984) and they are also treated by e.g. Leimer (1989). The first algorithm labels all the vertices in a simple undirected graph and is called the maximum cardinality search (MCS) algorithm.

Algorithm 5.1 Maximum cardinality search
Let $\mathcal{G} = (V, E)$ be a simple undirected graph. Label the vertices in $\mathcal{G}$ from $|V|$ to 1 in decreasing order by the following algorithm.

1. *Choose a vertex arbitrarily and label this by $|V|$.*
2. *Let $j := |V| - 1$.*
3. *As the next vertex to be labeled, select the unlabeled vertex adjacent to the largest number of labeled vertices. If more than one vertex fulfills this demand, choose arbitrarily between them. Label the vertex by j.*
4. *Let $j := j - 1$.*
5. *Repeat from step 3 until all the vertices are labeled.*

After the vertices have been labeled by the MCS-algorithm we are able to determine whether or not the graph is decomposable. This is done by the algorithm below. If the graph is decomposable we get in addition a RIP-ordering of the cliques in the graph.

Algorithm 5.2
Let $\mathcal{G} = (V, E)$ be a simple undirected graph with the vertices labeled by the MCS-algorithm stated in Algorithm 5.1 page 92. Let $\mathcal{C}$ be given and let $m = |\mathcal{C}|$. The algorithm to test decomposability of $\mathcal{G}$ is given as follows.

1. *Let $j := 1$, $k := m$ and $\mathcal{R} := \mathcal{C}$.*
2. *Consider the vertex labeled j. If $j \in C'$ and $j \in C''$ for $C' \neq C'' \in \mathcal{R}$, then stop and $\mathcal{G}$ is not decomposable. If $j \in C$ for $C \in \mathcal{R}$, then let $C_k := C$, $\mathcal{R} := \mathcal{R} \setminus C$ and $k := k - 1$.*
3. *If $\mathcal{R} = \emptyset$ then stop and $\mathcal{G}$ is decomposable. Otherwise let $j := j + 1$ and repeat from step 2.*

The ordering $C_1, C_2, \ldots, C_m$ is a RIP-ordering of the cliques in $\mathcal{G}$.

Note that there are at least $|\mathcal{C}|$ RIP-orderings of the cliques in a decomposable graph.

Example 5.10
Let $\mathcal{G} = (V, E)$ be the simple undirected graph illustrated in Figure 5.8a. Let the vertices of $\mathcal{G}$ be numbered by the MCS-algorithm. Then $\mathcal{G}$ can be illustrated e.g. as in Figure 5.8b.

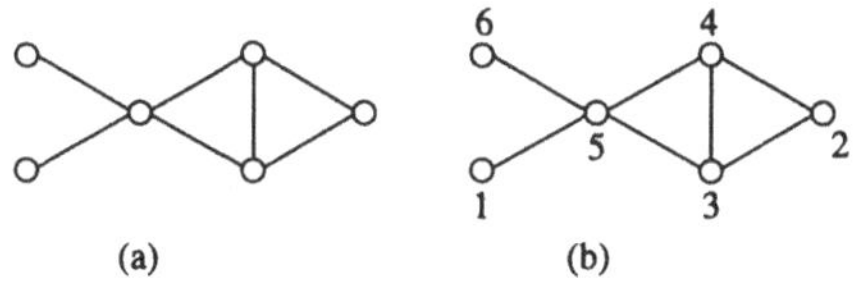

Figure 5.8: *In (a) an illustration of a simple undirected graph is given. In (b) we see an illustration of the same graph with the vertices numbered by the MCS-algorithm.*

We observe that $\mathcal{C} = \{\{5,6\}, \{1,5\}, \{3,4,5\}, \{2,3,4\}\}$. By Algorithm 5.2 we get that $\mathcal{G}$ is decomposable and that $C_1 = \{5,6\}$, $C_2 = \{3,4,5\}$, $C_3 = \{2,3,4\}$ and $C_4 = \{1,5\}$ is a RIP-ordering of the cliques in $\mathcal{G}$. Observe with $D_j = \bigcup_{k<j} C_k$ for $j = 2,3,4$ that

$$\begin{aligned} C_2 \cap D_2 &= \{5\} \subseteq C_1 \\ C_3 \cap D_3 &= \{3,4\} \subseteq C_2 \\ C_4 \cap D_4 &= \{5\} \subseteq C_1 \,. \end{aligned}$$

This points out that C_1, C_2, C_3 and C_4 in fact is a RIP-ordering of the cliques in $\mathcal{G}$. ■

A useful concept of a simple undirected graph in relation to graphical models is stated below.

Definition 5.19 Collapsibility
Let $\mathcal{G} = (V, E)$ be a simple undirected graph and let $A \subseteq V$. Furthermore let $\mathcal{G}_{B_j} = \left(B_j, E_{B_j}\right)$ for $j = 1, 2, \ldots, k$ be the k connectivity components of $\mathcal{G}_{V\setminus A} = \left(V \setminus A, E_{V\setminus A}\right)$, where it holds that $B_j \subseteq V \setminus A$ and $E_{B_j} = E_{V\setminus A} \cap (B_j \times B_j)$. If $\mathrm{bd}\,(B_j)$ is complete for all $j = 1, 2, \ldots k$, then $\mathcal{G}$ is collapsible onto A.

Example 5.11
Let $\mathcal{G} = (V, E)$ be the simple undirected graph with $V = \{1, 2, 3, 4, 5, 6\}$ and $E = \{\{1, 2\}, \{1, 4\}, \{2, 3\}, \{2, 4\}, \{2, 5\}, \{3, 5\}, \{4, 5\}, \{5, 6\}\}$. Then $\mathcal{G}$ can be illustrated as in Figure 5.9. Further let A be the subset of V given by $A = \{2, 3, 5\}$. We observe that $\mathcal{G}$ is collapsible onto A.

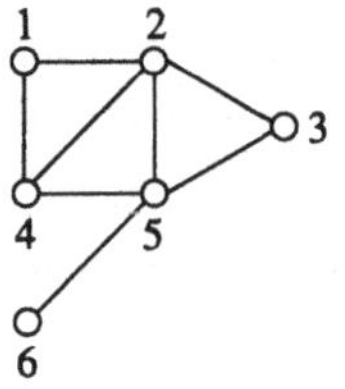

Figure 5.9: *Illustration of the simple undirected graph from Example 5.11. The graph is collapsible onto* $A = \{2, 3, 5\}$.

■

Remark that a simple undirected graph always is collapsible onto a clique. If we remove an edge from a clique, but still want the graph to be collapsible onto the same set of vertices, the removed edge must be regular. A regular edge is defined below.

Definition 5.20 Regular edge
Let $\mathcal{G} = (V, E)$ be a simple undirected graph with $\{\alpha, \beta\} \in E$ and let $C = \mathrm{bd}\,(\alpha) \cap \mathrm{bd}\,(\beta)$. If the following properties hold

1. *C is complete,*
2. *$\mathrm{bd}\,(\alpha)$ and $\mathrm{bd}\,(\beta)$ are separated by $C \cup \{\alpha, \beta\}$,*

then $\{\alpha, \beta\}$ is a regular edge.

The theorem below follows immediately from Definition 5.20.

Theorem 5.4
Let $\mathcal{G} = (V, E)$ be a simple undirected graph and let $\{\alpha, \beta\} \in E$ be a regular edge. Further let $C = \text{bd}(\alpha) \cap \text{bd}(\beta)$. Then $\{\alpha, \beta\}$ is contained in exactly one clique, namely $C \cup \{\alpha, \beta\}$.

Example 5.12
This example illustrates the concept of a regular edge. Let $\mathcal{G} = (V, E)$ be the simple undirected graph with $V = \{1, 2, 3, 4, 5, 6, 7, 8, 9, 10\}$ and $E = \{\{1,2\}, \{2,3\}, \{2,4\}, \{2,6\}, \{2,7\}, \{3,4\}, \{4,5\}, \{4,6\}, \{4,7\}, \{6,7\}, \{6,10\}, \{7,8\}, \{7,9\}, \{8,9\}, \{9,10\}\}$. In Figure 5.10 is an illustration of $\mathcal{G}$.

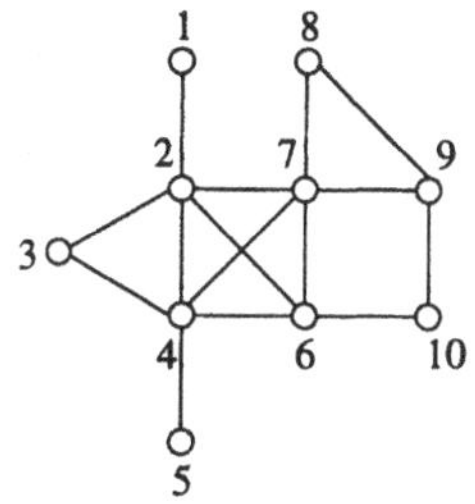

Figure 5.10: *Illustration of the simple undirected graph from Example 5.12. In this graph e.g. $\{4, 7\}$ is a regular edge.*

Observe that $\text{bd}(4) = \{2, 3, 5, 6, 7\}$ and $\text{bd}(7) = \{2, 4, 6, 8, 9\}$, i.e. $\text{bd}(4) \cap \text{bd}(7) = \{2, 6\}$, which is complete. Further $\text{bd}(4)$ and $\text{bd}(7)$ are separated by $\{2, 4, 6, 7\}$. Therefore we conclude that $\{4, 7\} \in E$ is a regular edge. According to Theorem 5.4 this implies that $\{4, 7\}$ is contained in exactly one clique, namely $\{2, 4, 6, 7\}$. We see in Figure 5.10 that this is true. ■

The concept regular edge is also useful when we consider simple undirected decomposable graphs. In the following we show that when two simple undirected decomposable graphs differ by exactly one edge, this particular edge is regular.

Theorem 5.5
Let $\mathcal{G} = (V, E)$ be a simple undirected decomposable graph with $\{\alpha, \beta\} \in E$ and further let $\mathcal{G}' = (V, E \setminus \{\alpha, \beta\})$. If $\mathcal{G}'$ is decomposable, then $\{\alpha, \beta\}$ is a regular edge.

Proof:
Let $\mathcal{G} = (V, E)$ be a simple undirected decomposable graph with $\{\alpha, \beta\} \in E$ and further let $\mathcal{G}' = (V, E \setminus \{\alpha, \beta\})$. Assume that $\mathcal{G}'$ is decomposable and let $C = \text{bd}(\alpha) \cap \text{bd}(\beta)$ in $\mathcal{G}$. We seek to show that

1. C is complete in $\mathcal{G}$.
2. bd (α) and bd (β) are separated by $C \cup \{\alpha, \beta\}$ in $\mathcal{G}$.

This is shown by contradiction.

Re 1:
Assume that C is not complete, which implies that C contains at least two elements. Moreover $\{\alpha, \beta\}$ is a member of at least two cliques in $\mathcal{G}$ and hereby $\mathcal{G}'$ contains a chordless 4-cycle, when $\{\alpha, \beta\}$ is removed from $\mathcal{G}$. Then according to Theorem 5.2 page 91 we have that $\mathcal{G}'$ is not decomposable. This is a contradiction, whereby we conclude that C is complete.

Re 2:
Assume that bd (α) and bd (β) are not separated by $C \cup \{\alpha, \beta\}$. Hence there exists a path in $\mathcal{G}'$ from α to β which do not intercept $C \cup \{\alpha, \beta\}$. Choose the shortest path from α to β which do not coincide with $C \cup \{\alpha, \beta\}$, i.e.

$$\alpha \sim \gamma_1 \sim \gamma_2 \sim \cdots \sim \gamma_p \sim \beta\,, \tag{5.1}$$

where $\gamma_1 \in \text{bd}\,(\alpha) \setminus \text{bd}\,(\beta)$ and $\gamma_p \in \text{bd}\,(\beta) \setminus \text{bd}\,(\alpha)$. The path in (5.1) combined with the edge $\{\alpha, \beta\}$ is a $(p+2)$-cycle in $\mathcal{G}$. Since $\mathcal{G}$ is decomposable it contains no chordless cycle of length greater than 3, i.e. $p = 1$. Hereby $\gamma_1 \in C$, which is a contradiction. Hence bd (α) and bd (β) are separated by $C \cup \{\alpha, \beta\}$. ■

From Lauritzen & Frydenberg (1989) and Theorem 5.5 we deduce the following important result.

Theorem 5.6
Let $\mathcal{G} = (V, E)$ be a simple undirected decomposable graph and let $\mathcal{G}' = (V, E')$ be a decomposable subgraph of $\mathcal{G}$ such that $|E| - |E'| = k$. Then there exists a sequence of decomposable graphs $\mathcal{G}' = \mathcal{G}_0 \subset \mathcal{G}_1 \subset \cdots \subset \mathcal{G}_k = \mathcal{G}$ such that $\mathcal{G}_{j-1}$ and $\mathcal{G}_j$ differ by exactly one regular edge for $j = 1, 2, \ldots, k$.

In the theorem above $\mathcal{G}_{j-1} \subset \mathcal{G}_j$ means that $E_{j-1} \subset E_j$ for $j = 1, 2, \ldots, k$.

A simple undirected graph containing a regular edge implies certain decompositions of subgraphs. These are stated in the next theorem.

Theorem 5.7
Let $\mathcal{G} = (V, E)$ be a simple undirected graph and let $\{\alpha, \beta\} \in E$ be a regular edge. Further let $C = \text{bd}\,(\alpha) \cap \text{bd}\,(\beta)$, $A = \{\gamma \in V \,|\, \gamma$ and β are separated by $C \cup \alpha\}$ and $B = (V \setminus A) \cup C$. The following properties hold.

1. $\mathcal{G} = \mathcal{G}_A \dot{\cup} \mathcal{G}_{B \cup \alpha}$ and $A \cap (B \cup \alpha) = C \cup \alpha$.

2. $\mathcal{G}_{A\cup\beta} = \mathcal{G}_A \dot{\cup} \mathcal{G}_{C\cup\{\alpha,\beta\}}$ *and* $A \cap (C \cup \{\alpha, \beta\}) = C \cup \alpha$.

3. $\mathcal{G}_{B\cup\alpha} = \mathcal{G}_B \dot{\cup} \mathcal{G}_{C\cup\{\alpha,\beta\}}$ *and* $B \cap (C \cup \{\alpha, \beta\}) = C \cup \beta$.

Proof:
Let $\mathcal{G} = (V, E)$ be a simple undirected graph and let $\{\alpha, \beta\} \in E$ be a regular edge. Further let A, B and C be given as in the theorem.

Since $\{\alpha, \beta\} \in E$ is a regular edge we deduce from Theorem 5.4 page 95 that $C \cup \{\alpha, \beta\}$ is a clique.

Re 1:
From the definitions of A and B we observe that $V = A \cup (B \cup \alpha)$. Further A and $B \cup \alpha$ are separated by $C \cup \alpha$. Hence $E = E_A \cup E_{B\cup\alpha}$. Next we see that $A \cap B = C$, therefore $A \cap (B \cup \alpha) = C \cup \alpha$ as $\alpha \in A$. This set is complete since $C \cup \{\alpha, \beta\}$ is a clique in $\mathcal{G}$. Hereby we conclude that $\mathcal{G} = \mathcal{G}_A \dot{\cup} \mathcal{G}_{B\cup\alpha}$.

Re 2:
Considering the induced graph $\mathcal{G}_{A\cup\beta} = (A \cup \beta, E_{A\cup\beta})$ we observe that the set given by $\{\gamma \in A \cup \beta |\, \gamma$ and β are separated by $C \cup \alpha\}$ is equal to A. However since $(A \cup \beta) \setminus A = \beta$ the set B in $\mathcal{G}_{A\cup\beta}$ is $B = C \cup \beta$. Applying part 1 on these subsets in $\mathcal{G}_{A\cup\beta}$ we observe $\mathcal{G}_{A\cup\beta} = \mathcal{G}_A \dot{\cup} \mathcal{G}_{C\cup\{\alpha,\beta\}}$ and $A \cap (C \cup \{\alpha, \beta\}) = C \cup \alpha$.

Re 3:
Consider the graph $\mathcal{G}_{B\cup\alpha} = (B \cup \alpha, E_{B\cup\alpha})$. In this graph the sets become $A = C \cup \alpha$ and $B = \{\gamma \in B \cup \alpha |\, \gamma$ and α are separated by $C \cup \beta\}$. This situation is similar to the one in part 2 and we deduce that $\mathcal{G}_{B\cup\alpha} = \mathcal{G}_B \dot{\cup} \mathcal{G}_{C\cup\{\alpha,\beta\}}$ and $B \cap (C \cup \{\alpha, \beta\}) = C \cup \beta$. ■

Example 5.13
Let $\mathcal{G}$ be the simple undirected graph from Example 5.12 page 95, where $\{4, 7\} \in E$ is a regular edge and $C = \{2, 6\}$. From Figure 5.10 page 95 we observe that the set A is given by $A = \{\gamma \in V |\, \gamma$ and 7 are separated by $\{2, 4, 6\}\} = \{1, 2, 3, 4, 5, 6\}$ and that the set B is given by $B = (V \setminus A) \cup C = \{2, 6, 7, 8, 9, 10\}$. Figure 5.11 page 98 illustrates $\mathcal{G}_A, \mathcal{G}_B$ and $\mathcal{G}_{C\cup\{\alpha,\beta\}}$. We see that $\mathcal{G} = \mathcal{G}_A \dot{\cup} \left(\mathcal{G}_B \dot{\cup} \mathcal{G}_{C\cup\{\alpha,\beta\}}\right)$, which must be fulfilled according to Theorem 5.7 page 96.

■

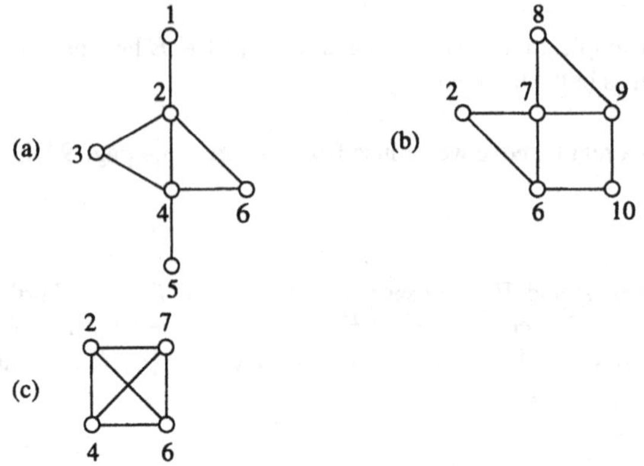

Figure 5.11: *(a), (b) and (c) illustrate the induced subgraphs $\mathcal{G}_A$, $\mathcal{G}_B$ and $\mathcal{G}_{C\cup\{\alpha,\beta\}}$ from Example 5.13.*

6

Conditional Independence and Markov Properties

When we consider graphical models for the multivariate complex normal distribution, we formulate the models in terms of simple undirected graphs, which illustrate conditional independence of complex random vectors. Therefore this chapter concentrates on conditional independence of complex random vectors. Conditional independence is studied formally by Dawid (1979), but has also been explored by others see e.g. Pearl (1988). We have chosen only to consider complex random vectors with continuous density function w.r.t. Lebesgue measure, since we only consider such complex random vectors in this book. Besides an acquaintance with Lebesgue measure no further measure theory is used. We define the conditional density function and the conditional distribution for complex random vectors. Then the conditional distribution of a measurable transformation of a complex random vector is defined. Using these definitions we are able to state the law of total probability and further to define conditional independence of two measurable transformations of a complex random vector given a third complex random vector. We establish some properties, which are equivalent to the definition of conditional independence, and we find properties which can be used to deduce conditional independences from others. Furthermore we study conditional independence in the special case, where a complex random vector is partitioned and we give useful theorems in this case. Next conditional independence in relation to simple undirected graphs is studied. Three different ways are used to specify conditional independence by means of a graph. This means that the distributions fulfilling these different conditional independence criteria may have different properties. These are the so called Markov properties. In Speed (1979) basic aspects in conditional independence are treated together with Markov properties. We examine the Markov properties and especially their equivalence. Furthermore the conditional independence graph of a complex random vector is defined. This graph gives a picture of the pattern of conditional independence of the complex random variables in the complex random vector.

6.1 Conditional Independence

We begin by defining the conditional density function and the conditional distribution.

Definition 6.1 The conditional density function and distribution
Let $\boldsymbol{X} = (\boldsymbol{X}_1, \boldsymbol{X}_2)$ *be a complex random vector with continuous density function w.r.t. Lebesgue*

measure. The conditional density function of $\boldsymbol{X}_1$ given $\boldsymbol{X}_2$ is defined as

$$f_{\boldsymbol{X}_1|\boldsymbol{X}_2}(\boldsymbol{x}_1|\,\boldsymbol{x}_2) = \frac{f_{\boldsymbol{X}_1,\boldsymbol{X}_2}(\boldsymbol{x}_1,\boldsymbol{x}_2)}{f_{\boldsymbol{X}_2}(\boldsymbol{x}_2)}$$

for all values of $\boldsymbol{x}_1$ and for all $\boldsymbol{x}_2$, where $f_{\boldsymbol{X}_2}(\boldsymbol{x}_2) > 0$.

The conditional distribution of $\boldsymbol{X}_1$ given $\boldsymbol{X}_2$ is defined as

$$P(\boldsymbol{X}_1 \in A_1|\,\boldsymbol{X}_2 = \boldsymbol{x}_2) = \int_{A_1} f_{\boldsymbol{X}_1|\boldsymbol{X}_2}(\boldsymbol{x}_1|\,\boldsymbol{x}_2)\,d\boldsymbol{x}_1$$

for all measurable sets A_1 in the sample space of $\boldsymbol{X}_1$ and for all $\boldsymbol{x}_2$, where $f_{\boldsymbol{X}_2}(\boldsymbol{x}_2) > 0$.

Next we define the conditional distribution of a measurable transformation of a complex random vector.

Definition 6.2 The conditional distribution of a transformation
Let $\boldsymbol{X} = (\boldsymbol{X}_1, \boldsymbol{X}_2)$ be a complex random vector with continuous density function w.r.t. Lebesgue measure and let $U(\boldsymbol{X})$ be a measurable transformation. The conditional distribution of $U(\boldsymbol{X})$ given $\boldsymbol{X}_2$ is defined as

$$P(U(\boldsymbol{X}) \in A|\,\boldsymbol{X}_2 = \boldsymbol{x}_2) = \int_{\{\boldsymbol{x}_1|U(\boldsymbol{x})\in A\}} f_{\boldsymbol{X}_1|\boldsymbol{X}_2}(\boldsymbol{x}_1|\,\boldsymbol{x}_2)\,d\boldsymbol{x}_1$$

for all measurable sets A in the sample space of $U(\boldsymbol{X})$ and for all $\boldsymbol{x}_2$, where $f_{\boldsymbol{X}_2}(\boldsymbol{x}_2) > 0$.

From the definitions above the law of total probability follows immediately.

Theorem 6.1 The law of total probability
Let $\boldsymbol{X} = (\boldsymbol{X}_1, \boldsymbol{X}_2)$ be a complex random vector with continuous density function w.r.t. Lebesgue measure and let $U(\boldsymbol{X})$ be a measurable transformation. It holds that

$$P(U(\boldsymbol{X}) \in A) = \int P(U(\boldsymbol{X}) \in A|\,\boldsymbol{X}_2 = \boldsymbol{x}_2)\,f_{\boldsymbol{X}_2}(\boldsymbol{x}_2)\,d\boldsymbol{x}_2$$

for all measurable sets A in the sample space of $U(\boldsymbol{X})$.

With the distributions above as a base, we are able to define conditional independence of two measurable transformations of a complex random vector given a part of the complex random vector.

Definition 6.3 Conditional independence
Let $\boldsymbol{X} = (\boldsymbol{X}_1, \boldsymbol{X}_2)$ be a complex random vector with continuous density function w.r.t. Lebesgue measure and let $U(\boldsymbol{X})$ and $V(\boldsymbol{X})$ be measurable transformations. If, for all measurable sets A

and B in the sample spaces of U$(\boldsymbol{X})$ *and V*$(\boldsymbol{X})$*, respectively, and for all* $\boldsymbol{x}_2$*, where* $f_{\boldsymbol{X}_2}(\boldsymbol{x}_2) > 0$*, it holds that*

$$P\left(U(\boldsymbol{X}) \in A, V(\boldsymbol{X}) \in B \middle| \boldsymbol{X}_2 = \boldsymbol{x}_2\right) = P\left(U(\boldsymbol{X}) \in A \middle| \boldsymbol{X}_2 = \boldsymbol{x}_2\right) P\left(V(\boldsymbol{X}) \in B \middle| \boldsymbol{X}_2 = \boldsymbol{x}_2\right),$$

then $U(\boldsymbol{X})$ *and* $V(\boldsymbol{X})$ *are conditionally independent given* $\boldsymbol{X}_2$*. This is denoted by*

$$U(\boldsymbol{X}) \perp\!\!\!\perp V(\boldsymbol{X}) \mid \boldsymbol{X}_2 \,.$$

Remark that considering $\boldsymbol{X}_2$ as trivial Definition 6.3 gives the definition of independence of two measurable transformations of a complex random vector. Obviously the definition of conditional independence is symmetric in $U(\boldsymbol{X})$ and $V(\boldsymbol{X})$. An extension of the definition to n measurable transformations of $\boldsymbol{X}$ is readily made. Furthermore Definition 6.3 includes the case where $\boldsymbol{X}$ is a complex random variable or a complex random matrix, since a vector can consist of only one element and a matrix can be interpreted as a vector.

Theorem 6.2 below states properties, which are equivalent to the definition of conditional independence. For this theorem we need the following definition.

Definition 6.4
Let $\boldsymbol{X} = (\boldsymbol{X}_1, \boldsymbol{X}_2)$ *be a complex random vector with continuous density function w.r.t. Lebesgue measure and let* $U(\boldsymbol{X})$ *be a measurable transformation. The joint distribution of* $U(\boldsymbol{X})$ *and* $\boldsymbol{X}_2$ *is defined as*

$$P\left(U(\boldsymbol{X}) \in A, \boldsymbol{X}_2 = \boldsymbol{x}_2\right) = \int_{\{\boldsymbol{x}_1 | U(\boldsymbol{x}) \in A\}} f_{\boldsymbol{X}_1,\boldsymbol{X}_2}(\boldsymbol{x}_1, \boldsymbol{x}_2)\, d\boldsymbol{x}_1$$

for all measurable sets A in the sample space of $U(\boldsymbol{X})$ *and for all* $\boldsymbol{x}_2$*.*

Further let $V(\boldsymbol{X})$ *be a measurable transformation. The conditional distribution of* $U(\boldsymbol{X})$ *given* $V(\boldsymbol{X})$ *and* $\boldsymbol{X}_2$ *is defined as*

$$P\left(U(\boldsymbol{X}) \in A \middle| V(\boldsymbol{X}) \in B, \boldsymbol{X}_2 = \boldsymbol{x}_2\right) = \frac{P\left(U(\boldsymbol{X}) \in A, V(\boldsymbol{X}) \in B, \boldsymbol{X}_2 = \boldsymbol{x}_2\right)}{P\left(V(\boldsymbol{X}) \in B, \boldsymbol{X}_2 = \boldsymbol{x}_2\right)}$$

for all measurable sets A in the sample space of $U(\boldsymbol{X})$*, all measurable sets B in the sample space of* $V(\boldsymbol{X})$ *and all* $\boldsymbol{x}_2$*, where* $P\left(V(\boldsymbol{X}) \in B, \boldsymbol{X}_2 = \boldsymbol{x}_2\right) > 0$*.*

If we in the definition above let $U(\boldsymbol{X})$ be trivial in the joint distribution of $U(\boldsymbol{X})$ and $\boldsymbol{X}_2$, we observe that $P\left(\boldsymbol{X}_2 = \boldsymbol{x}_2\right) = f_{\boldsymbol{X}_2}(\boldsymbol{x}_2)$. Hereby considering $V(\boldsymbol{X})$ as trivial in the conditional distribution of $U(\boldsymbol{X})$ given $V(\boldsymbol{X})$ and $\boldsymbol{X}_2$ we get a definition of $P\left(U(\boldsymbol{X}) \in A \middle| \boldsymbol{X}_2 = \boldsymbol{x}_2\right)$, which is in agreement with Definition 6.2 page 100. As promised the theorem about conditional independence is given below.

Theorem 6.2 Properties of conditional independence
Let $\boldsymbol{X} = (\boldsymbol{X}_1, \boldsymbol{X}_2)$ *be a complex random vector with continuous density function w.r.t. Lebesgue*

measure and let $U(\boldsymbol{X})$ and $V(\boldsymbol{X})$ be measurable transformations. Further let A and B be measurable sets in the sample spaces of $U(\boldsymbol{X})$ and $V(\boldsymbol{X})$, respectively. The following properties are equivalent.

(a) $U(\boldsymbol{X}) \perp\!\!\!\perp V(\boldsymbol{X}) \mid \boldsymbol{X}_2$.

(b) $P(U(\boldsymbol{X})\in A, V(\boldsymbol{X})\in B \mid \boldsymbol{X}_2 = \boldsymbol{x}_2) = P(U(\boldsymbol{X})\in A \mid \boldsymbol{X}_2 = \boldsymbol{x}_2) P(V(\boldsymbol{X})\in B \mid \boldsymbol{X}_2 = \boldsymbol{x}_2)$.

(c) $P(U(\boldsymbol{X})\in A, V(\boldsymbol{X})\in B, \boldsymbol{X}_2 = \boldsymbol{x}_2) = \frac{P(U(\boldsymbol{X})\in A, \boldsymbol{X}_2 = \boldsymbol{x}_2) P(V(\boldsymbol{X})\in B, \boldsymbol{X}_2 = \boldsymbol{x}_2)}{f_{\boldsymbol{X}_2}(\boldsymbol{x}_2)}$.

(d) $P(U(\boldsymbol{X})\in A, V(\boldsymbol{X})\in B, \boldsymbol{X}_2 = \boldsymbol{x}_2) = P(U(\boldsymbol{X})\in A \mid \boldsymbol{X}_2 = \boldsymbol{x}_2) P(V(\boldsymbol{X})\in B, \boldsymbol{X}_2 = \boldsymbol{x}_2)$.

(e) $P(U(\boldsymbol{X})\in A \mid V(\boldsymbol{X})\in B, \boldsymbol{X}_2 = \boldsymbol{x}_2) = P(U(\boldsymbol{X})\in A \mid \boldsymbol{X}_2 = \boldsymbol{x}_2)$.

(f) *There exist functions h and k such that*
$P(U(\boldsymbol{X})\in A, V(\boldsymbol{X})\in B, \boldsymbol{X}_2 = \boldsymbol{x}_2) = h(A, \boldsymbol{x}_2)\, k(B, \boldsymbol{x}_2)$.

(g) *There exists a function h such that*
$P(U(\boldsymbol{X})\in A \mid V(\boldsymbol{X})\in B, \boldsymbol{X}_2 = \boldsymbol{x}_2) = h(A, \boldsymbol{x}_2)$.

The equalities hold whenever all the quantities in question are well defined, i.e. when the distributions of all the conditioning variables are positive.

Proof:
The equivalence between (a) and (b) is given directly in Definition 6.3 page 100. The properties left are easily shown by starting with (b) and using Definition 6.4 page 101. ■

The functions h and k in (f) do not have to be uniquely determined, only the factorization into functions must exist. The property (f) is called the *factorization criterion* for conditional independence. The factorization criterion can also be formulated in an additive form as

(h) There exist functions h and k such that
$\log P(U(\boldsymbol{X})\in A, V(\boldsymbol{X})\in B, \boldsymbol{X}_2 = \boldsymbol{x}_2) = \log h(A, \boldsymbol{x}_2) + \log k(B, \boldsymbol{x}_2)$.

Note that whenever a relation includes the possibility of zero probabilities the multiplicative form is needed since the logarithm of zero is not defined.

The theorem below establishes some properties for deducing conditional independence from others. These properties are given for measurable transformations of a complex random vector.

Theorem 6.3
Let $\boldsymbol{X} = (\boldsymbol{X}_1, \boldsymbol{X}_2)$ be a complex random vector with continuous density function w.r.t. Lebesgue measure. Further let $U(\boldsymbol{X})$, $V(\boldsymbol{X})$ and $W(\boldsymbol{X})$ be measurable transformations and let $Y(U(\boldsymbol{X}), \boldsymbol{X}_2)$ and $Z(V(\boldsymbol{X}), \boldsymbol{X}_2)$ be measurable transformations. The following properties hold.

1. $U(\boldsymbol{X}) \perp\!\!\!\perp V(\boldsymbol{X}) \mid \boldsymbol{X}_2$ *and* $U(\boldsymbol{X}) \perp\!\!\!\perp \boldsymbol{X}_2 \Rightarrow U(\boldsymbol{X}) \perp\!\!\!\perp (V(\boldsymbol{X}), \boldsymbol{X}_2)$.

2. $U(\boldsymbol{X}) \perp\!\!\!\perp V(\boldsymbol{X}) \mid \boldsymbol{X}_2 \Rightarrow Y(U(\boldsymbol{X}), \boldsymbol{X}_2) \perp\!\!\!\perp Z(V(\boldsymbol{X}), \boldsymbol{X}_2) \mid \boldsymbol{X}_2$.

3. $U(\boldsymbol{X}) \perp\!\!\!\perp \boldsymbol{X}_2$, $V(\boldsymbol{X}) \perp\!\!\!\perp \boldsymbol{X}_2$, $W(\boldsymbol{X}) \perp\!\!\!\perp \boldsymbol{X}_2$ *and* $U(\boldsymbol{X})$, $V(\boldsymbol{X})$ *and* $W(\boldsymbol{X})$ *are mutually independent given* $\boldsymbol{X}_2 \Rightarrow U(\boldsymbol{X})$, $V(\boldsymbol{X})$, $W(\boldsymbol{X})$ *and* $\boldsymbol{X}_2$ *are mutually independent.*

Proof:
Apply Definition 6.3 page 100, Definition 6.4 and Theorem 6.2 page 101 to prove the theorem.

■

Note from part 2 that $\boldsymbol{X}_2$ can be considered as trivial, whereby we get a property for deducing independence from another independence. The results discussed until this point can easily be applied on any partition of $\boldsymbol{X}$. Considering e.g. a complex random vector $\boldsymbol{X}$ partitioned as $\boldsymbol{X} = (\boldsymbol{X}_1, \boldsymbol{X}_2, \boldsymbol{X}_3)$ and the measurable transformations $U(\boldsymbol{X}) = \boldsymbol{X}_1$ and $V(\boldsymbol{X}) = \boldsymbol{X}_2$ we get from Definition 6.3 a condition for the conditional independence of $\boldsymbol{X}_1$ and $\boldsymbol{X}_2$ given $\boldsymbol{X}_3$.

Definition 6.5 Conditional independence
Let $(\boldsymbol{X}_1, \boldsymbol{X}_2, \boldsymbol{X}_3)$ *be a partitioned complex random vector with continuous density function w.r.t. Lebesgue measure. If, for all measurable sets A and B in the sample spaces of* $\boldsymbol{X}_1$ *and* $\boldsymbol{X}_2$*, respectively, and for all* $\boldsymbol{x}_3$*, where* $f_{\boldsymbol{X}_3}(\boldsymbol{x}_3) > 0$*, it holds that*

$$P(\boldsymbol{X}_1 \in A, \boldsymbol{X}_2 \in B | \boldsymbol{X}_3 = \boldsymbol{x}_3) = P(\boldsymbol{X}_1 \in A | \boldsymbol{X}_3 = \boldsymbol{x}_3) P(\boldsymbol{X}_2 \in B | \boldsymbol{X}_3 = \boldsymbol{x}_3) ,$$

then $\boldsymbol{X}_1$ *and* $\boldsymbol{X}_2$ *are conditionally independent given* $\boldsymbol{X}_3$*. This is denoted by*

$$\boldsymbol{X}_1 \perp\!\!\!\perp \boldsymbol{X}_2 \mid \boldsymbol{X}_3 .$$

Remark that Definition 6.5 is a special case of Definition 6.3, but we have chosen to state it as a definition anyway. Considering $\boldsymbol{X}_3$ as trivial in Definition 6.5 gives the definition of independence of $\boldsymbol{X}_1$ and $\boldsymbol{X}_2$. The extension to mutual independence of n random vectors is defined earlier in Definition 1.21 page 13. Some properties which are equivalent to conditional independence of $\boldsymbol{X}_1$ and $\boldsymbol{X}_2$ given $\boldsymbol{X}_3$ are established in the following theorem.

Theorem 6.4 Properties of conditional independence
Let $(\boldsymbol{X}_1, \boldsymbol{X}_2, \boldsymbol{X}_3)$ *be a partitioned complex random vector with continuous density function w.r.t. Lebesgue measure. The following properties are equivalent.*

(a) $\boldsymbol{X}_1 \perp\!\!\!\perp \boldsymbol{X}_2 \mid \boldsymbol{X}_3$.

(b) $f_{\boldsymbol{X}_1,\boldsymbol{X}_2|\boldsymbol{X}_3}(\boldsymbol{x}_1, \boldsymbol{x}_2 | \boldsymbol{x}_3) = f_{\boldsymbol{X}_1|\boldsymbol{X}_3}(\boldsymbol{x}_1 | \boldsymbol{x}_3) f_{\boldsymbol{X}_2|\boldsymbol{X}_3}(\boldsymbol{x}_2 | \boldsymbol{x}_3)$.

(c) $f_{\boldsymbol{X}_1,\boldsymbol{X}_2,\boldsymbol{X}_3}(\boldsymbol{x}_1, \boldsymbol{x}_2, \boldsymbol{x}_3) = \frac{f_{\boldsymbol{X}_1,\boldsymbol{X}_3}(\boldsymbol{x}_1,\boldsymbol{x}_3) f_{\boldsymbol{X}_2,\boldsymbol{X}_3}(\boldsymbol{x}_2,\boldsymbol{x}_3)}{f_{\boldsymbol{X}_3}(\boldsymbol{x}_3)}$.

(d) $f_{\boldsymbol{X}_1,\boldsymbol{X}_2,\boldsymbol{X}_3}(\boldsymbol{x}_1, \boldsymbol{x}_2, \boldsymbol{x}_3) = f_{\boldsymbol{X}_1|\boldsymbol{X}_3}(\boldsymbol{x}_1 | \boldsymbol{x}_3) f_{\boldsymbol{X}_2,\boldsymbol{X}_3}(\boldsymbol{x}_2, \boldsymbol{x}_3)$.

(e) $f_{X_1|X_2,X_3}(\boldsymbol{x}_1|\boldsymbol{x}_2,\boldsymbol{x}_3) = f_{X_1|X_3}(\boldsymbol{x}_1|\boldsymbol{x}_3)$.

(f) *There exist functions h and k such that*
$f_{X_1,X_2,X_3}(\boldsymbol{x}_1,\boldsymbol{x}_2,\boldsymbol{x}_3) = h(\boldsymbol{x}_1,\boldsymbol{x}_3)\,k(\boldsymbol{x}_2,\boldsymbol{x}_3)$.

(g) *There exists a function h such that*
$f_{X_1|X_2,X_3}(\boldsymbol{x}_1|\boldsymbol{x}_2,\boldsymbol{x}_3) = h(\boldsymbol{x}_1,\boldsymbol{x}_3)$.

The equalities hold whenever all the quantities in question are well defined, i.e. when the density functions of all the conditioning variables are positive.

Proof:
Use Definition 6.5 page 103 and Definition 6.1 page 99 to show that (a) is equivalent to (b). The properties left are shown similarly as the corresponding properties of Theorem 6.2 page 101 by use of Bayes' rule. ■

As in the general case the factorization criterion (f) is also given in an additive form as given below.

(h) *There exist functions h and k such that*
$\log f_{X_1,X_2,X_3}(\boldsymbol{x}_1,\boldsymbol{x}_2,\boldsymbol{x}_3) = \log h(\boldsymbol{x}_1,\boldsymbol{x}_3) + \log k(\boldsymbol{x}_2,\boldsymbol{x}_3)$.

Properties of conditional independence of parts of a complex random vector can be deduced from others by using the theorem below.

Theorem 6.5
Let $(\boldsymbol{X}_1,\boldsymbol{X}_2,\boldsymbol{X}_3)$ be a partitioned complex random vector with continuous density function w.r.t. Lebesgue measure. Further let $U(\boldsymbol{X}_1,\boldsymbol{X}_3)$ and $V(\boldsymbol{X}_2,\boldsymbol{X}_3)$ be measurable transformations. The following properties hold.

1. $\boldsymbol{X}_1 \perp\!\!\!\perp \boldsymbol{X}_2 \mid \boldsymbol{X}_3$ *and* $\boldsymbol{X}_1 \perp\!\!\!\perp \boldsymbol{X}_3 \Rightarrow \boldsymbol{X}_1 \perp\!\!\!\perp (\boldsymbol{X}_2,\boldsymbol{X}_3)$.
2. $\boldsymbol{X}_1 \perp\!\!\!\perp \boldsymbol{X}_2 \mid \boldsymbol{X}_3 \Rightarrow U(\boldsymbol{X}_1,\boldsymbol{X}_3) \perp\!\!\!\perp V(\boldsymbol{X}_2,\boldsymbol{X}_3) \mid \boldsymbol{X}_3$.

Let $(\boldsymbol{X}_1,\boldsymbol{X}_2,\boldsymbol{X}_3,\boldsymbol{X}_4)$ be a partitioned complex random vector with continuous density function w.r.t. Lebesgue measure. The following property holds.

3. $\boldsymbol{X}_1 \perp\!\!\!\perp (\boldsymbol{X}_2,\boldsymbol{X}_3) \mid \boldsymbol{X}_4 \Rightarrow \boldsymbol{X}_1 \perp\!\!\!\perp \boldsymbol{X}_2 \mid (\boldsymbol{X}_3,\boldsymbol{X}_4)$.

Proof:
Use Theorem 6.3 page 102 to show part 1 and 2. Part 3 is shown by use of Theorem 6.4 page 103, Bayes' rule and part 2. ■

Again some properties concerning independence can be obtained by considering the conditions as trivial. In addition to Theorem 6.5 we can show another important property of conditional independence. It does not hold universally but only under the additional assumption of positive density function w.r.t. Lebesgue measure.

Theorem 6.6
Let $(\boldsymbol{X}_1, \boldsymbol{X}_2, \boldsymbol{X}_3, \boldsymbol{X}_4)$ be a partitioned complex random vector with positive and continuous density function w.r.t. Lebesgue measure. It holds that

$$\begin{aligned} &\boldsymbol{X}_1 \perp\!\!\!\perp \boldsymbol{X}_2 \mid (\boldsymbol{X}_3, \boldsymbol{X}_4) \text{ and } \boldsymbol{X}_1 \perp\!\!\!\perp \boldsymbol{X}_3 \mid (\boldsymbol{X}_2, \boldsymbol{X}_4) \\ \Downarrow \\ &\boldsymbol{X}_1 \perp\!\!\!\perp (\boldsymbol{X}_2, \boldsymbol{X}_3) \mid \boldsymbol{X}_4 . \end{aligned}$$

Proof:
Let $(\boldsymbol{X}_1, \boldsymbol{X}_2, \boldsymbol{X}_3, \boldsymbol{X}_4)$ be a partitioned complex random vector with positive and continuous density function w.r.t. Lebesgue measure. Moreover let

$$\boldsymbol{X}_1 \perp\!\!\!\perp \boldsymbol{X}_2 \mid (\boldsymbol{X}_3, \boldsymbol{X}_4) \text{ and } \boldsymbol{X}_1 \perp\!\!\!\perp \boldsymbol{X}_3 \mid (\boldsymbol{X}_2, \boldsymbol{X}_4) .$$

By using Theorem 6.4 page 103 this is equivalent to

$$\begin{aligned} f_{\boldsymbol{X}_1,\boldsymbol{X}_2,\boldsymbol{X}_3,\boldsymbol{X}_4}(\boldsymbol{x}_1, \boldsymbol{x}_2, \boldsymbol{x}_3, \boldsymbol{x}_4) &= h_1(\boldsymbol{x}_1, \boldsymbol{x}_3, \boldsymbol{x}_4)\, k_1(\boldsymbol{x}_2, \boldsymbol{x}_3, \boldsymbol{x}_4) \\ &= h_2(\boldsymbol{x}_1, \boldsymbol{x}_2, \boldsymbol{x}_4)\, k_2(\boldsymbol{x}_2, \boldsymbol{x}_3, \boldsymbol{x}_4) , \end{aligned} \tag{6.1}$$

where h_1, h_2, k_1 and k_2 are suitable positive functions as it holds that the density function $f_{\boldsymbol{X}_1,\boldsymbol{X}_2,\boldsymbol{X}_3,\boldsymbol{X}_4}(\boldsymbol{x}_1, \boldsymbol{x}_2, \boldsymbol{x}_3, \boldsymbol{x}_4)$ is positive. Thus we are allowed to perform division in (6.1) and for all $\boldsymbol{x}_3$ we get

$$h_2(\boldsymbol{x}_1, \boldsymbol{x}_2, \boldsymbol{x}_4) = \frac{h_1(\boldsymbol{x}_1, \boldsymbol{x}_3, \boldsymbol{x}_4)\, k_1(\boldsymbol{x}_2, \boldsymbol{x}_3, \boldsymbol{x}_4)}{k_2(\boldsymbol{x}_2, \boldsymbol{x}_3, \boldsymbol{x}_4)} .$$

Now let $\boldsymbol{x}_3$ be fixed at $\boldsymbol{a}$. Then

$$h_2(\boldsymbol{x}_1, \boldsymbol{x}_2, \boldsymbol{x}_4) = \pi(\boldsymbol{x}_1, \boldsymbol{x}_4)\, \rho(\boldsymbol{x}_2, \boldsymbol{x}_4) ,$$

where $\pi(\boldsymbol{x}_1, \boldsymbol{x}_4) = h_1(\boldsymbol{x}_1, \boldsymbol{a}, \boldsymbol{x}_4)$ and $\rho(\boldsymbol{x}_2, \boldsymbol{x}_4) = \frac{k_1(\boldsymbol{x}_2, \boldsymbol{a}, \boldsymbol{x}_4)}{k_2(\boldsymbol{x}_2, \boldsymbol{a}, \boldsymbol{x}_4)}$. Using this in (6.1) we get

$$f_{\boldsymbol{X}_1,\boldsymbol{X}_2,\boldsymbol{X}_3,\boldsymbol{X}_4}(\boldsymbol{x}_1, \boldsymbol{x}_2, \boldsymbol{x}_3, \boldsymbol{x}_4) = \pi(\boldsymbol{x}_1, \boldsymbol{x}_4)\, \rho(\boldsymbol{x}_2, \boldsymbol{x}_4)\, k_2(\boldsymbol{x}_2, \boldsymbol{x}_3, \boldsymbol{x}_4) .$$

From Theorem 6.4 this is equivalent to

$$\boldsymbol{X}_1 \perp\!\!\!\perp (\boldsymbol{X}_2, \boldsymbol{X}_3) \mid \boldsymbol{X}_4 .$$

■

Notice that the converse of the theorem holds even if the joint density function of $\boldsymbol{X}_1, \boldsymbol{X}_2, \boldsymbol{X}_3$ and $\boldsymbol{X}_4$ w.r.t. Lebesgue measure is not everywhere positive. This is due to Theorem 6.5 page 104.

Other useful results on connections between conditional independence are stated in the following corollaries. Theorem 6.6 gives the first corollary by considering $\boldsymbol{X}_4$ as trivial.

Corollary 6.1
Let $(\boldsymbol{X}_1, \boldsymbol{X}_2, \boldsymbol{X}_3)$ be a partitioned complex random vector with positive and continuous density function w.r.t. Lebesgue measure. It holds that

$$\boldsymbol{X}_1 \perp\!\!\!\perp \boldsymbol{X}_2 \mid \boldsymbol{X}_3 \text{ and } \boldsymbol{X}_1 \perp\!\!\!\perp \boldsymbol{X}_3 \mid \boldsymbol{X}_2$$
$$\Updownarrow$$
$$\boldsymbol{X}_1 \perp\!\!\!\perp (\boldsymbol{X}_2, \boldsymbol{X}_3) \ .$$

Using Corollary 6.1 and Theorem 6.5 page 104 we get the following corollary.

Corollary 6.2
Let $(\boldsymbol{X}_1, \boldsymbol{X}_2, \boldsymbol{X}_3)$ be a partitioned complex random vector with positive and continuous density function w.r.t. Lebesgue measure. It holds that

$$\boldsymbol{X}_1 \perp\!\!\!\perp \boldsymbol{X}_2 \mid \boldsymbol{X}_3 \ , \ \boldsymbol{X}_1 \perp\!\!\!\perp \boldsymbol{X}_3 \mid \boldsymbol{X}_2 \text{ and } \boldsymbol{X}_2 \perp\!\!\!\perp \boldsymbol{X}_3 \mid \boldsymbol{X}_1$$
$$\Downarrow$$
$$\boldsymbol{X}_1, \boldsymbol{X}_2 \text{ and } \boldsymbol{X}_3 \text{ are mutually independent.}$$

Moreover from Theorem 6.6 page 105 and Theorem 6.5 page 104 we get the corollary below.

Corollary 6.3
Let $(\boldsymbol{X}_1, \boldsymbol{X}_2, \boldsymbol{X}_3, \boldsymbol{X}_4)$ be a partitioned complex random vector with positive and continuous density function w.r.t. Lebesgue measure. It holds that

$$\boldsymbol{X}_1 \perp\!\!\!\perp \boldsymbol{X}_2 \mid (\boldsymbol{X}_3, \boldsymbol{X}_4) \text{ and } \boldsymbol{X}_1 \perp\!\!\!\perp \boldsymbol{X}_3 \mid (\boldsymbol{X}_2, \boldsymbol{X}_4)$$
$$\Downarrow$$
$$\boldsymbol{X}_1 \perp\!\!\!\perp \boldsymbol{X}_2 \mid \boldsymbol{X}_4 \text{ and } \boldsymbol{X}_1 \perp\!\!\!\perp \boldsymbol{X}_3 \mid \boldsymbol{X}_4 \ .$$

6.2 Markov Properties in Relation to Simple Undirected Graphs

Let $\mathcal{G} = (V, E)$ be a simple undirected graph with V as a finite set of vertices and E as a set of edges. Let the set of vertices V also be an index set of a $|V|$-dimensional complex random vector given by $\boldsymbol{X} = (X_v)_{v \in V}$, where $\boldsymbol{X}$ is taking values in the vector space $\mathbb{C}^{|V|}$. For a subset A of V we let $\boldsymbol{X}_A = (X_v)_{v \in A}$ be a $|A|$-dimensional random vector taking values in $\mathbb{C}^{|A|}$. Notice that $\boldsymbol{X} = \boldsymbol{X}_V$ and that the α'th element in $\boldsymbol{X}$ is referred to as $\boldsymbol{X}_\alpha$. The elements of $\mathbb{C}^{|A|}$ are denoted by $\boldsymbol{x}_A$ and the elements of $\mathbb{C}^{|V|}$ are denoted by $\boldsymbol{x} = \boldsymbol{x}_V$.

Definition 6.6 The factorization property
Let $\mathcal{G} = (V, E)$ be a simple undirected graph and let $\boldsymbol{X}$ be a $|V|$-dimensional complex random

vector with distribution P on $\mathbb{C}^{|V|}$. If, for all complete subsets A of V, there exist nonnegative functions

$$\phi_A : \mathbb{C}^{|A|} \mapsto \overline{\mathbb{R}}_+$$

and if $\boldsymbol{X}$ has density function $f_{\boldsymbol{X}}$ w.r.t. Lebesgue measure of the form

$$f_{\boldsymbol{X}}(\boldsymbol{x}) = \prod_A \phi_A(\boldsymbol{x}_A) \ ,$$

then P is said to factorize according to $\mathcal{G}$. We denote this property of P by (F).

Some of the functions ϕ_A can be multiplied together or split up in different ways, therefore the functions ϕ_A are not uniquely determined. Since a clique is a maximal complete subset we can choose only to look at the cliques and then (F) is rewritten as

$$f_{\boldsymbol{X}}(\boldsymbol{x}) = \prod_{C \in \mathcal{C}} \phi_C(\boldsymbol{x}_C) \ ,$$

where $\mathcal{C}$ denotes the set of cliques in $\mathcal{G}$.

We examine three Markov properties. Formally they are different but under a certain assumption they are equivalent. This is treated in Theorem 6.8 page 110.

Definition 6.7 The pairwise Markov property
Let $\mathcal{G} = (V, E)$ be a simple undirected graph and let $\boldsymbol{X}$ be a $|V|$-dimensional complex random vector with distribution P on $\mathbb{C}^{|V|}$. If, for all pairs α and β of nonadjacent vertices, it holds that

$$\boldsymbol{X}_\alpha \perp\!\!\!\perp \boldsymbol{X}_\beta \mid \boldsymbol{X}_{V \setminus \{\alpha,\beta\}} \ ,$$

then P has the pairwise Markov property w.r.t. $\mathcal{G}$. We denote this property of P by (P).

Definition 6.8 The local Markov property
Let $\mathcal{G} = (V, E)$ be a simple undirected graph and let $\boldsymbol{X}$ be a $|V|$-dimensional complex random vector with distribution P on $\mathbb{C}^{|V|}$. If, for all $\alpha \in V$, it holds that

$$\boldsymbol{X}_\alpha \perp\!\!\!\perp \boldsymbol{X}_{V \setminus \mathrm{cl}(\alpha)} \mid \boldsymbol{X}_{\mathrm{bd}(\alpha)} \ ,$$

then P has the local Markov property w.r.t $\mathcal{G}$. We denote this property of P by (L).

Definition 6.9 The global Markov property
Let $\mathcal{G} = (V, E)$ be a simple undirected graph and let $\boldsymbol{X}$ be a $|V|$-dimensional complex random vector with distribution P on $\mathbb{C}^{|V|}$. If, for all triples (A, B, S) of disjoint subsets of V such that A and B are separated by S in $\mathcal{G}$, it holds that

$$\boldsymbol{X}_A \perp\!\!\!\perp \boldsymbol{X}_B \mid \boldsymbol{X}_S \ ,$$

then P has the global Markov property w.r.t. $\mathcal{G}$. We denote this property of P by (G).

Remark by Theorem 6.5 page 104, that the global Markov property also could have been defined for A, B and S not being disjoint subsets of V. The global Markov property is very useful, since it gives a method to infer, when two complex random vectors $\boldsymbol{X}_A$ and $\boldsymbol{X}_B$ are conditionally independent given another complex random vector $\boldsymbol{X}_S$ by means of a graph. Example 6.1 illustrates this.

Example 6.1
Let $\mathcal{G} = (V, E)$ be the simple undirected graph given by $V = \{1,2,3,4,5,6,7,8,9,10\}$ and $E = \{\{1,2\},\{2,3\},\{3,4\},\{3,5\},\{4,5\},\{4,8\},\{5,6\},\{5,7\},\{8,9\},\{8,10\}\}$. The graph is illustrated in Figure 6.1. Let A, B and S be subsets of V given by $A = \{2,3,4\}$, $B = \{7,8\}$ and $S = \{4,5,7\}$. We observe that $A \setminus S$ and $B \setminus S$ are separated by S and moreover $A \setminus S$, $B \setminus S$ and S are disjoint sets.

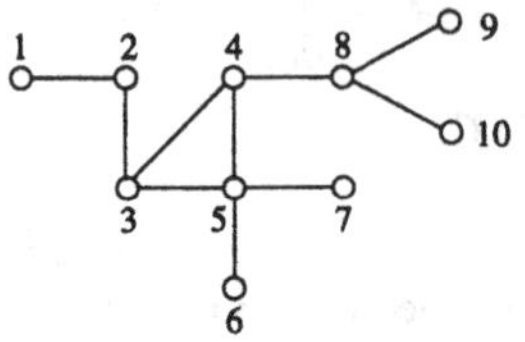

Figure 6.1: *Illustration of the simple undirected graph from Example 6.1*

Assume that the distribution of $\boldsymbol{X}$ has the global Markov property w.r.t. $\mathcal{G}$. Then we know from Definition 6.9 page 107 that $\boldsymbol{X}_{A\setminus S} \perp\!\!\!\perp \boldsymbol{X}_{B\setminus S} \mid \boldsymbol{X}_S$. According to Theorem 6.5 page 104 we see that $\boldsymbol{X}_A \perp\!\!\!\perp \boldsymbol{X}_B \mid \boldsymbol{X}_S$. ■

The following definition of the conditional independence graph is useful to visualize the pattern of conditional independence between random variables.

Definition 6.10 The conditional independence graph
Let $\boldsymbol{X}$ be a $|V|$-dimensional complex random vector with distribution P on $\mathbb{C}^{|V|}$. The conditional independence graph of $\boldsymbol{X}$ is the simple undirected graph $\mathcal{G} = (V, E)$, where for $\alpha \neq \beta \in V$ it holds that

$$\boldsymbol{X}_\alpha \perp\!\!\!\perp \boldsymbol{X}_\beta \mid \boldsymbol{X}_{V\setminus\{\alpha,\beta\}} \Leftrightarrow \{\alpha, \beta\} \notin E .$$

Remark that the conditional independence graph is the minimal graph for which the distribution of $\boldsymbol{X}$ has the pairwise Markov property .

In the following theorem we describe a relation between the factorization property and the Markov properties.

Theorem 6.7
Let $\mathcal{G} = (V, E)$ be a simple undirected graph and let $\boldsymbol{X}$ be a $|V|$-dimensional complex random vector with distribution P on $\mathbb{C}^{|V|}$. If $\boldsymbol{X}$ has continuous density function w.r.t. Lebesgue measure on $\mathbb{C}^{|V|}$, then for the distribution P it holds that

$$(F) \Rightarrow (G) \Rightarrow (L) \Rightarrow (P) .$$

Proof:
Let $\mathcal{G} = (V, E)$ be a simple undirected graph and let $\boldsymbol{X}$ be a $|V|$-dimensional complex random vector with distribution P on $\mathbb{C}^{|V|}$. Further assume that $\boldsymbol{X}$ has continuous density function w.r.t. Lebesgue measure on $\mathbb{C}^{|V|}$.

(F) $\Rightarrow$ (G):
Let the distribution P on $\mathbb{C}^{|V|}$ have property (F), then the continuous density function $f_{\boldsymbol{X}}$ has the form

$$f_{\boldsymbol{X}}(\boldsymbol{x}) = \prod_{T} \phi_T(\boldsymbol{x}_T) , \tag{6.2}$$

where ϕ_T are nonnegative functions defined on $\mathbb{C}^{|T|}$ and T is any complete subset of V.

Let (A, B, S) be an arbitrary triple of disjoint subsets of V such that A and B are separated by S in $\mathcal{G}$. Further let $C = \text{bd}(A) \setminus S$ and $D = V \setminus (A \cup B \cup C \cup S)$. From (6.2) this means

$$f_{\boldsymbol{X}}(\boldsymbol{x}) = h(\boldsymbol{x}_A, \boldsymbol{x}_C, \boldsymbol{x}_S)\, k(\boldsymbol{x}_B, \boldsymbol{x}_D, \boldsymbol{x}_S) .$$

Using Theorem 6.4 page 103 we see

$$(\boldsymbol{X}_A, \boldsymbol{X}_C) \perp\!\!\!\perp (\boldsymbol{X}_B, \boldsymbol{X}_D) \mid \boldsymbol{X}_S ,$$

which by Theorem 6.5 page 104 implies

$$\boldsymbol{X}_A \perp\!\!\!\perp \boldsymbol{X}_B \mid \boldsymbol{X}_S ,$$

i.e. the global Markov property is obtained.

(G) $\Rightarrow$ (L):
Let the distribution P on $\mathbb{C}^{|V|}$ have the global Markov property. For all $\alpha \in V$ we notice that $\text{bd}(\alpha)$ separates α from $V \setminus \text{cl}(\alpha)$ in $\mathcal{G}$, thus by the global Markov property

$$\boldsymbol{X}_\alpha \perp\!\!\!\perp \boldsymbol{X}_{V\setminus\text{cl}(\alpha)} \mid \boldsymbol{X}_{\text{bd}(\alpha)} ,$$

which is the local Markov property.

(L) $\Rightarrow$ (P):
Let the distribution P on $\mathbb{C}^{|V|}$ have the local Markov property, i.e. for all $\alpha \in V$

$$\boldsymbol{X}_\alpha \perp\!\!\!\perp \boldsymbol{X}_{V\setminus\text{cl}(\alpha)} \mid \boldsymbol{X}_{\text{bd}(\alpha)} . \tag{6.3}$$

Furthermore let $\beta \in V$ be a vertex with $\{\alpha,\beta\} \notin E$, i.e. $\beta \in V \setminus \mathrm{cl}(\alpha)$. Hereby (6.3) page 109 can be rewritten as

$$\boldsymbol{X}_\alpha \perp\!\!\!\perp \left(\boldsymbol{X}_\beta, \boldsymbol{X}_{V\setminus(\beta\cup\mathrm{cl}(\alpha))}\right) \mid \boldsymbol{X}_{\mathrm{bd}(\alpha)},$$

which by Theorem 6.5 implies

$$\boldsymbol{X}_\alpha \perp\!\!\!\perp \boldsymbol{X}_\beta \mid \boldsymbol{X}_{V\setminus\{\alpha,\beta\}}.$$

This is the pairwise Markov property. ■

The Markov properties (G), (L) and (P) are not equivalent in general. But if the density function of $\boldsymbol{X}$ w.r.t. Lebesgue measure is positive and continuous, we can show that all the properties (G), (L) and (P) are equivalent. This is a direct consequence of the theorem below, which first was shown by Pearl & Paz (1986). This proof can also be found in Pearl (1988).

Theorem 6.8 Equivalence of Markov properties
Let $\mathcal{G} = (V,E)$ be a simple undirected graph and let $\boldsymbol{X}$ be a $|V|$-dimensional complex random vector with distribution P on $\mathbb{C}^{|V|}$. If $\boldsymbol{X}$ has continuous density function w.r.t. Lebesgue measure on $\mathbb{C}^{|V|}$ and it holds that

$$\boldsymbol{X}_C \perp\!\!\!\perp \boldsymbol{X}_D \mid (\boldsymbol{X}_E, \boldsymbol{X}_F) \ \text{ and } \ \boldsymbol{X}_C \perp\!\!\!\perp \boldsymbol{X}_E \mid (\boldsymbol{X}_D, \boldsymbol{X}_F) \tag{6.4}$$
$$\Downarrow$$
$$\boldsymbol{X}_C \perp\!\!\!\perp (\boldsymbol{X}_D, \boldsymbol{X}_E) \mid \boldsymbol{X}_F$$

for all disjoint subsets C, D, E and F of V, then

$$(G) \Leftrightarrow (L) \Leftrightarrow (P).$$

Proof:
Let $\mathcal{G} = (V,E)$ be a simple undirected graph and let $\boldsymbol{X}$ be a $|V|$-dimensional complex random vector with distribution P on $\mathbb{C}^{|V|}$. Further assume that $\boldsymbol{X}$ has continuous density function w.r.t. Lebesgue measure on $\mathbb{C}^{|V|}$.

From Theorem 6.7 page 109 the only thing left to show is that (P) $\Rightarrow$ (G), when (6.4) holds. Therefore assume that (P) and (6.4) hold and let (A, B, S) be an arbitrary triple of disjoint subsets of V such that A and B are separated by S in $\mathcal{G}$. Without loss of generality we assume that A and B are nonempty. The theorem is proved by backward induction on the number of vertices in S. Let $n = |S|$.

Induction start:
For $n = |V| - 2$ we observe that both A and B only consist of one vertex each, and these vertices are nonadjacent. Since (P) holds, this implies that

$$\boldsymbol{X}_A \perp\!\!\!\perp \boldsymbol{X}_B \mid \boldsymbol{X}_{V\setminus(A\cup B)}.$$

Because $n = |V| - 2$ we observe that $V \setminus (A \cup B) = S$. Therefore

$$\boldsymbol{X}_A \perp\!\!\!\perp \boldsymbol{X}_B \mid \boldsymbol{X}_S ,$$

which is the global Markov property.

Induction step:
Assume that $n < |V| - 2$ and that (G) holds for all S with more than n elements.

First let $V = A \cup B \cup S$. This means that either A or B has more than one element. Let us say A. If $\alpha \in A$ then $A \setminus \alpha$ and B are separated by $S \cup \alpha$ and moreover α and B are separated by $S \cup A \setminus \alpha$. By the induction assumption this means that

$$\boldsymbol{X}_{A\setminus\alpha} \perp\!\!\!\perp \boldsymbol{X}_B \mid \boldsymbol{X}_{S\cup\alpha}$$

and

$$\boldsymbol{X}_\alpha \perp\!\!\!\perp \boldsymbol{X}_B \mid \boldsymbol{X}_{S\cup A\setminus\alpha} .$$

Since (6.4) holds for all disjoint subsets of V we get

$$\boldsymbol{X}_A \perp\!\!\!\perp \boldsymbol{X}_B \mid \boldsymbol{X}_S .$$

Hence the global Markov property is shown, when $V = A \cup B \cup S$.

Let $A \cup B \cup S \subset V$ and choose $\alpha \in V \setminus (A \cup B \cup S)$. We see that A and B are separated by $S \cup \alpha$, which by the induction assumption implies that

$$\boldsymbol{X}_A \perp\!\!\!\perp \boldsymbol{X}_B \mid \boldsymbol{X}_{S\cup\alpha} . \tag{6.5}$$

Furthermore either

B and α are separated by $A \cup S$ or A and α are separated by $B \cup S$.

In other words by the induction assumption either

$$\boldsymbol{X}_\alpha \perp\!\!\!\perp \boldsymbol{X}_B \mid \boldsymbol{X}_{A\cup S} \text{ or } \boldsymbol{X}_\alpha \perp\!\!\!\perp \boldsymbol{X}_A \mid \boldsymbol{X}_{B\cup S} .$$

Each case together with (6.5) imply by using (6.4) and Theorem 6.5 page 104 that

$$\boldsymbol{X}_A \perp\!\!\!\perp \boldsymbol{X}_B \mid \boldsymbol{X}_S ,$$

and the global Markov property is obtained. ■

If the density function of $\boldsymbol{X}$ w.r.t. Lebesgue measure is positive and continuous and if two subsets form a decomposition of a graph, then the factorization property is decomposed accordingly and hereby the other Markov properties are decomposed accordingly. The decomposition of the factorization property is stated in the following theorem.

Theorem 6.9 Decomposition of the factorization property
Let $\mathcal{G} = (V, E)$ be a simple undirected graph and let A and B be subsets of V forming a decomposition of $\mathcal{G}$. Further let $\boldsymbol{X}$ be a $|V|$-dimensional complex random vector with positive and continuous density function $f_{\boldsymbol{X}}$ w.r.t. Lebesgue measure on $\mathbb{C}^{|V|}$. The distribution of $\boldsymbol{X}$ factorizes according to $\mathcal{G}$ iff the distributions of $\boldsymbol{X}_A$ and $\boldsymbol{X}_B$ factorize according to $\mathcal{G}_A$ and $\mathcal{G}_B$, respectively, and the densities satisfy

$$f_{\boldsymbol{X}}(\boldsymbol{x}) = \frac{f_{\boldsymbol{X}_A}(\boldsymbol{x}_A)\, f_{\boldsymbol{X}_B}(\boldsymbol{x}_B)}{f_{\boldsymbol{X}_{A\cap B}}(\boldsymbol{x}_{A\cap B})}\,. \tag{6.6}$$

Proof:
Let $\mathcal{G} = (V, E)$ be a simple undirected graph and let A and B be subsets of V forming a decomposition of $\mathcal{G}$. Further let $\boldsymbol{X}$ be a $|V|$-dimensional complex random vector with positive and continuous density function $f_{\boldsymbol{X}}$ w.r.t. Lebesgue measure on $\mathbb{C}^{|V|}$ and let the distribution of $\boldsymbol{X}$ factorize according to $\mathcal{G}$ as

$$f_{\boldsymbol{X}}(\boldsymbol{x}) = \prod_{C\in\mathcal{C}} \phi_C(\boldsymbol{x}_C)\,, \tag{6.7}$$

where $\mathcal{C}$ denotes the set of cliques in $\mathcal{G}$. Since A and B form a decomposition of $\mathcal{G}$ we know that

- $V = A \cup B$.
- $A \cap B$ is a complete subset of V.
- $A \setminus B$ and $B \setminus A$ are separated by $A \cap B$.

Thus all the cliques are either subsets of A or B, i.e. (6.7) becomes

$$f_{\boldsymbol{X}}(\boldsymbol{x}) = \prod_{C\in\mathcal{A}} \phi_C(\boldsymbol{x}_C) \prod_{C\in\mathcal{B}} \phi_C(\boldsymbol{x}_C)\,, \tag{6.8}$$

where $\mathcal{A}$ is the set of cliques in $\mathcal{G}_A$ and $\mathcal{B}$ is the set of cliques in $\mathcal{G}_B$. Hereby we see that

$$f_{\boldsymbol{X}}(\boldsymbol{x}) = h(\boldsymbol{x}_A)\, k(\boldsymbol{x}_B) \tag{6.9}$$

and by integration we find

$$\begin{aligned} f_{\boldsymbol{X}_A}(\boldsymbol{x}_A) &= \int f_{\boldsymbol{X}}(\boldsymbol{x})\, d\boldsymbol{x}_{B\setminus A} \\ &= \int h(\boldsymbol{x}_A)\, k(\boldsymbol{x}_B)\, d\boldsymbol{x}_{B\setminus A} \\ &= h(\boldsymbol{x}_A)\, k_1(\boldsymbol{x}_{A\cap B})\,, \end{aligned}$$

where $k_1(\boldsymbol{x}_{A\cap B}) = \int k(\boldsymbol{x}_B)\, d\boldsymbol{x}_{B\setminus A}$. Similarly

$$f_{\boldsymbol{X}_B}(\boldsymbol{x}_B) = k(\boldsymbol{x}_B)\, h_1(\boldsymbol{x}_{A\cap B})\,,$$

where $h_1(\boldsymbol{x}_{A\cap B}) = \int h(\boldsymbol{x}_A)\, d\boldsymbol{x}_{A\setminus B}$, and

$$\begin{aligned} f_{\boldsymbol{X}_{A\cap B}}(\boldsymbol{x}_{A\cap B}) &= \int f_{\boldsymbol{X}}(\boldsymbol{x})\, d\boldsymbol{x}_{V\setminus(A\cap B)} \\ &= \int\int h(\boldsymbol{x}_A)\, k(\boldsymbol{x}_B)\, d\boldsymbol{x}_{A\setminus B} d\boldsymbol{x}_{B\setminus A} \\ &= h_1(\boldsymbol{x}_{A\cap B})\, k_1(\boldsymbol{x}_{A\cap B}) \ . \end{aligned}$$

Thus by (6.9) page 112 we get

$$f_{\boldsymbol{X}}(\boldsymbol{x}) = \frac{f_{\boldsymbol{X}_A}(\boldsymbol{x}_A)\, f_{\boldsymbol{X}_B}(\boldsymbol{x}_B)}{f_{\boldsymbol{X}_{A\cap B}}(\boldsymbol{x}_{A\cap B})}, \tag{6.10}$$

which was requested. From (6.8) page 112 and (6.10) we also see that the distributions of $\boldsymbol{X}_A$ and $\boldsymbol{X}_B$ factorize according to $\mathcal{G}_A$ and $\mathcal{G}_B$. The converse is immediately seen from (6.6) page 112 and the fact that the distributions of $\boldsymbol{X}_A$ and $\boldsymbol{X}_B$ factorize according to $\mathcal{G}_A$ and $\mathcal{G}_B$, respectively. ■

7

Complex Normal Graphical Models

Graphical models are used to examine conditional independence among random variables. In this chapter we take graphical models for the multivariate complex normal distribution w.r.t. simple undirected graphs into consideration. This is the first published presentation of these models. Graphical models for the multivariate real normal distribution, also called covariance selection models, have already been studied in the literature. The initial work on covariance selection models is done by Dempster (1972) and Wermuth (1976) and a presentation of these models is given in Eriksen (1992). Graphical models for contingency tables are introduced in statistics by Darroch et al. (1980) and further these are well-described in Lauritzen (1989). Graphical association models are treated in general in Whittaker (1990) and Lauritzen (1993). The complex normal graphical models are quite similar to the covariance selection models. We have chosen to develop this chapter without use of exponential families. We study definition of the model, maximum likelihood estimation and hypothesis testing. To verify some of the results in the chapter we use results from mathematical analysis. These can be found in e.g. Rudin (1987). In graphical models one uses the concentration matrix instead of the variance matrix as it is more advantageous. Therefore we define this matrix and derive a relation which is basic for complex normal graphical models. Afterwards we formally define a complex normal graphical model w.r.t. a simple undirected graph. As these models are used to examine conditional independence of selected pairs of variables given the remaining ones we are mainly interested in inference on the concentration matrix. It is possible to base the maximum likelihood estimation of the concentration matrix on a complex random matrix with mean zero. The maximum likelihood estimate of the concentration matrix is determined by the likelihood equations and if it exists it is uniquely determined. The likelihood equations do not give an explicit expression for the estimate, so often one may use an iterative algorithm to determine it. We examine the iterative proportional scaling (IPS) algorithm. In Speed & Kiiveri (1986) the IPS-algorithm for covariance selection models is described together with an alternative algorithm. We have transferred these ideas to complex normal graphical models. If the simple undirected graph in the model has a decomposition the estimation problem is decomposed accordingly. In fact if the graph is decomposable we are able to find an explicit expression of the the maximum likelihood estimate of the concentration matrix. This expression only involves submatrices which can be found directly from the empirical variance matrix, whereby the IPS-algorithm is not needed. Finally hypothesis testing in complex normal graphical models is performed. Generally the likelihood ratio test statistic is asymptotically chi-square distributed under the null hypothesis. The hypothesis testing problem may be reduced if the graph is collapsible onto a subset of the vertices and it makes the number of observations larger compared to the number of parameters to estimate in the test. We see that in the test of removal of a regular edge the likelihood ratio test statistic has a beta distribution. Further in test the distribution of the likelihood ratio test statistic between two decomposable models is proved to be equal to the distribution of a product of mutually independent beta distributed random variables. First we introduce the notation used.

7.1 Notation

Let V be an index set and let $\boldsymbol{X}$ be a $|V|$-dimensional complex random vector given by $\boldsymbol{X} = (X_v)_{v\in V}$. Then $\boldsymbol{X}$ takes values in the vector space $\mathbb{C}^{|V|}$. For an arbitrary subset $A \subseteq V$, let $\boldsymbol{X}_A = (X_v)_{v\in A}$ be an $|A|$-dimensional complex random vector taking values in $\mathbb{C}^{|A|}$. The vector $\boldsymbol{X}_A$ is called a subvector of $\boldsymbol{X}$. It is obvious that $\boldsymbol{X} = \boldsymbol{X}_V$.

A $|V| \times |V|$ complex matrix with entries indexed by V is defined as

$$\boldsymbol{C} = (c_{\alpha\beta})_{\alpha,\beta\in V} .$$

For two arbitrary subsets A and B of V the $|A| \times |B|$ complex matrix $\boldsymbol{C}_{AB}$ defined by

$$\boldsymbol{C}_{AB} = (c_{\alpha\beta})_{\alpha\in A,\beta\in B} ,$$

is called a submatrix of $\boldsymbol{C}$. Note that $\boldsymbol{C} = \boldsymbol{C}_{VV}$. This notation of a subvector and a submatrix infers that the α'th element of a complex random vector $\boldsymbol{X}$ is referred to as $\boldsymbol{X}_\alpha$ and the $\alpha\beta$'th element of a complex matrix $\boldsymbol{C}$ is referred to as $\boldsymbol{C}_{\alpha\beta}$. Furthermore note that one must be cautious to distinguish between $\boldsymbol{C}_{AB}^{-1}$ and $\left(\boldsymbol{C}^{-1}\right)_{AB}$.

In subsequent considerations we need the matrix obtained from a submatrix by filling in missing entries with zeros to get full dimension. For a submatrix $\boldsymbol{C}_{AB}$ of $\boldsymbol{C}$ we let $[\boldsymbol{C}_{AB}]$ denote the $|V| \times |V|$ complex matrix with the $\alpha\beta$'th element defined by

$$[\boldsymbol{C}_{AB}]_{\alpha\beta} = \begin{cases} \boldsymbol{C}_{\alpha\beta} & \text{if } \alpha \in A \text{ and } \beta \in B \\ 0 & \text{otherwise} \end{cases} .$$

If we partition V into two disjoint subsets A and B with $V = A \cup B$, we get that $\boldsymbol{X}$ and $\boldsymbol{C}$ can be partitioned accordingly by subvectors and submatrices e.g. as

$$\boldsymbol{X} = \begin{pmatrix} \boldsymbol{X}_A \\ \boldsymbol{X}_B \end{pmatrix} \text{ and } \boldsymbol{C} = \begin{pmatrix} \boldsymbol{C}_{AA} & \boldsymbol{C}_{AB} \\ \boldsymbol{C}_{BA} & \boldsymbol{C}_{BB} \end{pmatrix} .$$

7.2 The Concentration Matrix

Let $\boldsymbol{X}$ be a $|V|$-dimensional complex random vector with $\mathcal{L}(\boldsymbol{X}) = \mathbb{C}\mathcal{N}_{|V|}(\boldsymbol{\theta}, \boldsymbol{H})$, where $\boldsymbol{\theta} \in \mathbb{C}^{|V|}$ and $\boldsymbol{H} \in \mathbb{C}_+^{|V|\times|V|}$. The variance matrix being positive definite ensures that $\boldsymbol{H}^{-1}$ exists. The inverse of this variance matrix is called the *concentration matrix*. We recall from Theorem 2.12 page 28 that a marginal independence in the multivariate complex normal distribution is equivalent to a zero entry in the variance matrix. Next we show that a pairwise conditional independence is equivalent to a zero entry in the concentration matrix. This is similar to the real case and is basic for graphical models for the multivariate normal distribution.

Theorem 7.1
Let X be a $|V|$-dimensional complex random vector with $\mathcal{L}(X) = \mathbb{C}\mathcal{N}_{|V|}(\theta, H)$, where $\theta \in \mathbb{C}^{|V|}$ and $H \in \mathbb{C}_+^{|V|\times|V|}$. For $\alpha \neq \beta \in V$ it holds that

$$X_\alpha \perp\!\!\!\perp X_\beta \mid X_{V\setminus\{\alpha,\beta\}} \Leftrightarrow K_{\alpha\beta} = 0\,,$$

where $H^{-1} = K$.

Proof:
Let X be a $|V|$-dimensional complex random vector with $\mathcal{L}(X) = \mathbb{C}\mathcal{N}_{|V|}(\theta, H)$ and concentration matrix $K = H^{-1}$, where $\theta \in \mathbb{C}^{|V|}$ and $H \in \mathbb{C}_+^{|V|\times|V|}$. Further let $\alpha \neq \beta \in V$ be arbitrary but fixed vertices and let $X_{\{\alpha,\beta\}}$ be the subvector of X given by

$$X_{\{\alpha,\beta\}} = \begin{pmatrix} X_\alpha \\ X_\beta \end{pmatrix}.$$

Moreover let X, H and K be partitioned as

$$X = \begin{pmatrix} X_{\{\alpha,\beta\}} \\ X_{V\setminus\{\alpha,\beta\}} \end{pmatrix}, \quad H = \begin{pmatrix} H_{\{\alpha,\beta\}\{\alpha,\beta\}} & H_{\{\alpha,\beta\},V\setminus\{\alpha,\beta\}} \\ H_{V\setminus\{\alpha,\beta\},\{\alpha,\beta\}} & H_{V\setminus\{\alpha,\beta\},V\setminus\{\alpha,\beta\}} \end{pmatrix}$$

and

$$K = \begin{pmatrix} K_{\{\alpha,\beta\}\{\alpha,\beta\}} & K_{\{\alpha,\beta\},V\setminus\{\alpha,\beta\}} \\ K_{V\setminus\{\alpha,\beta\},\{\alpha,\beta\}} & K_{V\setminus\{\alpha,\beta\},V\setminus\{\alpha,\beta\}} \end{pmatrix}.$$

The matrix $H_{V\setminus\{\alpha,\beta\},V\setminus\{\alpha,\beta\}}$ is nonsingular since $H > O$, then according to Theorem 2.16 page 31 the variance matrix of $X_{\{\alpha,\beta\}}$ given $X_{V\setminus\{\alpha,\beta\}}$ is

$$(7.1)\ \mathbb{V}\big(X_{\{\alpha,\beta\}} \big| X_{V\setminus\{\alpha,\beta\}}\big) = H_{\{\alpha,\beta\}\{\alpha,\beta\}} - H_{\{\alpha,\beta\},V\setminus\{\alpha,\beta\}} H^{-1}_{V\setminus\{\alpha,\beta\},V\setminus\{\alpha,\beta\}} H_{V\setminus\{\alpha,\beta\},\{\alpha,\beta\}}.$$

From the partition of H we get $K_{\{\alpha,\beta\}\{\alpha,\beta\}}$ as

$$(7.2)\quad K_{\{\alpha,\beta\}\{\alpha,\beta\}} = \big(H_{\{\alpha,\beta\}\{\alpha,\beta\}} - H_{\{\alpha,\beta\},V\setminus\{\alpha,\beta\}} H^{-1}_{V\setminus\{\alpha,\beta\},V\setminus\{\alpha,\beta\}} H_{V\setminus\{\alpha,\beta\},\{\alpha,\beta\}}\big)^{-1},$$

where $K_{\{\alpha,\beta\}\{\alpha,\beta\}}$ exists since $H > O$.

Combining (7.1) and (7.2) implies $K_{\{\alpha,\beta\}\{\alpha,\beta\}} = \Big(\mathbb{V}\big(X_{\{\alpha,\beta\}} \big| X_{V\setminus\{\alpha,\beta\}}\big)\Big)^{-1}$. Hereby we get that $K_{\{\alpha,\beta\}\{\alpha,\beta\}}$ is the concentration matrix of $X_{\{\alpha,\beta\}}$ given $X_{V\setminus\{\alpha,\beta\}}$. Let

$$K_{\{\alpha,\beta\}\{\alpha,\beta\}} = \begin{pmatrix} K_{\alpha\alpha} & K_{\alpha\beta} \\ K_{\beta\alpha} & K_{\beta\beta} \end{pmatrix}.$$

By matrix inversion we find the variance matrix of $X_{\{\alpha,\beta\}}$ given $X_{V\setminus\{\alpha,\beta\}}$ as

$$\mathbb{V}\big(X_{\{\alpha,\beta\}} \big| X_{V\setminus\{\alpha,\beta\}}\big) = \frac{1}{\det\big(K_{\{\alpha,\beta\}\{\alpha,\beta\}}\big)} \begin{pmatrix} K_{\beta\beta} & -K_{\alpha\beta} \\ -K_{\beta\alpha} & K_{\alpha\alpha} \end{pmatrix}.$$

Theorem 2.12 page 28 states that $X_\alpha \perp\!\!\!\perp X_\beta \mid X_{V\setminus\{\alpha,\beta\}}$ is equivalent to $\frac{-K_{\alpha\beta}}{\det(K_{\{\alpha,\beta\}\{\alpha,\beta\}})} = 0$. Since $K_{\{\alpha,\beta\}\{\alpha,\beta\}} > O$ we deduce that $\det\big(K_{\{\alpha,\beta\}\{\alpha,\beta\}}\big) > 0$, thus $X_\alpha \perp\!\!\!\perp X_\beta \mid X_{V\setminus\{\alpha,\beta\}}$ iff $K_{\alpha\beta} = 0$. ■

7.3 Complex Normal Graphical Models

Consider a $|V|$-dimensional complex normally distributed random vector $\boldsymbol{X}$ and assume conditional independence of selected pairs of variables given the remaining variables in $\boldsymbol{X}$. Let $\mathcal{G} = (V, E)$ be a simple undirected graph satisfying that the distribution of $\boldsymbol{X}$ has the pairwise Markov property w.r.t. $\mathcal{G}$. Hence from Definition 6.7 page 107 it holds for $\alpha \neq \beta \in V$ that

$$\{\alpha, \beta\} \notin E \Rightarrow \boldsymbol{X}_\alpha \perp\!\!\!\perp \boldsymbol{X}_\beta \mid \boldsymbol{X}_{V \setminus \{\alpha,\beta\}} . \tag{7.3}$$

Hereby a model specifying the assumed pairwise conditional independence can be formulated in terms of $\mathcal{G}$. This model is called a complex normal graphical model of $\boldsymbol{X}$ w.r.t. $\mathcal{G}$.

Using Theorem 7.1 page 117 we observe that assuming pairwise conditional independences is equivalent to assuming zero entries in the concentration matrix. Then for $\alpha \neq \beta \in V$ we get that (7.3) is equivalent to

$$\{\alpha, \beta\} \notin E \Rightarrow \boldsymbol{K}_{\alpha\beta} = 0 , \tag{7.4}$$

where $\boldsymbol{K}_{\alpha\beta}$ is the $\alpha\beta$'th element of the concentration matrix of $\boldsymbol{X}$.

To give a formal definition of the complex normal graphical model of $\boldsymbol{X}$ w.r.t. $\mathcal{G}$ we need the set given by

$$\mathbb{C}_+(\mathcal{G}) = \left\{ \boldsymbol{K} \in \mathbb{C}_+^{|V|\times|V|} \;\middle|\; \forall\, \alpha \neq \beta \in V : \{\alpha, \beta\} \notin E \Rightarrow \boldsymbol{K}_{\alpha\beta} = 0 \right\} . \tag{7.5}$$

Hence $\mathbb{C}_+(\mathcal{G})$ is the set of all $|V| \times |V|$ positive definite matrices containing a zero entry, when the corresponding vertices are nonadjacent in $\mathcal{G}$.

Definition 7.1 The complex normal graphical model
Let $\mathcal{G} = (V, E)$ be a simple undirected graph and furthermore let $\boldsymbol{X}$ be a $|V|$-dimensional complex random vector. The complex normal graphical model of $\boldsymbol{X}$ w.r.t. $\mathcal{G}$ is described by $\mathcal{L}(\boldsymbol{X}) = \mathbb{C}\mathcal{N}_{|V|}\left(\boldsymbol{\theta}, \boldsymbol{K}(\mathcal{G})^{-1}\right)$, where $\boldsymbol{\theta} \in \mathbb{C}^{|V|}$ and $\boldsymbol{K}(\mathcal{G}) \in \mathbb{C}_+(\mathcal{G})$.

Note that the concentration matrix of $\boldsymbol{X}$ depends on $\mathcal{G}$ as specified in (7.4). This is indicated by the notation $\boldsymbol{K}(\mathcal{G})$. When $\boldsymbol{X}$ is a complex random vector with $\mathcal{L}(\boldsymbol{X}) = \mathbb{C}\mathcal{N}_{|V|}\left(\boldsymbol{\theta}, \boldsymbol{K}(\mathcal{G})^{-1}\right)$, where $\boldsymbol{\theta} \in \mathbb{C}^{|V|}$ and $\boldsymbol{K}(\mathcal{G}) \in \mathbb{C}_+(\mathcal{G})$, we shortly say that $\boldsymbol{X}$ is described by a complex normal graphical model w.r.t. $\mathcal{G}$.

In the following example we see how complex normal graphical models are related to covariance selection models.

Example 7.1
Let $\mathcal{G} = (V, E)$ be a simple undirected graph and let $\boldsymbol{X}$ be a $|V|$-dimensional complex random vector with $\mathcal{L}(\boldsymbol{X}) = \mathbb{C}\mathcal{N}_{|V|}\left(\boldsymbol{\theta}, \boldsymbol{K}(\mathcal{G})^{-1}\right)$, where $\boldsymbol{\theta} \in \mathbb{C}^{|V|}$ and $\boldsymbol{K}(\mathcal{G}) \in \mathbb{C}_+(\mathcal{G})$.

Assume that $\boldsymbol{K}(\mathcal{G})$ can be written as $\boldsymbol{K}(\mathcal{G}) = \boldsymbol{C} + i\boldsymbol{D}$, whereby we deduce that if $\boldsymbol{K}(\mathcal{G})_{\alpha\beta} = 0$ then $\boldsymbol{C}_{\alpha\beta} = 0$ and $\boldsymbol{D}_{\alpha\beta} = 0$.

According to Theorem 2.9 page 25 we know that the distribution of $[\boldsymbol{X}]$ is given as

$$\begin{aligned} \mathcal{L}([\boldsymbol{X}]) &= \mathcal{N}_{2|V|}\left([\boldsymbol{\theta}], \frac{1}{2}\left\{\boldsymbol{K}(\mathcal{G})^{-1}\right\}\right) \\ &= \mathcal{N}_{2|V|}\left([\boldsymbol{\theta}], \frac{1}{2}\left\{\boldsymbol{K}(\mathcal{G})\right\}^{-1}\right) . \end{aligned}$$

This means that the concentration matrix of $[\boldsymbol{X}]$ is

$$\begin{aligned} \mathbb{V}([\boldsymbol{X}])^{-1} &= 2\left\{\boldsymbol{K}(\mathcal{G})\right\} \\ &= 2\begin{pmatrix} \boldsymbol{C} & -\boldsymbol{D} \\ \boldsymbol{D} & \boldsymbol{C} \end{pmatrix} , \end{aligned}$$

where $\boldsymbol{C}$ is symmetric and $\boldsymbol{D}$ is skew symmetric, since $\mathbb{V}([\boldsymbol{X}])^{-1}$ is symmetric.

As $\boldsymbol{D}$ is skew symmetric we have that the diagonal elements of $\boldsymbol{D}$ are zero. Furthermore the zero entries of $\boldsymbol{K}(\mathcal{G})$ are inherited both in $\boldsymbol{C}$ and $\boldsymbol{D}$ as described above. Hereby we see that a complex normal graphical model can be represented as a covariance selection model with additional structure on the concentration matrix.

We illustrate the above considerations by a concrete example. Therefore let $V = \{1, 2, 3\}$ and $E = \{\{1, 2\}, \{2, 3\}\}$. This particular graph is visualized in Figure 7.1.

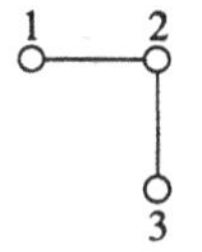

Figure 7.1: *Illustration of the simple undirected graph from Example 7.1.*

We observe that $\boldsymbol{X}$ is the 3-dimensional complex random vector with $\mathcal{L}(\boldsymbol{X}) = \mathbb{C}\mathcal{N}_3\left(\boldsymbol{\theta}, \boldsymbol{K}(\mathcal{G})^{-1}\right)$. Furthermore $\mathcal{L}([\boldsymbol{X}]) = \mathcal{N}_6\left([\boldsymbol{\theta}], \frac{1}{2}\left\{\boldsymbol{K}(\mathcal{G})\right\}^{-1}\right)$ and the concentration matrix of $[\boldsymbol{X}]$ is given as

$$\begin{aligned} \mathbb{V}([\boldsymbol{X}])^{-1} &= 2\begin{pmatrix} \boldsymbol{C} & -\boldsymbol{D} \\ \boldsymbol{D} & \boldsymbol{C} \end{pmatrix} \\ &= 2\begin{pmatrix} c_{11} & c_{12} & 0 & 0 & -d_{12} & 0 \\ c_{12} & c_{22} & c_{23} & d_{12} & 0 & -d_{23} \\ 0 & c_{23} & c_{33} & 0 & d_{23} & 0 \\ 0 & d_{12} & 0 & c_{11} & c_{12} & 0 \\ -d_{12} & 0 & d_{23} & c_{12} & c_{22} & c_{23} \\ 0 & -d_{23} & 0 & 0 & c_{23} & c_{33} \end{pmatrix} , \end{aligned}$$

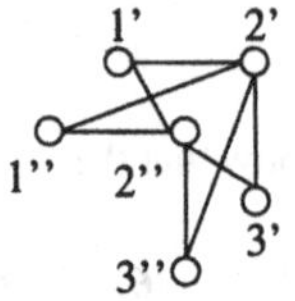

Figure 7.2: *A graph which illustrates the conditional independence in the distribution of* $[X]$ *from Example 7.1. The vertices j' and j" represent real and imaginary parts of vertex j from Figure 7.1, respectively.*

where $c_{\alpha\beta} = C_{\alpha\beta}$ and $d_{\alpha\beta} = D_{\alpha\beta}$ for notational convenience. A graph which illustrates the conditional independence in the distribution of $[X]$ is illustrated in Figure 7.2.

Considering Figure 7.1 and Figure 7.2 we observe that the complex normal graphical model has a decomposable graph, whereas the graph of the covariance selection model is not decomposable. ∎

Recall that the distribution of X has the pairwise Markov property w.r.t. $\mathcal{G}$. From Theorem 2.10 page 26 we know that the density function of X w.r.t. Lebesgue measure on $\mathbb{C}^{|V|}$ is positive and continuous and it is given as

$$\begin{aligned} f_X(x) &= \pi^{-p} \det(K(\mathcal{G})) \exp\left(-(x-\theta)^* K(\mathcal{G})(x-\theta)\right) \\ &= \pi^{-p} \det(K(\mathcal{G})) \exp\left(-\sum_{\alpha,\beta}(x-\theta)^*_\alpha K(\mathcal{G})_{\alpha\beta}(x-\theta)_\beta\right) \\ &= \pi^{-p} \det(K(\mathcal{G})) \exp\left(-\sum_{\alpha\sim\beta}(x-\theta)^*_\alpha K(\mathcal{G})_{\alpha\beta}(x-\theta)_\beta\right) \\ &= \pi^{-p} \det(K(\mathcal{G})) \prod_{\alpha\sim\beta} \exp\left(-(x-\theta)^*_\alpha K(\mathcal{G})_{\alpha\beta}(x-\theta)_\beta\right) . \end{aligned}$$

According to Definition 6.6 page 106 this means that the distribution of X factorizes according to $\mathcal{G}$. Using Theorem 6.7 page 109 we hereby observe that the distribution of X also has the local and the global Markov property w.r.t. $\mathcal{G}$.

The following example illustrates that complex graphical models are relevant when predicting multivariate time series.

Example 7.2
Let $\{X_t\}_{t\in\mathbb{Z}} = \left\{(X_{tv})_{v\in V}\right\}_{t\in\mathbb{Z}}$ be a $|V|$-dimensional Gaussian time series with spectral density matrix $\Sigma(\omega)$ for $\omega \in [0,\pi[$ and let $X_\alpha = \{X_{t\alpha}\}_{t\in\mathbb{Z}}$ denote the α-coordinate of the time series and let $X_a = \left\{(X_{t\alpha})_{\alpha\in a}\right\}_{t\in\mathbb{Z}}$, $a \subseteq V$, denote the a-subset of the time series.

Consider a simple undirected graph $\mathcal{G} = (V, E)$ and a model specifying that

$$\boldsymbol{X}_\alpha \perp\!\!\!\perp \boldsymbol{X}_\beta | \boldsymbol{X}_{V \setminus \{\alpha,\beta\}} \quad \text{if } \{\alpha, \beta\} \notin E \, .$$

In terms of the spectral density matrix this is equivalent to

$$\boldsymbol{\Sigma}(\omega)^{-1} \in \mathbb{C}_+(\mathcal{G}) \, , \; \omega \in]0, \pi[$$

and similarly for $\boldsymbol{\Sigma}(0)$ except that we are now in the real case.

The model is particularly relevant for prediction purpose, e.g. when predicting $X_{\alpha t}$ based on the history until time t of the time series. In that respect, it resembles the issues of canonical analysis of time series as discussed in Chapter 10 of Brillinger (1975). The advantage of the present model is that when predicting e.g. $X_{\alpha t}$, this is only based on the neighbouring part of the time series, thereby limiting the need for collecting data.

It may be so that different models apply to different parts of the frequency domain, in which case a predictor should filter out frequencies for irrelevant part of the series and thereby increase the accuracy of the predictor. ■

7.4 Maximum Likelihood Estimation of the Concentration Matrix

In this section we determine a maximum likelihood estimate of the concentration matrix in a complex normal graphical model. We show that it is uniquely determined if it exists. From the unique maximum likelihood estimate of the concentration matrix we can obtain the unique maximum likelihood estimate of the variance matrix by matrix inversion.

Let $\mathcal{G} = (V, E)$ be a simple undirected graph. Normally making inference on parameters in a model one considers a sample from the model. This is a set of observations corresponding to independent identically distributed random variables described by the model. This means that we normally consider observations from e.g. l independent complex random vectors $\boldsymbol{Y}_1, \boldsymbol{Y}_2, \ldots, \boldsymbol{Y}_l$ with $\mathcal{L}(\boldsymbol{Y}_j) = \mathbb{C}\mathcal{N}_{|V|}\left(\boldsymbol{\theta}, \boldsymbol{K}(\mathcal{G})^{-1}\right)$, where $\boldsymbol{\theta} \in \mathbb{C}^{|V|}$ and $\boldsymbol{K}(\mathcal{G}) \in \mathbb{C}_+(\mathcal{G})$ for $j = 1, 2, \ldots, l$. In Chapter 4 page 67 we have examined the complex MANOVA model, where the means of the $\boldsymbol{Y}_j$'s not necessarily are equal. Therefore we do not want to put the restriction of equal mean onto our complex random vectors, but we only require that $\boldsymbol{Y} = (\boldsymbol{Y}_1, \boldsymbol{Y}_2, \ldots, \boldsymbol{Y}_l)^*$ is an $l \times |V|$ complex random matrix described by a complex MANOVA model with parameter set $\left(N \otimes \mathbb{C}^{|V|}\right) \times \mathbb{C}_+(\mathcal{G})$, where N is a k-dimensional subspace of $\mathbb{C}^l$. This means that $\mathcal{L}(\boldsymbol{Y}) = \mathbb{C}\mathcal{N}_{l \times |V|}\left(\Theta, \boldsymbol{I}_l \otimes \boldsymbol{K}(\mathcal{G})^{-1}\right)$, where $\Theta = (\boldsymbol{\theta}_1, \boldsymbol{\theta}_2, \ldots, \boldsymbol{\theta}_l)^* \in N \otimes \mathbb{C}^{|V|}$ and $\boldsymbol{K}(\mathcal{G}) \in \mathbb{C}_+(\mathcal{G})$. Doing this we require that each column of $\boldsymbol{Y}^*$ is described by a complex normal graphical model w.r.t. $\mathcal{G}$, but do not necessarily have equal means.

From Theorem 4.1 page 70 we deduce that $\left(\boldsymbol{P}_N \boldsymbol{Y}, \boldsymbol{Y}^* \boldsymbol{P}_N^\perp \boldsymbol{Y}\right)$ is a sufficient statistic for $(\Theta, \boldsymbol{K}(\mathcal{G}))$ and that $\boldsymbol{P}_N \boldsymbol{Y}$ is the maximum likelihood estimator of Θ, where $\boldsymbol{P}_N$ is an $l \times l$

complex matrix representing the orthogonal projection of $\mathbb{C}^l$ onto N. Notice that we do not need to assume $l - k \geq |V|$ to ensure that $\boldsymbol{P}_N \boldsymbol{Y}$ is the maximum likelihood estimator of Θ. According to Theorem 4.3 page 73 it holds that $\boldsymbol{P}_N \boldsymbol{Y}$ and $\boldsymbol{Y}^* \boldsymbol{P}_N^{\perp} \boldsymbol{Y}$ are independent. Thus inference on $\boldsymbol{K}(\mathcal{G})$ is naturally based only on $\boldsymbol{Y}^* \boldsymbol{P}_N^{\perp} \boldsymbol{Y}$. Furthermore from Theorem 4.3 we have $\mathcal{L}\left(\boldsymbol{Y}^* \boldsymbol{P}_N^{\perp} \boldsymbol{Y}\right) = \mathbb{C}\mathcal{W}_{|V|}\left(\boldsymbol{K}(\mathcal{G})^{-1}, n\right)$, where $n = l - k$. Hence there exists an $n \times |V|$ complex random matrix $\boldsymbol{X}$ with $\mathcal{L}(\boldsymbol{X}) = \mathbb{C}\mathcal{N}_{n\times|V|}\left(\boldsymbol{O}, \boldsymbol{I}_n \otimes \boldsymbol{K}(\mathcal{G})^{-1}\right)$ such that $\mathcal{L}\left(\boldsymbol{Y}^* \boldsymbol{P}_N^{\perp} \boldsymbol{Y}\right) = \mathcal{L}(\boldsymbol{X}^* \boldsymbol{X})$. Therefore the situation can be considered as having a sample of size n from $\mathbb{C}\mathcal{N}_{|V|}\left(\boldsymbol{0}, \boldsymbol{K}(\mathcal{G})^{-1}\right)$, where $\boldsymbol{K}(\mathcal{G}) \in \mathbb{C}_+(\mathcal{G})$, or equivalently having a sample of size n from a complex normal graphical model w.r.t. $\mathcal{G}$ and with mean zero. As stated in (4.2) page 74 we have that $\boldsymbol{S} = \frac{1}{n}\boldsymbol{X}^* \boldsymbol{X}$ is an unbiased estimator of $\boldsymbol{K}(\mathcal{G})^{-1}$, but the inverse of this estimator does not necessarily contain zero entries at the right places as we require an estimator of $\boldsymbol{K}(\mathcal{G})$ to do. In the following we seek an estimator which fulfills this demand. We define by means of $\boldsymbol{S}$ the $|V| \times |V|$ Hermitian matrix $\boldsymbol{S}(\mathcal{G})$ as

$$\boldsymbol{S}(\mathcal{G})_{\alpha\beta} = \begin{cases} 0 & \text{if } \{\alpha, \beta\} \notin E \text{ for } \alpha \neq \beta \in V \\ \boldsymbol{S}_{\alpha\beta} & \text{otherwise} \end{cases} . \tag{7.6}$$

Let $\boldsymbol{x}$ be an observation matrix of $\boldsymbol{X}$, then the corresponding observation matrix of $\boldsymbol{S}$ is given by $\boldsymbol{s} = \frac{1}{n}\boldsymbol{x}^*\boldsymbol{x}$ and the corresponding observation matrix of $\boldsymbol{S}(\mathcal{G})$ is given by

$$\boldsymbol{s}(\mathcal{G})_{\alpha\beta} = \begin{cases} 0 & \text{if } \{\alpha, \beta\} \notin E \text{ for } \alpha \neq \beta \in V \\ \boldsymbol{s}_{\alpha\beta} & \text{otherwise} \end{cases} . \tag{7.7}$$

The density function of $\boldsymbol{X}$ w.r.t. Lebesgue measure on $\mathbb{C}^{n\times|V|}$ is from Theorem 2.17 page 33 given as

$$f_{\boldsymbol{X}}(\boldsymbol{x}) = \pi^{-n|V|} \det\left(\boldsymbol{K}(\mathcal{G})\right)^n \exp\left(-\operatorname{tr}\left(\boldsymbol{K}(\mathcal{G})\, \boldsymbol{x}^*\boldsymbol{x}\right)\right) . \tag{7.8}$$

It holds that

$$\begin{aligned}
\operatorname{tr}\left(\boldsymbol{K}(\mathcal{G})\, \boldsymbol{x}^*\boldsymbol{x}\right) &= n \operatorname{tr}\left(\boldsymbol{K}(\mathcal{G})\, \boldsymbol{s}\right) \\
&= n \sum_{\alpha} \left(\boldsymbol{K}(\mathcal{G})\, \boldsymbol{s}\right)_{\alpha\alpha} \\
&= n \sum_{\alpha,\beta} \boldsymbol{K}(\mathcal{G})_{\alpha\beta}\, \boldsymbol{s}_{\beta\alpha} \\
&= n \left(\sum_{\alpha\sim\beta} \boldsymbol{K}(\mathcal{G})_{\alpha\beta}\, \boldsymbol{s}_{\beta\alpha} + \sum_{\alpha\not\sim\beta} \boldsymbol{K}(\mathcal{G})_{\alpha\beta}\, \boldsymbol{s}_{\beta\alpha} \right) \\
&= n \sum_{\alpha\sim\beta} \boldsymbol{K}(\mathcal{G})_{\alpha\beta}\, \boldsymbol{s}_{\beta\alpha} \\
&= n \sum_{\alpha,\beta} \boldsymbol{K}(\mathcal{G})_{\alpha\beta}\, \boldsymbol{s}(\mathcal{G})_{\beta\alpha} \\
&= n \operatorname{tr}\left(\boldsymbol{K}(\mathcal{G})\, \boldsymbol{s}(\mathcal{G})\right) .
\end{aligned}$$

Hence (7.8) can be written as

$$f_{\boldsymbol{X}}(\boldsymbol{x}) = \pi^{-n|V|} \det\left(\boldsymbol{K}(\mathcal{G})\right)^n \exp\left(-n \operatorname{tr}\left(\boldsymbol{K}(\mathcal{G})\, \boldsymbol{s}(\mathcal{G})\right)\right) .$$

Similar to the proof of Theorem 4.1 page 70 we observe that $\boldsymbol{S}(\mathcal{G})$ is a sufficient statistic for $\boldsymbol{K}(\mathcal{G})$. The likelihood function of $\boldsymbol{K}(\mathcal{G})$ is determined as

$$L\left(\boldsymbol{K}(\mathcal{G})|\,\boldsymbol{x}\right) = \pi^{-n|V|} \det\left(\boldsymbol{K}(\mathcal{G})\right)^n \exp\left(-n \operatorname{tr}\left(\boldsymbol{K}(\mathcal{G})\,\boldsymbol{s}(\mathcal{G})\right)\right) ,$$

thus the log-likelihood function is given by

$$l\left(\boldsymbol{K}(\mathcal{G})|\,\boldsymbol{x}\right) = -n|V| \log \pi + n \log\left(\det\left(\boldsymbol{K}(\mathcal{G})\right)\right) - n \operatorname{tr}\left(\boldsymbol{K}(\mathcal{G})\,\boldsymbol{s}(\mathcal{G})\right) . \tag{7.9}$$

By means of the log-likelihood function we are able to derive the likelihood equations. For an observation matrix a maximum likelihood estimate of $\boldsymbol{K}(\mathcal{G})$ can be determined from the likelihood equations. If there exists an estimate, it is unique. The likelihood equations are given in Theorem 7.2 page 128, but to prove this theorem the following lemmas are needed.

Lemma 7.1
Let $F(t), t \in]t_0, t_1[\subseteq \mathbb{R}$, be a differentiable curve in $\mathbb{C}_+^{p\times p}$, so that $F(t) = (f_{jk}(t)) = (a_{jk}(t) + ib_{jk}(t))$, where $a_{jk}(t)$ and $b_{jk}(t)$, $j, k = 1, 2, \ldots, p$, are differentiable real valued functions. For $t \in]t_0, t_1[$ it holds that

1. $\frac{d}{dt} \log\left(\det\left(F(t)\right)\right) = \operatorname{tr}\left(\left(F(t)\right)^{-1} \frac{d}{dt} F(t)\right)$
2. $\frac{d}{dt} \left(F(t)\right)^{-1} = -\left(F(t)\right)^{-1} \left(\frac{d}{dt} F(t)\right) \left(F(t)\right)^{-1}$,

where the jk'th element of $\frac{d}{dt}F(t)$ is given by $\left(\frac{d}{dt}F(t)\right)_{jk} = \frac{d}{dt} f_{jk}(t) = \frac{d}{dt} a_{jk}(t) + i \frac{d}{dt} b_{jk}(t)$.

Proof:
Let $F(t)$ be a differentiable curve in the set of all nonsingular $p \times p$ complex matrices.

Re 1:
It is a well known fact that

$$\frac{d}{dt} \log\left(\det\left(F(t)\right)\right) = \frac{1}{\det\left(F(t)\right)} \frac{d}{dt} \det\left(F(t)\right) .$$

Therefore we seek to prove

$$\frac{d}{dt} \det\left(F(t)\right) = \det\left(F(t)\right) \operatorname{tr}\left(\left(F(t)\right)^{-1} \frac{d}{dt} F(t)\right)$$

or equivalently

$$\lim_{h \to 0} \frac{\det\left(F(t+h)\right) - \det\left(F(t)\right)}{h} = \det\left(F(t)\right) \operatorname{tr}\left(\left(F(t)\right)^{-1} \frac{d}{dt} F(t)\right) .$$

To begin the proof we define

$$D_t(h) = \frac{F(t+h) - F(t)}{h} ,$$

which yields that

$$\begin{aligned}\det(F(t+h)) &= \det(F(t)+D_t(h)\,h)\\ &= \det(F(t))\det\left(\boldsymbol{I}_p+(F(t))^{-1}D_t(h)\,h\right).\end{aligned}$$

Using Definition A.7 page 168 we obtain

$$\det(F(t+h)) = \det(F(t))\left(1+\operatorname{tr}\left((F(t))^{-1}D_t(h)\right)h+\sum_{j=2}^{p}G_{tj}(h)\,h^j\right),$$

where $G_{tj}(h)$ is a product of j entries of $(F(t))^{-1}D_t(h)$ for $j=2,3,\dots,p$. This leads to

(7.10)
$$\frac{\det(F(t+h))-\det(F(t))}{h}=\det(F(t))\left(\operatorname{tr}\left((F(t))^{-1}D_t(h)\right)+\sum_{j=2}^{p}G_{tj}(h)\,h^{j-1}\right).$$

As $F(t)$ is differentiable in the set of all nonsingular $p\times p$ complex matrices we have

$$\lim_{h\to 0}D_t(h)=\frac{d}{dt}F(t),$$

which infers

$$\lim_{h\to 0}(F(t))^{-1}D_t(h)=(F(t))^{-1}\frac{d}{dt}F(t). \tag{7.11}$$

Using (7.10) we get

$$\lim_{h\to 0}\frac{\det(F(t+h))-\det(F(t))}{h}=\det(F(t))\lim_{h\to 0}\left(\operatorname{tr}\left((F(t))^{-1}D_t(h)\right)+\sum_{j=2}^{p}G_{tj}(h)\,h^{j-1}\right).$$

This expression contains only finite sums and therefore it becomes

$$\lim_{h\to 0}\frac{\det(F(t+h))-\det(F(t))}{h}=\det(F(t))\left(\operatorname{tr}\left(\lim_{h\to 0}(F(t))^{-1}D_t(h)\right)+\sum_{j=2}^{p}\lim_{h\to 0}G_{tj}(h)\,h^{j-1}\right).$$

Because $G_{tj}(h)$ is a product of entries of the matrix $(F(t))^{-1}D_t(h)$ for which (7.11) holds, we get $\lim_{h\to 0}G_{tj}(h)\,h^{j-1}=0$. Hence

$$\lim_{h\to 0}\frac{\det(F(t+h))-\det(F(t))}{h}=\det(F(t))\operatorname{tr}\left((F(t))^{-1}\frac{d}{dt}F(t)\right).$$

Re 2:
To verify 2 we use

$$F(t)(F(t))^{-1}=\boldsymbol{I}_p.$$

By differentiation this yields

$$\left(\frac{d}{dt}F(t)\right)(F(t))^{-1}+F(t)\left(\frac{d}{dt}(F(t))^{-1}\right)=\boldsymbol{O}$$

or equivalently

$$\frac{d}{dt}(F(t))^{-1} = -(F(t))^{-1}\left(\frac{d}{dt}F(t)\right)(F(t))^{-1}.$$

■

In order to perform further calculations we need the set defined by

$$\mathbb{C}_H(\mathcal{G}) = \left\{ \boldsymbol{K} \in \mathbb{C}_H^{|V|\times|V|} \;\middle|\; \forall\, \alpha \neq \beta \in V : \{\alpha,\beta\} \notin E \Rightarrow \boldsymbol{K}_{\alpha\beta} = 0 \right\}. \tag{7.12}$$

This set is a vector space over $\mathbb{R}$ as it satisfies the axioms for a vector space over this field. The dimension of $\mathbb{C}_H(\mathcal{G})$ is $|V| + 2|E|$. Such a vector space is exemplified in Example 7.3.

Example 7.3
Let $\mathcal{G} = (V, E)$ be the simple undirected graph with $V = \{1,2,3\}$ and $E = \{\{1,2\},\{2,3\}\}$. A matrix in $\mathbb{C}_H(\mathcal{G})$ can be written as

$$\boldsymbol{K} = \begin{pmatrix} k_{11} & k_{12} & 0 \\ \overline{k}_{12} & k_{22} & k_{23} \\ 0 & \overline{k}_{23} & k_{33} \end{pmatrix},$$

where $k_{rr} \in \mathbb{R}$ and $k_{rs} \in \mathbb{C}$ for $r \neq s$ and $r, s = 1,2,3$. A basis for the real vector space $\mathbb{C}_H(\mathcal{G})$ is e.g. given by the following matrices

$$\boldsymbol{E}_1 = \begin{pmatrix} 1 & 0 & 0 \\ 0 & 0 & 0 \\ 0 & 0 & 0 \end{pmatrix}, \quad \boldsymbol{E}_2 = \begin{pmatrix} 0 & 0 & 0 \\ 0 & 1 & 0 \\ 0 & 0 & 0 \end{pmatrix}, \quad \boldsymbol{E}_3 = \begin{pmatrix} 0 & 0 & 0 \\ 0 & 0 & 0 \\ 0 & 0 & 1 \end{pmatrix},$$

$$\boldsymbol{E}_4 = \begin{pmatrix} 0 & 1 & 0 \\ 1 & 0 & 0 \\ 0 & 0 & 0 \end{pmatrix}, \quad \boldsymbol{E}_5 = \begin{pmatrix} 0 & 0 & 0 \\ 0 & 0 & 1 \\ 0 & 1 & 0 \end{pmatrix},$$

$$\boldsymbol{E}_6 = \begin{pmatrix} 0 & i & 0 \\ -i & 0 & 0 \\ 0 & 0 & 0 \end{pmatrix}, \quad \boldsymbol{E}_7 = \begin{pmatrix} 0 & 0 & 0 \\ 0 & 0 & i \\ 0 & -i & 0 \end{pmatrix}.$$

Every matrix in $\mathbb{C}_H(\mathcal{G})$ can hereby be expressed as a real linear combination of $\boldsymbol{E}_1, \boldsymbol{E}_2, \dots, \boldsymbol{E}_7$. There is one basis matrix for each element in V and two basis matrices for each element in E. The dimension of $\mathbb{C}_H(\mathcal{G})$ is $3 + 2 \cdot 2 = 7$, which corresponds to the number of basis matrices.

■

The following lemma examines the relationship between $\mathbb{C}_+(\mathcal{G})$ and $\mathbb{C}_H(\mathcal{G})$.

Lemma 7.2
Let $\mathcal{G} = (V, E)$ be a simple undirected graph. The set $\mathbb{C}_+(\mathcal{G})$ defined in (7.5) page 118 is an open and convex subset of $\mathbb{C}_H(\mathcal{G})$ defined in (7.12) page 125.

Proof:
Let $\mathcal{G} = (V, E)$ be a simple undirected graph.

The set $\mathbb{C}_+(\mathcal{G})$ is obviously a subset of $\mathbb{C}_H(\mathcal{G})$. Furthermore for $\boldsymbol{K}_1, \boldsymbol{K}_2 \in \mathbb{C}_+(\mathcal{G})$ it holds for all $\boldsymbol{k} \in \mathbb{C}^{|V|} \setminus \{0\}$ and all $\lambda \in [0,1]$ that

$$\begin{aligned} \boldsymbol{k}^* (\lambda \boldsymbol{K}_1 + (1-\lambda) \boldsymbol{K}_2) \boldsymbol{k} &= \lambda \boldsymbol{k}^* \boldsymbol{K}_1 \boldsymbol{k} + (1-\lambda) \boldsymbol{k}^* \boldsymbol{K}_2 \boldsymbol{k} \\ &> 0 . \end{aligned}$$

Thus $\lambda \boldsymbol{K}_1 + (1-\lambda) \boldsymbol{K}_2 \in \mathbb{C}_+(\mathcal{G})$ for all $\lambda \in [0,1]$, i.e. $\mathbb{C}_+(\mathcal{G})$ is a convex subset of $\mathbb{C}_H(\mathcal{G})$.

To see that it is an open subset of $\mathbb{C}_H(\mathcal{G})$ we show that $\mathbb{C}_H(\mathcal{G}) \setminus \mathbb{C}_+(\mathcal{G})$ is closed. Therefore let $\{\boldsymbol{K}_j\}$ be any convergent sequence in $\mathbb{C}_H(\mathcal{G}) \setminus \mathbb{C}_+(\mathcal{G})$ which converges to $\boldsymbol{K}'$. If we can show that $\boldsymbol{K}' \in \mathbb{C}_H(\mathcal{G}) \setminus \mathbb{C}_+(\mathcal{G})$, then the set is closed. Therefore we seek a $\boldsymbol{k} \in \mathbb{C}^{|V|} \setminus \{0\}$ such that $\boldsymbol{k}^* \boldsymbol{K}' \boldsymbol{k} \leq 0$.

Since $\boldsymbol{K}_j \in \mathbb{C}_H(\mathcal{G}) \setminus \mathbb{C}_+(\mathcal{G})$ we know that there exists a $\boldsymbol{k}_j \in \mathbb{C}^{|V|} \setminus \{0\}$ such that

$$\boldsymbol{k}_j^* \boldsymbol{K}_j \boldsymbol{k}_j \leq 0 ,$$

or equivalently

$$\frac{\boldsymbol{k}_j^*}{||\boldsymbol{k}_j||} \boldsymbol{K}_j \frac{\boldsymbol{k}_j}{||\boldsymbol{k}_j||} \leq 0 , \tag{7.13}$$

where $||\boldsymbol{k}_j||^2 = \boldsymbol{k}_j^* \boldsymbol{k}_j$.

Let $\boldsymbol{u}_j = \frac{\boldsymbol{k}_j}{||\boldsymbol{k}_j||}$, then $||\boldsymbol{u}_j|| = 1$, thus the $\boldsymbol{u}_j$'s belong to the $|2V|$-dimensional unit-sphere, U, which is closed and bounded. Hence U is a compact metric space. Therefore a subsequence $\{\boldsymbol{u}_{j_r}\}$ of $\{\boldsymbol{u}_j\}$ converges to an element in U, let us call it $\boldsymbol{u}$. Since $\{\boldsymbol{K}_j\}$ converges to $\boldsymbol{K}'$ any subsequence $\{\boldsymbol{K}_{j_r}\}$ also converges to $\boldsymbol{K}'$. Altogether we have

$$\lim_{r \to \infty} \boldsymbol{u}_{j_r}^* \boldsymbol{K}_{j_r} \boldsymbol{u}_{j_r} = \boldsymbol{u}^* \boldsymbol{K}' \boldsymbol{u} .$$

By using (7.13) we conclude

$$\boldsymbol{u}^* \boldsymbol{K}' \boldsymbol{u} \leq 0 ,$$

i.e. $\mathbb{C}_+(\mathcal{G})$ is an open subset of $\mathbb{C}_H(\mathcal{G})$. ∎

The lemma below tells us that the log-likelihood function of $\boldsymbol{K}(\mathcal{G})$ is strictly concave, which means that the log-likelihood function of $\boldsymbol{K}(\mathcal{G})$ has at most one maximum.

Lemma 7.3
Let $\mathcal{G} = (V, E)$ be a simple undirected graph and let $\boldsymbol{X}$ be an $n \times |V|$ complex random matrix with $\mathcal{L}(\boldsymbol{X}) = \mathbb{C}\mathcal{N}_{n \times |V|}\left(\boldsymbol{O}, \boldsymbol{I}_n \otimes \boldsymbol{K}(\mathcal{G})^{-1}\right)$, where $\boldsymbol{K}(\mathcal{G}) \in \mathbb{C}_+(\mathcal{G})$. The log-likelihood function of $\boldsymbol{K}(\mathcal{G})$ is strictly concave.

Proof:
Let $\mathcal{G} = (V, E)$ be a simple undirected graph and let $\boldsymbol{X}$ be an $n \times |V|$ complex random matrix with $\mathcal{L}(\boldsymbol{X}) = \mathbb{C}\mathcal{N}_{n\times|V|}\left(\boldsymbol{O}, \boldsymbol{I}_n \otimes \boldsymbol{K}(\mathcal{G})^{-1}\right)$, where $\boldsymbol{K}(\mathcal{G}) \in \mathbb{C}_+(\mathcal{G})$.

Let $\boldsymbol{K}$ be any matrix in $\mathbb{C}_H(\mathcal{G}) \setminus \{\boldsymbol{O}\}$. Since $\mathbb{C}_+(\mathcal{G})$ is an open subset of $\mathbb{C}_H(\mathcal{G})$ it holds that $\boldsymbol{K}(\mathcal{G}) + t\boldsymbol{K} \in \mathbb{C}_+(\mathcal{G})$, where $t \in [-\varepsilon, \varepsilon]$ for some $\varepsilon > 0$. Thus the log-likelihood function of $\boldsymbol{K}(\mathcal{G}) + t\boldsymbol{K}$ is well defined. To prove that $l\,(\boldsymbol{K}(\mathcal{G})|\,\boldsymbol{x})$ is strictly concave we have to prove that

$$\frac{d^2}{dt^2} l\,(\boldsymbol{K}(\mathcal{G}) + t\boldsymbol{K}|\,\boldsymbol{x})\Big|_{t=0} < 0$$

for all $\boldsymbol{K}(\mathcal{G}) \in \mathbb{C}_+(\mathcal{G})$ and all $\boldsymbol{K} \in \mathbb{C}_H(\mathcal{G}) \setminus \{\boldsymbol{O}\}$.

From (7.9) page 123 the log-likelihood function of $\boldsymbol{K}(\mathcal{G}) + t\boldsymbol{K}$ is given by

$$l\,(\boldsymbol{K}(\mathcal{G}) + t\boldsymbol{K}|\,\boldsymbol{x}) = -n|V|\log\pi + n\log\left(\det\left(\boldsymbol{K}(\mathcal{G}) + t\boldsymbol{K}\right)\right) - n\,\mathrm{tr}\left(\left(\boldsymbol{K}(\mathcal{G}) + t\boldsymbol{K}\right)\boldsymbol{s}(\mathcal{G})\right) .$$

By using Lemma 7.1 page 123 we get the first derivative of the log-likelihood function w.r.t. t in direction of $\boldsymbol{K}$ as

$$\frac{d}{dt} l\,(\boldsymbol{K}(\mathcal{G}) + t\boldsymbol{K}|\,\boldsymbol{x}) = n\,\mathrm{tr}\left(\left(\boldsymbol{K}(\mathcal{G}) + t\boldsymbol{K}\right)^{-1}\boldsymbol{K}\right) - n\,\mathrm{tr}\left(\boldsymbol{K}\boldsymbol{s}(\mathcal{G})\right) ,$$

and the second derivative of the log-likelihood function w.r.t. t in direction of $\boldsymbol{K}$ as

$$\frac{d^2}{dt^2} l\,(\boldsymbol{K}(\mathcal{G}) + t\boldsymbol{K}|\,\boldsymbol{x}) = -n\,\mathrm{tr}\left(\left(\boldsymbol{K}(\mathcal{G}) + t\boldsymbol{K}\right)^{-1}\boldsymbol{K}\left(\boldsymbol{K}(\mathcal{G}) + t\boldsymbol{K}\right)^{-1}\boldsymbol{K}\right) .$$

Hereby the second derivative of the log-likelihood function w.r.t. t in direction of $\boldsymbol{K}$ at $t = 0$ is

$$\frac{d^2}{dt^2} l\,(\boldsymbol{K}(\mathcal{G}) + t\boldsymbol{K}|\,\boldsymbol{x})\Big|_{t=0} = -n\,\mathrm{tr}\left(\boldsymbol{K}(\mathcal{G})^{-1}\boldsymbol{K}\boldsymbol{K}(\mathcal{G})^{-1}\boldsymbol{K}\right) .$$

Since it holds that $\boldsymbol{K}(\mathcal{G})^{-1} > \boldsymbol{O}$ there exists a nonsingular $|V| \times |V|$ complex matrix $\boldsymbol{D}$ such that $\boldsymbol{K}(\mathcal{G})^{-1} = \boldsymbol{D}^*\boldsymbol{D}$. Then we have

$$\begin{aligned}\frac{d^2}{dt^2} l\,(\boldsymbol{K}(\mathcal{G}) + t\boldsymbol{K}|\,\boldsymbol{x})\Big|_{t=0} &= -n\,\mathrm{tr}\left(\boldsymbol{D}^*\boldsymbol{D}\boldsymbol{K}\boldsymbol{D}^*\boldsymbol{D}\boldsymbol{K}\right)\\ &= -n\,\mathrm{tr}\left(\boldsymbol{D}\boldsymbol{K}\boldsymbol{D}^*\left(\boldsymbol{D}\boldsymbol{K}\boldsymbol{D}^*\right)^*\right) .\end{aligned}$$

Let $\boldsymbol{D}\boldsymbol{K}\boldsymbol{D}^*$ be written as $\boldsymbol{D}\boldsymbol{K}\boldsymbol{D}^* = \left(\boldsymbol{c}_1, \boldsymbol{c}_2, \ldots, \boldsymbol{c}_{|V|}\right)^*$, where $\boldsymbol{c}_j \in \mathbb{C}^{|V|}$ for $j = 1, 2, \ldots, |V|$. The diagonal elements in $\boldsymbol{D}\boldsymbol{K}\boldsymbol{D}^*\left(\boldsymbol{D}\boldsymbol{K}\boldsymbol{D}^*\right)^*$ are $\boldsymbol{c}_j^*\boldsymbol{c}_j$. These elements are positive unless $\boldsymbol{D}\boldsymbol{K}\boldsymbol{D}^* = \boldsymbol{O}$. Since $\boldsymbol{D}$ is nonsingular this means unless $\boldsymbol{K} = \boldsymbol{O}$. As $\boldsymbol{K}$ is assumed not to be $\boldsymbol{O}$ we hereby have $\mathrm{tr}\left(\boldsymbol{D}\boldsymbol{K}\boldsymbol{D}^*\left(\boldsymbol{D}\boldsymbol{K}\boldsymbol{D}^*\right)^*\right) > 0$, which implies

$$\frac{d^2}{dt^2} l\,(\boldsymbol{K}(\mathcal{G}) + t\boldsymbol{K}|\,\boldsymbol{x})\Big|_{t=0} < 0 .$$

■

We are now able to deduce the likelihood equations from which a maximum likelihood estimate of $\boldsymbol{K}(\mathcal{G})$ is determined. If a solution exists, it is unique. This is due to Lemma 7.3.

Theorem 7.2 The likelihood equations
Let $\mathcal{G} = (V, E)$ be a simple undirected graph. Let $\boldsymbol{X}$ be an $n \times |V|$ complex random matrix with $\mathcal{L}(\boldsymbol{X}) = \mathbb{C}\mathcal{N}_{n\times|V|}\left(\boldsymbol{O}, \boldsymbol{I}_n \otimes \boldsymbol{K}(\mathcal{G})^{-1}\right)$, where $\boldsymbol{K}(\mathcal{G}) \in \mathbb{C}_+(\mathcal{G})$, and let $\boldsymbol{x}$ be an observation matrix of $\boldsymbol{X}$. The maximum likelihood estimate of $\boldsymbol{K}(\mathcal{G})$ is determined by the likelihood equations

$$\left(\boldsymbol{K}(\mathcal{G})^{-1}\right)_{CC} = \boldsymbol{s}_{CC} \quad \forall\, C \in \mathcal{C},$$

where $\boldsymbol{s} = \frac{1}{n}\boldsymbol{x}^*\boldsymbol{x}$ and $\mathcal{C}$ is the set of cliques in $\mathcal{G}$.

Proof:
Let $\mathcal{G} = (V, E)$ be a simple undirected graph. Let $\boldsymbol{X}$ be an $n \times |V|$ complex random matrix with $\mathcal{L}(\boldsymbol{X}) = \mathbb{C}\mathcal{N}_{n\times|V|}\left(\boldsymbol{O}, \boldsymbol{I}_n \otimes \boldsymbol{K}(\mathcal{G})^{-1}\right)$, where $\boldsymbol{K}(\mathcal{G}) \in \mathbb{C}_+(\mathcal{G})$, and let $\boldsymbol{x}$ be an observation matrix of $\boldsymbol{X}$.

From Lemma 7.3 page 126 the log-likelihood function is strictly concave, therefore it has at most one maximum. To determine this possible maximum we consider for $\boldsymbol{K}(\mathcal{G}) \in \mathbb{C}_+(\mathcal{G})$ the equation

$$\text{(7.14)} \qquad \frac{d}{dt} l\left(\boldsymbol{K}(\mathcal{G}) + t\boldsymbol{K} \,|\, \boldsymbol{x}\right)\Big|_{t=0} = 0$$

for all $\boldsymbol{K} \in \mathbb{C}_H(\mathcal{G}) \setminus \{\boldsymbol{O}\}$.

Using Lemma 7.1 page 123 we find the first derivative of the log-likelihood function w.r.t. t in direction of $\boldsymbol{K}$ at $t = 0$ as

$$\frac{d}{dt} l\left(\boldsymbol{K}(\mathcal{G}) + t\boldsymbol{K} \,|\, \boldsymbol{x}\right)\Big|_{t=0} = n \operatorname{tr}\left(\boldsymbol{K}(\mathcal{G})^{-1}\boldsymbol{K}\right) - n \operatorname{tr}\left(\boldsymbol{K}\boldsymbol{s}(\mathcal{G})\right) .$$

Hereby (7.14) becomes

$$\begin{array}{rcll} \operatorname{tr}\left(\boldsymbol{K}\boldsymbol{K}(\mathcal{G})^{-1}\right) & = & \operatorname{tr}\left(\boldsymbol{K}\boldsymbol{s}(\mathcal{G})\right) & \forall\, \boldsymbol{K} \in \mathbb{C}_H(\mathcal{G}) \setminus \{\boldsymbol{O}\} \\ \Updownarrow & & & \\ \operatorname{tr}\left([\boldsymbol{K}_{CC}]\boldsymbol{K}(\mathcal{G})^{-1}\right) & = & \operatorname{tr}\left([\boldsymbol{K}_{CC}]\boldsymbol{s}(\mathcal{G})\right) & \forall\, \boldsymbol{K}_{CC} \in \mathbb{C}_H^{|C|\times|C|},\ \forall\, C \in \mathcal{C} \\ \Updownarrow & & & \\ \operatorname{tr}\left(\boldsymbol{K}_{CC}\left(\boldsymbol{K}(\mathcal{G})^{-1}\right)_{CC}\right) & = & \operatorname{tr}\left(\boldsymbol{K}_{CC}\boldsymbol{s}(\mathcal{G})_{CC}\right) & \forall\, \boldsymbol{K}_{CC} \in \mathbb{C}_H^{|C|\times|C|},\ \forall\, C \in \mathcal{C} \\ \Updownarrow & & & \\ \left(\boldsymbol{K}(\mathcal{G})^{-1}\right)_{CC} & = & \boldsymbol{s}(\mathcal{G})_{CC} & \forall\, C \in \mathcal{C} \\ \Updownarrow & & & \\ \left(\boldsymbol{K}(\mathcal{G})^{-1}\right)_{CC} & = & \boldsymbol{s}_{CC} & \forall\, C \in \mathcal{C}, \end{array}$$

where $\mathcal{C}$ denotes the set of cliques in $\mathcal{G}$. Hence we conclude that $l\left(\boldsymbol{K}(\mathcal{G}) | \boldsymbol{x}\right)$ is maximized, when $\boldsymbol{K}(\mathcal{G}) \in \mathbb{C}_+(\mathcal{G})$ fulfills the equations $\left(\boldsymbol{K}(\mathcal{G})^{-1}\right)_{CC} = \boldsymbol{s}_{CC}$ for all $C \in \mathcal{C}$. These equations are called the likelihood equations. ■

Given a clique C we can partition $\boldsymbol{x}$ as $\boldsymbol{x} = \left(\boldsymbol{x}_C, \boldsymbol{x}_{V \setminus C}\right)$, where $\boldsymbol{x}_C$ and $\boldsymbol{x}_{V\setminus C}$ have dimensions $n \times |C|$ and $n \times |V \setminus C|$, respectively. This implies that $\boldsymbol{s}_{CC}$ in Theorem 7.2 can be expressed as $\boldsymbol{s}_{CC} = \frac{1}{n}\boldsymbol{x}_C^* \boldsymbol{x}_C$. Moreover it is interesting to observe that the unique maximum likelihood estimate also satisfies the likelihood equations on every complete subset of V. Besides if no pairwise conditional independence is assumed, i.e. $\mathcal{G}$ is complete, the maximum likelihood estimate of the variance matrix is determined as $\boldsymbol{s} = \frac{1}{n}\boldsymbol{x}^*\boldsymbol{x}$.

7.4.1 Iterative Proportional Scaling

The likelihood equations for the maximum likelihood estimate of the concentration matrix are given in Theorem 7.2 page 128. However this theorem does not give a description of how to solve these equations. In general there is no explicit solution and it is necessary to use an iterative method. There exist several methods and we have chosen to describe the method called the iterative proportional scaling algorithm (IPS-algorithm), which iteratively adjusts a matrix to satisfy the likelihood equations for all the cliques.

To ensure that the likelihood function is bounded from above, i.e. the maximum likelihood estimate exists and is unique, we need the concept of $\boldsymbol{s}(\mathcal{G})$ being $\mathcal{G}$-regular. Loosely spoken $\boldsymbol{s}(\mathcal{G})$ is $\mathcal{G}$-regular, if the zero entries inherited from $\mathcal{G}$ can be changed to some other value, such that the matrix obtained is positive definite. Formally we define this in Definition 7.2, where we need the set given by

$$\mathbb{C}_S(\mathcal{G}) = \left\{ \boldsymbol{K} \in \mathbb{C}_S^{|V|\times|V|} \;\middle|\; \forall \alpha \neq \beta \in V : \{\alpha, \beta\} \notin E \;\Rightarrow\; \boldsymbol{K}_{\alpha\beta} = 0 \right\} .$$

Definition 7.2 $\mathcal{G}$-regularity
The complex matrix $\boldsymbol{s}(\mathcal{G})$ is said to be $\mathcal{G}$-regular, if $\operatorname{tr}(\boldsymbol{K}\boldsymbol{s}(\mathcal{G})) > 0$ *for all* $\boldsymbol{K} \in \mathbb{C}_S(\mathcal{G}) \setminus \{\boldsymbol{O}\}$.

On page 134 we show that a necessary and sufficient condition to ensure that the likelihood function is bounded from above is that $\boldsymbol{s}(\mathcal{G})$ is $\mathcal{G}$-regular.

When $\boldsymbol{s}(\mathcal{G})$ is $\mathcal{G}$-regular it holds that $\operatorname{tr}(\boldsymbol{K}\boldsymbol{s}(\mathcal{G})) > 0$ for all $\boldsymbol{K} \in \mathbb{C}_S(\mathcal{G}) \setminus \{\boldsymbol{O}\}$. Therefore for an arbitrary complete subset A of V we can choose $\boldsymbol{K} = [\boldsymbol{d}_A \boldsymbol{d}_A^*]$ for all $\boldsymbol{d}_A \in \mathbb{C}^{|A|} \setminus \{0\}$ and get

$$\begin{aligned} \operatorname{tr}(\boldsymbol{K}\boldsymbol{s}(\mathcal{G})) &= \operatorname{tr}(\boldsymbol{d}_A \boldsymbol{d}_A^* \boldsymbol{s}_{AA}) \\ &= \operatorname{tr}(\boldsymbol{d}_A^* \boldsymbol{s}_{AA} \boldsymbol{d}_A) \\ &= \boldsymbol{d}_A^* \boldsymbol{s}_{AA} \boldsymbol{d}_A \\ &> 0 . \end{aligned}$$

Hence for a'l complete subsets A of V we have $\boldsymbol{s}_{AA} > \boldsymbol{O}$, when $\boldsymbol{s}(\mathcal{G})$ is $\mathcal{G}$-regular.

We are now able to define an operator for each clique, which conforms a concentration matrix to the likelihood equations for the particular clique.

Definition 7.3 The C-marginal adjusting operator
Let $\mathcal{G} = (V, E)$ be a simple undirected graph. Let $\boldsymbol{X}$ be an $n \times |V|$ complex random matrix with $\mathcal{L}(\boldsymbol{X}) = \mathbb{C}\mathcal{N}_{n\times|V|}\left(\boldsymbol{O}, \boldsymbol{I}_n \otimes \boldsymbol{K}(\mathcal{G})^{-1}\right)$, where $\boldsymbol{K}(\mathcal{G}) \in \mathbb{C}_+(\mathcal{G})$, and let $\boldsymbol{x}$ be an observation matrix of $\boldsymbol{X}$. Furthermore let $\boldsymbol{s} = \frac{1}{n}\boldsymbol{x}^*\boldsymbol{x}$ and let $\boldsymbol{s}(\mathcal{G})$ defined in (7.7) page 122 be $\mathcal{G}$-regular. The operator $T_C : \mathbb{C}_+(\mathcal{G}) \mapsto \mathbb{C}_+(\mathcal{G})$ given by

$$T_C\boldsymbol{K} = \boldsymbol{K} + \left[\boldsymbol{s}_{CC}^{-1} - \left(\boldsymbol{K}^{-1}\right)_{CC}^{-1}\right], \tag{7.15}$$

where $\boldsymbol{K} \in \mathbb{C}_+(\mathcal{G})$ and $C \in \mathcal{C}$, is called the C-marginal adjusting operator.

Observe that $\boldsymbol{s}_{CC} > \boldsymbol{O}$, since $\boldsymbol{s}(\mathcal{G})$ is $\mathcal{G}$-regular, and that $\left(\boldsymbol{K}^{-1}\right)_{CC} > \boldsymbol{O}$, since $\boldsymbol{K}^{-1} > \boldsymbol{O}$. Hereby $\boldsymbol{s}_{CC}^{-1}$ and $\left(\boldsymbol{K}^{-1}\right)_{CC}^{-1}$ in the C-marginal adjusting operator exist.

Let $\boldsymbol{K} \in \mathbb{C}_+(\mathcal{G})$ be partitioned as

$$\boldsymbol{K} = \begin{pmatrix} \boldsymbol{K}_{CC} & \boldsymbol{K}_{C,V\setminus C} \\ \boldsymbol{K}_{V\setminus C,C} & \boldsymbol{K}_{V\setminus C,V\setminus C} \end{pmatrix},$$

then we get

$$\left(\boldsymbol{K}^{-1}\right)_{CC} = \left(\boldsymbol{K}_{CC} - \boldsymbol{K}_{C,V\setminus C}\boldsymbol{K}_{V\setminus C,V\setminus C}^{-1}\boldsymbol{K}_{V\setminus C,C}\right)^{-1}.$$

Hereby the C-marginal adjusting operator is also given by

$$T_C\boldsymbol{K} = \begin{pmatrix} \boldsymbol{s}_{CC}^{-1} + \boldsymbol{K}_{C,V\setminus C}\boldsymbol{K}_{V\setminus C,V\setminus C}^{-1}\boldsymbol{K}_{V\setminus C,C} & \boldsymbol{K}_{C,V\setminus C} \\ \boldsymbol{K}_{V\setminus C,C} & \boldsymbol{K}_{V\setminus C,V\setminus C} \end{pmatrix}. \tag{7.16}$$

Using this alternative expression of $T_C\boldsymbol{K}$ we see that $T_C\boldsymbol{K} > \boldsymbol{O}$, since $\boldsymbol{K}_{V\setminus C,V\setminus C} > \boldsymbol{O}$ and $\boldsymbol{s}_{CC}^{-1} + \boldsymbol{K}_{C,V\setminus C}\boldsymbol{K}_{V\setminus C,V\setminus C}^{-1}\boldsymbol{K}_{V\setminus C,C} - \boldsymbol{K}_{C,V\setminus C}\boldsymbol{K}_{V\setminus C,V\setminus C}^{-1}\boldsymbol{K}_{V\setminus C,C} = \boldsymbol{s}_{CC}^{-1} > \boldsymbol{O}$. From (7.15) the pattern of zeros determined by $\mathcal{G}$ in $\boldsymbol{K}$ is preserved. Hence $T_C\boldsymbol{K} \in \mathbb{C}_+(\mathcal{G})$, if $\boldsymbol{K} \in \mathbb{C}_+(\mathcal{G})$, i.e. the adjusting operator is well defined.

Observe that

$$\left((T_C\boldsymbol{K})^{-1}\right)_{CC} = \boldsymbol{s}_{CC}. \tag{7.17}$$

Thus the adjusted matrix $T_C\boldsymbol{K}$ fulfills the likelihood equation for the clique C, which is chosen arbitrarily among the cliques in $\mathcal{G}$. We seek a matrix $\boldsymbol{K} \in \mathbb{C}_+(\mathcal{G})$, which fulfills the likelihood equations for all $C \in \mathcal{C}$, therefore we introduce the IPS-algorithm.

Algorithm 7.1 The IPS-algorithm
Let $\mathcal{G} = (V, E)$ be a simple undirected graph. Let $\boldsymbol{X}$ be an $n \times |V|$ complex random matrix with $\mathcal{L}(\boldsymbol{X}) = \mathbb{C}\mathcal{N}_{n\times|V|}\left(\boldsymbol{O}, \boldsymbol{I}_n \otimes \boldsymbol{K}(\mathcal{G})^{-1}\right)$, where $\boldsymbol{K}(\mathcal{G}) \in \mathbb{C}_+(\mathcal{G})$, and let $\boldsymbol{x}$ be an observation matrix of $\boldsymbol{X}$. Furthermore let $\boldsymbol{s} = \frac{1}{n}\boldsymbol{x}^*\boldsymbol{x}$ and let $\boldsymbol{s}(\mathcal{G})$ defined in (7.7) page 122 be $\mathcal{G}$-regular. The IPS-algorithm is given as follows.

1. *Choose an ordering* $C_1, C_2, \ldots C_m$ *of the cliques in* $\mathcal{G}$.

2. *Choose an arbitrary starting point* $\boldsymbol{K}_0 \in \mathbb{C}_+(\mathcal{G})$.

3. *Define the operator* $T : \mathbb{C}_+(\mathcal{G}) \mapsto \mathbb{C}_+(\mathcal{G})$ *as* $T = T_{C_1} T_{C_2} \cdots T_{C_m}$, *where* T_{C_j} *is defined in Definition 7.3 page 130.*

4. *Define recursively for* $s = 0, 1, 2, \ldots$ *the IPS-updating equation as*

$$\boldsymbol{K}_{s+1} = T\boldsymbol{K}_s \,. \tag{7.18}$$

The operator T is well defined as T_{C_j} for $j = 1, 2, \ldots, m$ maps $\mathbb{C}_+(\mathcal{G})$ onto itself. It is used to make adjustments to each marginal in turn. The IPS-updating equation is used to get closer and closer to a matrix $\boldsymbol{K} \in \mathbb{C}_+(\mathcal{G})$, which fulfills the likelihood equations for all $C \in \mathcal{C}$. The theorem below tells us that, when the unique maximum likelihood estimate of $\boldsymbol{K}(\mathcal{G})$ exists, it can be found by repeated use of the IPS-updating equation.

Theorem 7.3 The maximum likelihood estimate of the concentration matrix
Let $\mathcal{G} = (V, E)$ *be a simple undirected graph. Let* $\boldsymbol{X}$ *be an* $n \times |V|$ *complex random matrix with* $\mathcal{L}(\boldsymbol{X}) = \mathbb{C}\mathcal{N}_{n \times |V|}\left(\boldsymbol{O}, \boldsymbol{I}_n \otimes \boldsymbol{K}(\mathcal{G})^{-1}\right)$, *where* $\boldsymbol{K}(\mathcal{G}) \in \mathbb{C}_+(\mathcal{G})$, *and let* $\boldsymbol{x}$ *be an observation matrix of* $\boldsymbol{X}$. *Furthermore let* $\boldsymbol{s} = \frac{1}{n}\boldsymbol{x}^*\boldsymbol{x}$ *and let* $\boldsymbol{s}(\mathcal{G})$ *defined in (7.7) page 122 be* $\mathcal{G}$*-regular. The unique maximum likelihood estimate of* $\boldsymbol{K}(\mathcal{G})$ *is given by*

$$\widehat{\boldsymbol{K}}(\mathcal{G}) = \lim_{s \to \infty} \boldsymbol{K}_s \,,$$

where $\boldsymbol{K}_s$ *is defined in (7.18) page 131.*

Proof:
Let the assumptions in the theorem be satisfied and choose arbitrarily $\boldsymbol{K}_0 \in \mathbb{C}_+(\mathcal{G})$.

Define the set

$$\mathcal{K} = \{\boldsymbol{K} \in \mathbb{C}_+(\mathcal{G}) \mid l(\boldsymbol{K}|\,\boldsymbol{x}) \geq l(\boldsymbol{K}_0|\,\boldsymbol{x})\} \,.$$

First we seek to show that $\mathcal{K}$ is compact. Obviously $\mathcal{K}$ is a subset of $\mathbb{C}_H(\mathcal{G})$, which can be regarded as a vector space over $\mathbb{R}$. Therefore if we can show that $\mathcal{K}$ is closed and bounded, then $\mathcal{K}$ is compact. We begin by showing that $\mathcal{K}$ is closed.

Let $\{\boldsymbol{K}_r\}$ be an arbitrary convergent sequence in $\mathcal{K}$ which converges to $\boldsymbol{K}'$. If $\boldsymbol{K}' \in \mathcal{K}$, then $\mathcal{K}$ is closed. Since the log-likelihood function is continuous we find

$$\begin{aligned} \lim_{r \to \infty} (l(\boldsymbol{K}_r|\,\boldsymbol{x})) &= l\left(\lim_{r \to \infty} \boldsymbol{K}_r \middle|\, \boldsymbol{x}\right) \\ &= l(\boldsymbol{K}'|\,\boldsymbol{x}) \,. \end{aligned}$$

Because $\boldsymbol{K}_r \in \mathcal{K}$ we know that

$$l(\boldsymbol{K}_r|\boldsymbol{x}) \geq l(\boldsymbol{K}_0|\boldsymbol{x}) ,$$

thus it holds that

$$l(\boldsymbol{K}'|\boldsymbol{x}) \geq l(\boldsymbol{K}_0|\boldsymbol{x}) .$$

Therefore $\boldsymbol{K}'$ is an element in $\mathcal{K}$, i.e. $\mathcal{K}$ is closed.

Next we show by contradiction that $\mathcal{K}$ is bounded. Therefore assume that $\mathcal{K}$ is unbounded.

From Lemma 7.3 page 126 we have that the log-likelihood function is strictly concave. Using this we find for $\boldsymbol{K}_r, \boldsymbol{K}_t \in \mathcal{K}$ and all $\lambda \in [0, 1]$ that

$$\begin{aligned} l(\lambda \boldsymbol{K}_r + (1-\lambda)\boldsymbol{K}_t|\boldsymbol{x}) &> \lambda l(\boldsymbol{K}_r|\boldsymbol{x}) + (1-\lambda) l(\boldsymbol{K}_t|\boldsymbol{x}) \\ &\geq \lambda l(\boldsymbol{K}_0|\boldsymbol{x}) + (1-\lambda) l(\boldsymbol{K}_0|\boldsymbol{x}) \\ &= l(\boldsymbol{K}_0|\boldsymbol{x}) . \end{aligned}$$

Hereby $\lambda \boldsymbol{K}_r + (1-\lambda)\boldsymbol{K}_t \in \mathcal{K}$ and $\mathcal{K}$ is convex.

Using that $\mathcal{K}$ is unbounded, closed and convex we know that $\mathcal{K}$ contains at least one point at infinity (Rockafellar 1970, p. 64), whereby there exists a $\boldsymbol{D} \in \mathbb{C}_H(\mathcal{G}) \setminus \{\boldsymbol{O}\}$, such that

$$\left\{\boldsymbol{K}_0 + t\boldsymbol{D} \mid t \in \overline{\mathbb{R}}_+\right\} \subseteq \mathcal{K} .$$

From (7.9) page 123 the log-likelihood function at $\boldsymbol{K}_0 + t\boldsymbol{D} \in \mathcal{K}$ is

$$\begin{aligned} l(\boldsymbol{K}_0 + t\boldsymbol{D}|\boldsymbol{x}) &= -n|V|\log\pi + n\log(\det(\boldsymbol{K}_0 + t\boldsymbol{D})) - n\,\mathrm{tr}((\boldsymbol{K}_0 + t\boldsymbol{D})\boldsymbol{s}(\mathcal{G})) \\ &= -n|V|\log\pi + n\log(\det(\boldsymbol{K}_0)) + n\log\left(\det\left(\boldsymbol{I}_{|V|} + t\boldsymbol{K}_0^{-1}\boldsymbol{D}\right)\right) \\ &\quad -n\,\mathrm{tr}(\boldsymbol{K}_0\boldsymbol{s}(\mathcal{G})) - nt\,\mathrm{tr}(\boldsymbol{D}\boldsymbol{s}(\mathcal{G})) \\ &= -n|V|\log\pi + n\log(\det(\boldsymbol{K}_0)) + n\sum_{j=1}^{|V|}\log(1 + td_j) \\ &\quad -n\,\mathrm{tr}(\boldsymbol{K}_0\boldsymbol{s}(\mathcal{G})) - nt\,\mathrm{tr}(\boldsymbol{D}\boldsymbol{s}(\mathcal{G})) \\ &= -n|V|\log\pi + n\log(\det(\boldsymbol{K}_0)) - n\,\mathrm{tr}(\boldsymbol{K}_0\boldsymbol{s}(\mathcal{G})) \\ &\quad -nt\,\mathrm{tr}(\boldsymbol{D}\boldsymbol{s}(\mathcal{G}))\left(1 - \frac{\sum_{j=1}^{|V|}\log(1 + td_j)}{t\,\mathrm{tr}(\boldsymbol{D}\boldsymbol{s}(\mathcal{G}))}\right) , \end{aligned}$$

where the d_j's are the eigenvalues of $\boldsymbol{K}_0^{-1}\boldsymbol{D}$. We have $\boldsymbol{K}_0 > \boldsymbol{O}$ and $\boldsymbol{K}_0 + t\boldsymbol{D} > \boldsymbol{O}$, which for $t \in \mathbb{R}_+$ implies $\boldsymbol{D} \geq \boldsymbol{O}$, i.e. $\boldsymbol{D} \in \mathbb{C}_S(\mathcal{G}) \setminus \{\boldsymbol{O}\}$. Therefore as $\boldsymbol{s}(\mathcal{G})$ is $\mathcal{G}$-regular we have for $t \in \mathbb{R}_+$ that $\mathrm{tr}(\boldsymbol{D}\boldsymbol{s}(\mathcal{G})) > 0$ and we conclude

$$\lim_{t\to\infty} l(\boldsymbol{K}_0 + t\boldsymbol{D}|\boldsymbol{x}) = -\infty .$$

This contradicts that $\boldsymbol{K}_0 + t\boldsymbol{D} \in \mathcal{K}$ for $t \in \overline{\mathbb{R}}_+$, whereby $\mathcal{K}$ is bounded. Altogether we have shown that $\mathcal{K}$ is compact.

Let C be an arbitrary clique in $\mathcal{G}$ and let $\boldsymbol{X}$ be partitioned as

$$\boldsymbol{X} = \left(\boldsymbol{X}_{V \setminus C}, \boldsymbol{X}_C\right) ,$$

where $\boldsymbol{X}_{V\setminus C}$ and $\boldsymbol{X}_C$ have dimensions $n \times |V \setminus C|$ and $n \times |C|$, respectively. By Bayes' rule we obtain

$$f_{\boldsymbol{X}_{V\setminus C}, \boldsymbol{X}_C}\left(\boldsymbol{x}_{V\setminus C}, \boldsymbol{x}_C\right) = f_{\boldsymbol{X}_{V\setminus C} | \boldsymbol{X}_C}\left(\boldsymbol{x}_{V\setminus C} \middle| \boldsymbol{x}_C\right) f_{\boldsymbol{X}_C}\left(\boldsymbol{x}_C\right) .$$

From Theorem 2.25 page 37 we observe

$$\begin{aligned} \mathcal{L}\left(\boldsymbol{X}_{V\setminus C} \middle| \boldsymbol{X}_C\right) &= \mathbb{C}\mathcal{N}_{n\times|V\setminus C|}\Big(\boldsymbol{X}_C \left(\boldsymbol{K}(\mathcal{G})^{-1}\right)_{CC}^{-1} \left(\boldsymbol{K}(\mathcal{G})^{-1}\right)_{C,V\setminus C}, \\ &\qquad \boldsymbol{I}_n \otimes \Big(\left(\boldsymbol{K}(\mathcal{G})^{-1}\right)_{V\setminus C,V\setminus C} - \left(\boldsymbol{K}(\mathcal{G})^{-1}\right)_{V\setminus C,C} \left(\boldsymbol{K}(\mathcal{G})^{-1}\right)_{CC}^{-1} \left(\boldsymbol{K}(\mathcal{G})^{-1}\right)_{C,V\setminus C}\Big)\Big) \\ &= \mathbb{C}\mathcal{N}_{n\times|V\setminus C|}\left(-\boldsymbol{X}_C \boldsymbol{K}(\mathcal{G})_{C,V\setminus C} \boldsymbol{K}(\mathcal{G})_{V\setminus C,V\setminus C}^{-1}, \boldsymbol{I}_n \otimes \boldsymbol{K}(\mathcal{G})_{V\setminus C,V\setminus C}^{-1}\right) . \end{aligned}$$

Further from Theorem 2.23 page 36 we have

$$\mathcal{L}\left(\boldsymbol{X}_C\right) = \mathbb{C}\mathcal{N}_{n\times|C|}\left(\boldsymbol{O}, \boldsymbol{I}_n \otimes \left(\boldsymbol{K}(\mathcal{G})^{-1}\right)_{CC}\right) .$$

Therefore the likelihood function of $\boldsymbol{K}(\mathcal{G}) \in \mathbb{C}_+(\mathcal{G})$ factorizes as

$$L\left(\boldsymbol{K}(\mathcal{G}) | \boldsymbol{x}\right) = L\left(\boldsymbol{K}(\mathcal{G})_{C,V\setminus C}, \boldsymbol{K}(\mathcal{G})_{V\setminus C,V\setminus C} \middle| \boldsymbol{x}\right) L\left(\left(\boldsymbol{K}(\mathcal{G})^{-1}\right)_{CC} \middle| \boldsymbol{x}_C\right) .$$

Using this on the likelihood function for $T_C\boldsymbol{K} \in \mathbb{C}_+(\mathcal{G})$ together with (7.16) and (7.17) page 130 we get

$$\begin{aligned} L\left(T_C\boldsymbol{K} | \boldsymbol{x}\right) &= L\left((T_C\boldsymbol{K})_{C,V\setminus C}, (T_C\boldsymbol{K})_{V\setminus C,V\setminus C} \middle| \boldsymbol{x}\right) L\left(\left((T_C\boldsymbol{K})^{-1}\right)_{CC} \middle| \boldsymbol{x}_C\right) \\ &= L\left(\boldsymbol{K}_{C,V\setminus C}, \boldsymbol{K}_{V\setminus C,V\setminus C} \middle| \boldsymbol{x}\right) L\left(\boldsymbol{s}_{CC} | \boldsymbol{x}_C\right) \\ &= \frac{L\left(\boldsymbol{K} | \boldsymbol{x}\right)}{L\left(\left(\boldsymbol{K}^{-1}\right)_{CC} \middle| \boldsymbol{x}_C\right)} L\left(\boldsymbol{s}_{CC} | \boldsymbol{x}_C\right) \\ &\geq L\left(\boldsymbol{K} | \boldsymbol{x}\right) . \end{aligned}$$

The inequality is obtained from $L\left(\boldsymbol{s}_{CC} | \boldsymbol{x}_C\right) \geq L\left(\left(\boldsymbol{K}^{-1}\right)_{CC} \middle| \boldsymbol{x}_C\right)$, since $\boldsymbol{s}_{CC} > \boldsymbol{O}$. Equality holds iff $\boldsymbol{s}_{CC} = \left(\boldsymbol{K}^{-1}\right)_{CC}$, i.e. from Definition 7.3 page 130 iff $T_C\boldsymbol{K} = \boldsymbol{K}$. It leads to the following inequality for the log-likelihood function

$$l\left(T_C\boldsymbol{K} | \boldsymbol{x}\right) \geq l\left(\boldsymbol{K} | \boldsymbol{x}\right) , \tag{7.19}$$

for $C \in \mathcal{C}$ and for $\boldsymbol{K} \in \mathbb{C}_+(\mathcal{G})$. Again equality is obtained iff $\boldsymbol{s}_{CC} = \left(\boldsymbol{K}^{-1}\right)_{CC}$ or equivalently iff $T_C\boldsymbol{K} = \boldsymbol{K}$.

Let $\{\boldsymbol{K}_s\}$ be the sequence produced by the IPS-algorithm with $\boldsymbol{K}_0$ as starting point. Applying (7.19) successively we get

$$l\left(T\boldsymbol{K} | \boldsymbol{x}\right) \geq l\left(T_{C_2} \cdots T_{C_m}\boldsymbol{K} | \boldsymbol{x}\right) \geq \cdots \geq l\left(T_{C_m}\boldsymbol{K} | \boldsymbol{x}\right) \geq l\left(\boldsymbol{K} | \boldsymbol{x}\right) \tag{7.20}$$

for $\boldsymbol{K} \in \mathbb{C}_+(\mathcal{G})$. Hereby we deduce that

$$l(\boldsymbol{K}_s|\boldsymbol{x}) \geq l(\boldsymbol{K}_{s'}|\boldsymbol{x}) ,$$

when $s \geq s'$. For $s \in \mathbb{N}_0$ we then have $\boldsymbol{K}_s \in \mathcal{K}$ and hereby $\{\boldsymbol{K}_s\}$ is a sequence in $\mathcal{K}$.

Since $\mathcal{K}$ is compact there exists a convergent subsequence $\{\boldsymbol{K}_{s_r}\}$ of $\{\boldsymbol{K}_s\}$ which converges to an element of $\mathcal{K}$. This element is denoted by $\boldsymbol{K}''$. Notice from (7.16) page 130 and step 3 of the IPS-algorithm page 130 that T is continuous. Using this together with the continuity of the log-likelihood function we get

$$\begin{aligned} l(\boldsymbol{K}''|\boldsymbol{x}) &= l\left(\lim_{r\to\infty} \boldsymbol{K}_{s_r}\middle|\boldsymbol{x}\right) \\ &= \lim_{r\to\infty} l(\boldsymbol{K}_{s_r}|\boldsymbol{x}) \\ &= \lim_{r\to\infty} l(T\boldsymbol{K}_{s_r-1}|\boldsymbol{x}) \\ &\geq \lim_{r\to\infty} l\left(T\boldsymbol{K}_{s_{r-1}}\middle|\boldsymbol{x}\right) \\ &= l\left(T\left(\lim_{r\to\infty} \boldsymbol{K}_{s_{r-1}}\right)\middle|\boldsymbol{x}\right) \\ &= l(T\boldsymbol{K}''|\boldsymbol{x}) \\ &\geq l(\boldsymbol{K}''|\boldsymbol{x}) . \end{aligned}$$

From the above we see

$$l(\boldsymbol{K}''|\boldsymbol{x}) \geq l(T\boldsymbol{K}''|\boldsymbol{x}) \geq l(\boldsymbol{K}''|\boldsymbol{x}) ,$$

which yields that

$$l(T\boldsymbol{K}''|\boldsymbol{x}) = l(\boldsymbol{K}''|\boldsymbol{x}) .$$

Using this in (7.20) page 133 we reach the following

$$\text{(7.21)} \qquad l(T\boldsymbol{K}''|\boldsymbol{x}) = l(T_{C_2}\cdots T_{C_m}\boldsymbol{K}''|\boldsymbol{x}) = \cdots = l(T_{C_m}\boldsymbol{K}''|\boldsymbol{x}) = l(\boldsymbol{K}''|\boldsymbol{x}) .$$

Now from (7.19) page 133 we obtain by reading (7.21) from right to left

$$\text{(7.22)} \qquad T_C\boldsymbol{K}'' = \boldsymbol{K}'' \quad \forall\, C \in \mathcal{C} ,$$

which means that $\boldsymbol{K}''$ is a fixpoint of T_C for all $C \in \mathcal{C}$. We know that (7.22) is equivalent to $\left(\boldsymbol{K}''^{-1}\right)_{CC} = \boldsymbol{s}_{CC}$ for all $C \in \mathcal{C}$, hence $\boldsymbol{K}''$ is a solution of the likelihood equations. Since the log-likelihood function is strictly concave we know that $\boldsymbol{K}''$ is the only solution, whereby $\boldsymbol{K}''$ must be the maximum likelihood estimate of $\boldsymbol{K}(\mathcal{G})$. We denote this by $\widehat{\boldsymbol{K}}(\mathcal{G})$, i.e. $\widehat{\boldsymbol{K}}(\mathcal{G}) = \boldsymbol{K}''$.

Thus we have shown that all convergent subsequences converge to the same point, namely $\widehat{\boldsymbol{K}}(\mathcal{G})$. This implies that $\{\boldsymbol{K}_s\}$ itself must converge to $\widehat{\boldsymbol{K}}(\mathcal{G})$, which completes the proof. ■

In the following we show that $\boldsymbol{s}(\mathcal{G})$ being $\mathcal{G}$-regular is a necessary and sufficient condition to ensure that the likelihood function is bounded from above.

First we assume that the likelihood function is bounded from above and we show by contradiction that $\boldsymbol{s}(\mathcal{G})$ is $\mathcal{G}$-regular. Therefore assume that $\boldsymbol{s}(\mathcal{G})$ is not $\mathcal{G}$-regular, whereby there exists a $\boldsymbol{D}_0 \in \mathbb{C}_S(\mathcal{G}) \setminus \{\boldsymbol{O}\}$ such that $\operatorname{tr}(\boldsymbol{D}_0\boldsymbol{s}(\mathcal{G})) \leq 0$. Obviously $\boldsymbol{K} + t\boldsymbol{D}_0$, where $\boldsymbol{K} \in \mathbb{C}_+(\mathcal{G})$ and $t \in \overline{\mathbb{R}}_+$, belongs to $\mathbb{C}_+(\mathcal{G})$. Similarly as in the proof above we observe

$$\begin{aligned} l(\boldsymbol{K} + t\boldsymbol{D}_0 | \boldsymbol{x}) &= -n|V|\log\pi + n\log(\det(\boldsymbol{K})) + n\sum_{j=1}^{|V|}\log(1 + td_j) \\ &\quad -n\operatorname{tr}(\boldsymbol{K}\boldsymbol{s}(\mathcal{G})) - nt\operatorname{tr}(\boldsymbol{D}_0\boldsymbol{s}(\mathcal{G})) \ , \end{aligned}$$

where the d_j's are the eigenvalues of $\boldsymbol{K}^{-1}\boldsymbol{D}_0$. Because $\boldsymbol{K}^{-1} > \boldsymbol{O}$ there exists a positive definite matrix $\boldsymbol{K}^{-\frac{1}{2}}$, such that $\boldsymbol{K}^{-1} = \left(\boldsymbol{K}^{-\frac{1}{2}}\right)^2$. Therefore the d_j's are also the eigenvalues of $\boldsymbol{K}^{-\frac{1}{2}}\boldsymbol{D}_0\boldsymbol{K}^{-\frac{1}{2}}$. Since $\boldsymbol{K}^{-\frac{1}{2}}\boldsymbol{D}_0\boldsymbol{K}^{-\frac{1}{2}}$ is Hermitian we know, that if all the d_j's are zero, then $\boldsymbol{K}^{-\frac{1}{2}}\boldsymbol{D}_0\boldsymbol{K}^{-\frac{1}{2}} = \boldsymbol{O}$. This is only fulfilled when $\boldsymbol{D}_0 = \boldsymbol{O}$, since $\boldsymbol{K}^{-\frac{1}{2}} > \boldsymbol{O}$. But it contradicts that $\boldsymbol{D}_0 \in \mathbb{C}_S(\mathcal{G}) \setminus \{\boldsymbol{O}\}$, whereby not all the d_j's are zero. Therefore we deduce that

$$\lim_{t\to\infty} l(\boldsymbol{K} + t\boldsymbol{D}_0 | \boldsymbol{x}) = \infty \ ,$$

which is a contradiction, i.e. $\boldsymbol{s}(\mathcal{G})$ is $\mathcal{G}$-regular.

Next we show that $\boldsymbol{s}(\mathcal{G})$ being $\mathcal{G}$-regular implies that the log-likelihood function is bounded from above. In the proof of Theorem 7.3 page 131 we have seen that if $\boldsymbol{s}(\mathcal{G})$ is $\mathcal{G}$-regular, then $\mathcal{K}$ is a compact set. From the definition of $\mathcal{K}$ we know that the log-likelihood function of $\boldsymbol{K} \in \mathbb{C}_+(\mathcal{G})$ attains the unique maximum in $\mathcal{K}$, if the maximum exists. Since the log-likelihood function is continuous and $\mathcal{K}$ is compact we deduce that the log-likelihood function attains the unique maximum, i.e. the log-likelihood function is bounded from above.

We conclude that $\boldsymbol{s}(\mathcal{G})$ being $\mathcal{G}$-regular is a necessary and sufficient condition to ensure the existence of the unique maximum likelihood estimate of $\boldsymbol{K}(\mathcal{G})$.

In practice when we solve the likelihood equations by the IPS-algorithm we stop the algorithm when the change from the elements in $\boldsymbol{K}_s$ to the elements in $\boldsymbol{K}_{s+1}$ is smaller than some predetermined constant. Besides we check if the inverse of $\boldsymbol{K}_{s+1}$ fulfills the likelihood equations on every clique. Furthermore one generally chooses $\boldsymbol{K}_0$ as $\boldsymbol{I}_{|V|}$, since it belongs to $\mathbb{C}_+(\mathcal{G})$ for all $\mathcal{G}$.

Example 7.4

Let $\mathcal{G} = (V, E)$ be the simple undirected graph with $V = \{1, 2, 3, 4\}$ and $E = \{\{1,2\}, \{1,4\}, \{2,3\}, \{3,4\}\}$. An illustration of $\mathcal{G}$ is shown in Figure 7.3 page 136.

Let $\boldsymbol{X}$ be an $n \times 4$ complex random matrix with $\mathcal{L}(\boldsymbol{X}) = \mathbb{C}\mathcal{N}_{n\times 4}\left(\boldsymbol{O}, \boldsymbol{I}_n \otimes \boldsymbol{K}(\mathcal{G})^{-1}\right)$, where $\boldsymbol{K}(\mathcal{G}) \in \mathbb{C}_+(\mathcal{G})$. Assume that we have an observation matrix of $\boldsymbol{X}$ which yields an observation matrix of $\boldsymbol{S}$ as

$$\boldsymbol{s} = \begin{pmatrix} 7 & 1+i2 & 8-i & 2 \\ 1-i2 & 5 & 5-i3 & 1-i \\ 8+i & 5+i3 & 17 & 4 \\ 2 & 1+i & 4 & 1 \end{pmatrix} . \tag{7.23}$$

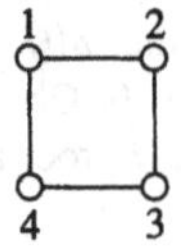

Figure 7.3: *Illustration of the simple undirected graph from Example 7.4.*

Note that $\boldsymbol{s}$ is positive definite which implies that $\boldsymbol{s}(\mathcal{G})$ is $\mathcal{G}$-regular, i.e. there exists a unique maximum likelihood estimate of $\boldsymbol{K}(\mathcal{G})$.

In this example we determine the unique maximum likelihood estimate, $\widehat{\boldsymbol{K}}(\mathcal{G})$, by means of the IPS-algorithm. The calculations have been performed by using the statistical package S-PLUS. Note that the values of the matrix entries have been rounded. According to the IPS-algorithm page 130 we let $C_1 = \{3,4\}$, $C_2 = \{2,3\}$, $C_3 = \{1,4\}$ and $C_4 = \{1,2\}$ be an ordering of the cliques in $\mathcal{G}$. We choose an arbitrary starting point as $K_0 = \boldsymbol{I}_4 \in \mathbb{C}_+(\mathcal{G})$ and we let $T = T_{C_1}T_{C_2}T_{C_3}T_{C_4}$. The IPS-updating equation for $s = 0$ is determined by

$$\boldsymbol{K}_1 = T\boldsymbol{K}_0 = T_{C_1}T_{C_2}T_{C_3}T_{C_4}\boldsymbol{I}_4 \,.$$

Using the C_4-marginal adjusting operator we get

$$T_{C_4}\boldsymbol{I}_4 = \begin{pmatrix} 0.17 & -0.033 - i0.067 & 0 & 0 \\ -0.033 + i0.067 & 0.23 & 0 & 0 \\ 0 & 0 & 1 & 0 \\ 0 & 0 & 0 & 1 \end{pmatrix}.$$

Note that this operator has adjusted $\boldsymbol{I}_4$ according to C_4. We continue by using the remaining C-marginal adjusting operators, whereby

$$T_{C_3}T_{C_4}\boldsymbol{I}_4 = \begin{pmatrix} 0.36 & -0.033 - i0.067 & 0 & -0.67 \\ -0.033 + i0.067 & 0.23 & 0 & 0 \\ 0 & 0 & 1 & 0 \\ -0.67 & 0 & 0 & 2.33 \end{pmatrix},$$

$$T_{C_2}T_{C_3}T_{C_4}\boldsymbol{I}_4 = \begin{pmatrix} 0.36 & -0.033 - i0.067 & 0 & -0.67 \\ -0.033 + i0.067 & 0.37 & -0.098 + i0.059 & 0 \\ 0 & -0.098 - i0.059 & 0.098 & 0 \\ -0.67 & 0 & 0 & 2.33 \end{pmatrix}$$

and finally

$$\boldsymbol{K}_1 = \begin{pmatrix} 0.36 & -0.033 - i0.067 & 0 & -0.67 \\ -0.033 + i0.067 & 0.37 & -0.098 + i0.059 & 0 \\ 0 & -0.098 - i0.059 & 1.04 & -3.96 - i0.024 \\ -0.67 & 0 & -3.96 + i0.024 & 18.3 \end{pmatrix}.$$

Thus after one iteration of the IPS-algorithm we get $\boldsymbol{K}_1$ as above. Note that $\boldsymbol{K}_1$ contains zero entries according to the missing edges in $\mathcal{G}$ as it is requested to. In order to be the unique maximum likelihood estimate of $\boldsymbol{K}(\mathcal{G})$ the complex matrix $\boldsymbol{K}_1$ must fulfill the likelihood equations. By matrix inversion we find

$$\boldsymbol{K}_1^{-1} = \begin{pmatrix} 7.70 & 2.95 + i2.97 & 8.77 + i0.62 & 2.18 + i0.15 \\ 2.95 - i2.97 & 5.68 & 5.46 - i4.27 & 1.29 - i1.02 \\ 8.77 - i0.62 & 5.46 + i4.27 & 17.0 & 4.0 \\ 2.18 - i0.15 & 1.29 + i1.02 & 4.0 & 1.0 \end{pmatrix}.$$

Obviously this matrix does not fulfill the likelihood equations on all cliques. After three iterations of the IPS-algorithm we get $\boldsymbol{K}_3$ as

$$\boldsymbol{K}_3 = \begin{pmatrix} 0.36 & 0.078 - i0.031 & 0 & -0.76 + i0.094 \\ 0.078 + i0.031 & 0.36 & -0.14 + i0.048 & 0 \\ 0 & -0.14 - i0.048 & 1.07 & -4.07 - i0.045 \\ -0.76 - i0.094 & 0 & -4.07 + i0.045 & 18.8 \end{pmatrix}$$

and further inverting $\boldsymbol{K}_3$ we observe

$$\boldsymbol{K}_3^{-1} = \begin{pmatrix} 6.81 & 1.55 + i1.66 & 7.74 + i0.043 & 1.96 - i0.006 \\ 1.55 - i1.66 & 5.00 & 5.00 - i2.99 & 1.15 - i0.71 \\ 7.74 - i0.043 & 5.00 + i2.99 & 17.0 & 4.0 \\ 1.96 + i0.006 & 1.15 + i0.71 & 4.0 & 1.0 \end{pmatrix}.$$

Normally one does not perform matrix inversion after each iteration. Here it is carried out to illustrate that $\boldsymbol{K}_3^{-1}$ does not fulfill the likelihood equations, and therefore further iterations are needed.

After 24 iterations of the IPS-algorithm the change of values of the matrix entries from $\boldsymbol{K}_{23}$ to $\boldsymbol{K}_{24}$ is on the fourth or fifth significant digit. Moreover by matrix inversion we get $\boldsymbol{K}_{24}^{-1}$ as

$$\boldsymbol{K}_{24}^{-1} = \begin{pmatrix} 7.00 & 1.00 + i2.00 & 7.81 + i0.14 & 2.00 \\ 1.00 - i2.00 & 5.00 & 5.00 - i3.00 & 1.11 - i0.72 \\ 7.81 - i0.14 & 5.00 + i3.00 & 17.00 & 4.00 \\ 2.00 & 1.11 + i0.72 & 4.00 & 1.00 \end{pmatrix}.$$

By comparing (7.23) page 135 with $\boldsymbol{K}_{24}^{-1}$ we observe that $\boldsymbol{K}_{24}^{-1}$ fulfills the likelihood equations for all $C \in \mathcal{C}$ at least to the third significant digit. Therefore we choose to accept $\boldsymbol{K}_{24}$ as the maximum likelihood estimate of $\boldsymbol{K}(\mathcal{G})$, i.e. $\widehat{\boldsymbol{K}}(\mathcal{G}) = \boldsymbol{K}_{24}$. ■

7.5 Decomposition of the Estimation Problem

Let $\mathcal{G} = (V, E)$ be a simple undirected graph and let $\boldsymbol{X}$ be an $n \times |V|$ complex random matrix with $\mathcal{L}(\boldsymbol{X}) = \mathbb{C}\mathcal{N}_{n\times|V|}\left(\boldsymbol{O}, \boldsymbol{I}_n \otimes \boldsymbol{K}(\mathcal{G})^{-1}\right)$, where $\boldsymbol{K}(\mathcal{G}) \in \mathbb{C}_+(\mathcal{G})$. Assume that A and B are subsets of V forming a decomposition of $\mathcal{G}$, i.e.

- $V = A \cup B$.
- $A \cap B$ is a complete subset of V.
- $A \setminus B$ and $B \setminus A$ are separated by $A \cap B$.

In this section we show that, when $\mathcal{G}$ has a decomposition, it is possible to decompose the estimation problem accordingly. Furthermore we consider the estimation problem, when $\mathcal{G}$ is decomposable.

Let $\boldsymbol{X}$ be partitioned according to the decomposition of $\mathcal{G}$ as

$$\boldsymbol{X} = \left(\boldsymbol{X}_{A\setminus B}, \boldsymbol{X}_{A\cap B}, \boldsymbol{X}_{B\setminus A}\right) ,$$

where $\boldsymbol{X}_{A\setminus B}, \boldsymbol{X}_{A\cap B}$ and $\boldsymbol{X}_{B\setminus A}$ have dimensions $n \times |A \setminus B|, n \times |A \cap B|$ and $n \times |B \setminus A|$, respectively. Further let $\boldsymbol{X}_A$ and $\boldsymbol{X}_B$ be the complex random matrices defined by means of $\boldsymbol{X}$ as

$$\boldsymbol{X}_A = \left(\boldsymbol{X}_{A\setminus B}, \boldsymbol{X}_{A\cap B}\right) \text{ and } \boldsymbol{X}_B = \left(\boldsymbol{X}_{A\cap B}, \boldsymbol{X}_{B\setminus A}\right) .$$

From Theorem 2.23 page 36 we have

$$\mathcal{L}\left(\boldsymbol{X}_A\right) = \mathbb{C}\mathcal{N}_{n\times|A|}\left(\boldsymbol{O}, \boldsymbol{I}_n \otimes \left(\boldsymbol{K}(\mathcal{G})^{-1}\right)_{AA}\right)$$

and

$$\mathcal{L}\left(\boldsymbol{X}_B\right) = \mathbb{C}\mathcal{N}_{n\times|B|}\left(\boldsymbol{O}, \boldsymbol{I}_n \otimes \left(\boldsymbol{K}(\mathcal{G})^{-1}\right)_{BB}\right) ,$$

where $\left(\boldsymbol{K}(\mathcal{G})^{-1}\right)_{AA}^{-1} \in \mathbb{C}_+(\mathcal{G}_A)$ and $\left(\boldsymbol{K}(\mathcal{G})^{-1}\right)_{BB}^{-1} \in \mathbb{C}_+(\mathcal{G}_B)$ according to the following lemma.

Lemma 7.4
Let $\mathcal{G} = (V, E)$ be a simple undirected graph and let $\boldsymbol{X}$ be an $n \times |V|$ complex random matrix with $\mathcal{L}\left(\boldsymbol{X}\right) = \mathbb{C}\mathcal{N}_{n\times|V|}\left(\boldsymbol{O}, \boldsymbol{I}_n \otimes \boldsymbol{K}(\mathcal{G})^{-1}\right)$, where $\boldsymbol{K}(\mathcal{G}) \in \mathbb{C}_+(\mathcal{G})$. Furthermore let A and B be subsets of V forming a decomposition of $\mathcal{G}$. The following properties hold.

1. $\boldsymbol{K}(\mathcal{G})_{A\setminus B, B\setminus A} = \boldsymbol{O}$.
2. $\left(\boldsymbol{K}(\mathcal{G})^{-1}\right)_{AA}^{-1} \in \mathbb{C}_+(\mathcal{G}_A)$.
3. $\left(\boldsymbol{K}(\mathcal{G})^{-1}\right)_{BB}^{-1} \in \mathbb{C}_+(\mathcal{G}_B)$.

Proof:
Let $\mathcal{G} = (V, E)$ be a simple undirected graph and let $\boldsymbol{X}$ be an $n \times |V|$ complex random matrix

with $\mathcal{L}(\boldsymbol{X}) = \mathbb{C}\mathcal{N}_{n\times|V|}\big(\boldsymbol{O}, \boldsymbol{I}_n \otimes \boldsymbol{K}(\mathcal{G})^{-1}\big)$, where $\boldsymbol{K}(\mathcal{G}) \in \mathbb{C}_+(\mathcal{G})$. Furthermore let A and B be subsets of V forming a decomposition of $\mathcal{G}$ and let $\boldsymbol{X}_A$ and $\boldsymbol{X}_B$ be defined as mentioned before the lemma.

Re 1:
Since A and B form a decomposition of $\mathcal{G}$ there are no edges between elements in $A \setminus B$ and elements in $B\setminus A$. Since $\boldsymbol{K}(\mathcal{G}) \in \mathbb{C}_+(\mathcal{G})$, we therefore have $\boldsymbol{K}(\mathcal{G})_{A\setminus B, B\setminus A} = \boldsymbol{K}(\mathcal{G})_{B\setminus A, A\setminus B} = \boldsymbol{O}$.

Re 2:
From page 120 we know that the density function of each column of $\boldsymbol{X}^*$ w.r.t. Lebesgue measure on $\mathbb{C}^{|V|}$ of each column of $\boldsymbol{X}^*$ is positive and continuous and that the distribution factorizes according to $\mathcal{G}$. Using Theorem 6.9 page 112 this means that the distribution of each column of $\boldsymbol{X}^*_A$ factorizes according to $\mathcal{G}_A$. Theorem 6.7 page 109 hereby tells us that the distribution of each column of $\boldsymbol{X}^*_A$ has the pairwise Markov property, which means that $\left(\boldsymbol{K}(\mathcal{G})^{-1}\right)^{-1}_{AA} \in \mathbb{C}_+(\mathcal{G}_A)$.

Re 3:
By similar arguments as in part 2 we find that $\left(\boldsymbol{K}(\mathcal{G})^{-1}\right)^{-1}_{BB} \in \mathbb{C}_+(\mathcal{G}_B)$. ■

It holds that $\boldsymbol{S}$ is given by

$$\boldsymbol{S} = \frac{1}{n}\boldsymbol{X}^*\boldsymbol{X} = \frac{1}{n}\begin{pmatrix} \boldsymbol{X}^*_{A\setminus B}\boldsymbol{X}_{A\setminus B} & \boldsymbol{X}^*_{A\setminus B}\boldsymbol{X}_{A\cap B} & \boldsymbol{X}^*_{A\setminus B}\boldsymbol{X}_{B\setminus A} \\ \boldsymbol{X}^*_{A\cap B}\boldsymbol{X}_{A\setminus B} & \boldsymbol{X}^*_{A\cap B}\boldsymbol{X}_{A\cap B} & \boldsymbol{X}^*_{A\cap B}\boldsymbol{X}_{B\setminus A} \\ \boldsymbol{X}^*_{B\setminus A}\boldsymbol{X}_{A\setminus B} & \boldsymbol{X}^*_{B\setminus A}\boldsymbol{X}_{A\cap B} & \boldsymbol{X}^*_{B\setminus A}\boldsymbol{X}_{B\setminus A} \end{pmatrix}.$$

From the definitions of $\boldsymbol{X}_A$ and $\boldsymbol{X}_B$ we get that

$$\boldsymbol{S}_{AA} = \frac{1}{n}\begin{pmatrix} \boldsymbol{X}^*_{A\setminus B}\boldsymbol{X}_{A\setminus B} & \boldsymbol{X}^*_{A\setminus B}\boldsymbol{X}_{A\cap B} \\ \boldsymbol{X}^*_{A\cap B}\boldsymbol{X}_{A\setminus B} & \boldsymbol{X}^*_{A\cap B}\boldsymbol{X}_{A\cap B} \end{pmatrix} = \frac{1}{n}\boldsymbol{X}^*_A\boldsymbol{X}_A$$

and

$$\boldsymbol{S}_{BB} = \frac{1}{n}\begin{pmatrix} \boldsymbol{X}^*_{A\cap B}\boldsymbol{X}_{A\cap B} & \boldsymbol{X}^*_{A\cap B}\boldsymbol{X}_{B\setminus A} \\ \boldsymbol{X}^*_{B\setminus A}\boldsymbol{X}_{A\cap B} & \boldsymbol{X}^*_{B\setminus A}\boldsymbol{X}_{B\setminus A} \end{pmatrix} = \frac{1}{n}\boldsymbol{X}^*_B\boldsymbol{X}_B.$$

Thus marginals in $\boldsymbol{S}$ can be found directly from marginals in $\boldsymbol{X}$.

Similar to the definition of $\boldsymbol{S}(\mathcal{G})$ in (7.6) page 122 we define by means of $\boldsymbol{S}_{AA}$ and $\boldsymbol{S}_{BB}$ the Hermitian matrices $\boldsymbol{S}(\mathcal{G}_A)$ and $\boldsymbol{S}(\mathcal{G}_B)$ as

$$\boldsymbol{S}(\mathcal{G}_A)_{\alpha\beta} = \begin{cases} 0 & \text{if } \{\alpha,\beta\} \notin E_A \text{ for } \alpha \neq \beta \in A \\ (\boldsymbol{S}_{AA})_{\alpha\beta} & \text{otherwise} \end{cases}$$

and

$$\boldsymbol{S}(\mathcal{G}_B)_{\alpha\beta} = \begin{cases} 0 & \text{if } \{\alpha,\beta\} \notin E_B \text{ for } \alpha \neq \beta \in B \\ (\boldsymbol{S}_{BB})_{\alpha\beta} & \text{otherwise} \end{cases},$$

where $\boldsymbol{S}(\mathcal{G}_A)$ and $\boldsymbol{S}(\mathcal{G}_B)$ have dimensions $|A| \times |A|$ and $|B| \times |B|$, respectively. Likewise the observation matrices $\boldsymbol{s}_{AA}$, $\boldsymbol{s}_{BB}$, $\boldsymbol{s}(\mathcal{G}_A)$ and $\boldsymbol{s}(\mathcal{G}_B)$ are defined.

Assume that $s(\mathcal{G})$ is $\mathcal{G}$-regular, then $s(\mathcal{G}_A)$ and $s(\mathcal{G}_B)$ are $\mathcal{G}_A$-regular and $\mathcal{G}_B$-regular, respectively, i.e. the unique marginal maximum likelihood estimates of $\boldsymbol{K}(\mathcal{G}_A) \in \mathbb{C}_+(\mathcal{G}_A)$ and $\boldsymbol{K}(\mathcal{G}_B) \in \mathbb{C}_+(\mathcal{G}_B)$ exist. Define

$$\widetilde{\boldsymbol{K}}(\mathcal{G}) = [\widehat{\boldsymbol{K}}(\mathcal{G}_A)] + [\widehat{\boldsymbol{K}}(\mathcal{G}_B)] - [\boldsymbol{s}_{A\cap B,A\cap B}^{-1}], \tag{7.24}$$

where $\widehat{\boldsymbol{K}}(\mathcal{G}_A)$ and $\widehat{\boldsymbol{K}}(\mathcal{G}_B)$ denote the unique marginal maximum likelihood estimates of $\boldsymbol{K}(\mathcal{G}_A) \in \mathbb{C}_+(\mathcal{G}_A)$ and $\boldsymbol{K}(\mathcal{G}_B) \in \mathbb{C}_+(\mathcal{G}_B)$, respectively. Then $\widehat{\boldsymbol{K}}(\mathcal{G}_A)$ satisfies the likelihood equations $\left(\boldsymbol{K}(\mathcal{G}_A)^{-1}\right)_{CC} = \boldsymbol{s}_{CC}$ for all cliques C in $\mathcal{G}_A$, and $\widehat{\boldsymbol{K}}(\mathcal{G}_B)$ satisfies the likelihood equations $\left(\boldsymbol{K}(\mathcal{G}_B)^{-1}\right)_{CC} = \boldsymbol{s}_{CC}$ for all cliques C in $\mathcal{G}_B$. Obviously $\widehat{\boldsymbol{K}}(\mathcal{G}_A) \in \mathbb{C}_+(\mathcal{G}_A)$ and $\widehat{\boldsymbol{K}}(\mathcal{G}_B) \in \mathbb{C}_+(\mathcal{G}_B)$. Further, since $s(\mathcal{G})$ is $\mathcal{G}$-regular and $A \cap B$ is complete, we know that $\boldsymbol{s}_{A\cap B,A\cap B} > \boldsymbol{O}$, therefore $\boldsymbol{s}_{A\cap B,A\cap B}^{-1}$ exists. In Theorem 7.4 page 142 we show that $\widetilde{\boldsymbol{K}}(\mathcal{G})$ is the unique maximum likelihood estimate of $\boldsymbol{K}(\mathcal{G})$, but in order to prove this we need the following lemma.

Lemma 7.5
Let $\mathcal{G} = (V, E)$ be a simple undirected graph. Let $\boldsymbol{X}$ be an $n \times |V|$ complex random matrix with $\mathcal{L}(\boldsymbol{X}) = \mathbb{C}\mathcal{N}_{n\times|V|}\left(\boldsymbol{O}, \boldsymbol{I}_n \otimes \boldsymbol{K}(\mathcal{G})^{-1}\right)$, where $\boldsymbol{K}(\mathcal{G}) \in \mathbb{C}_+(\mathcal{G})$, and let $\boldsymbol{x}$ be an observation matrix of $\boldsymbol{X}$. Furthermore let $\boldsymbol{s} = \frac{1}{n}\boldsymbol{x}^*\boldsymbol{x}$ and let $s(\mathcal{G})$ defined in (7.7) page 122 be $\mathcal{G}$-regular. Finally let A and B be subsets of V forming a decomposition of $\mathcal{G}$ and let $\widetilde{\boldsymbol{K}}(\mathcal{G})$ be given as in (7.24) page 140. The following properties hold.

1. $\widetilde{\boldsymbol{K}}(\mathcal{G}) \in \mathbb{C}_+(\mathcal{G})$.

2. $\det\left(\widetilde{\boldsymbol{K}}(\mathcal{G})\right) = \det\left(\widehat{\boldsymbol{K}}(\mathcal{G}_A)\right) \det\left(\widehat{\boldsymbol{K}}(\mathcal{G}_B)\right) \det\left(\boldsymbol{s}_{A\cap B,A\cap B}\right)$.

Proof:
Let the assumptions in the lemma be satisfied.

Re 1:
By the definition of $\widetilde{\boldsymbol{K}}(\mathcal{G})$ we observe that if $\{\alpha, \beta\} \notin E$, then $\widetilde{\boldsymbol{K}}(\mathcal{G})_{\alpha\beta} = 0$ for all $\alpha \neq \beta \in V$. Therefore we only have to prove that $\widetilde{\boldsymbol{K}}(\mathcal{G}) > \boldsymbol{O}$. Let $\widehat{\boldsymbol{K}}(\mathcal{G}_A)$ and $\widehat{\boldsymbol{K}}(\mathcal{G}_B)$ be partitioned as

$$\widehat{\boldsymbol{K}}(\mathcal{G}_A) = \begin{pmatrix} \widehat{\boldsymbol{K}}(\mathcal{G}_A)_{A\setminus B,A\setminus B} & \widehat{\boldsymbol{K}}(\mathcal{G}_A)_{A\setminus B,A\cap B} \\ \widehat{\boldsymbol{K}}(\mathcal{G}_A)_{A\cap B,A\setminus B} & \widehat{\boldsymbol{K}}(\mathcal{G}_A)_{A\cap B,A\cap B} \end{pmatrix} \tag{7.25}$$

and

$$\widehat{\boldsymbol{K}}(\mathcal{G}_B) = \begin{pmatrix} \widehat{\boldsymbol{K}}(\mathcal{G}_B)_{A\cap B,A\cap B} & \widehat{\boldsymbol{K}}(\mathcal{G}_B)_{A\cap B,B\setminus A} \\ \widehat{\boldsymbol{K}}(\mathcal{G}_B)_{B\setminus A,A\cap B} & \widehat{\boldsymbol{K}}(\mathcal{G}_B)_{B\setminus A,B\setminus A} \end{pmatrix}. \tag{7.26}$$

Rearranging the elements of $\widetilde{K}(\mathcal{G})$ we get

$$\widetilde{K}(\mathcal{G}) = \left(\begin{array}{cc|c} \widehat{K}(\mathcal{G}_A)_{A\setminus B,A\setminus B} & O & \widehat{K}(\mathcal{G}_A)_{A\setminus B,A\cap B} \\ O & \widehat{K}(\mathcal{G}_B)_{B\setminus A,B\setminus A} & \widehat{K}(\mathcal{G}_B)_{B\setminus A,A\cap B} \\ \hline \widehat{K}(\mathcal{G}_A)_{A\cap B,A\setminus B} & \widehat{K}(\mathcal{G}_B)_{A\cap B,B\setminus A} & \widehat{K}(\mathcal{G}_A)_{A\cap B,A\cap B} + \widehat{K}(\mathcal{G}_B)_{A\cap B,A\cap B} - s^{-1}_{A\cap B,A\cap B} \end{array}\right)$$

$$= \left(\begin{array}{c|c} \widetilde{K}(\mathcal{G})_{11} & \widetilde{K}(\mathcal{G})_{12} \\ \hline \widetilde{K}(\mathcal{G})_{21} & \widetilde{K}(\mathcal{G})_{22} \end{array}\right).$$

If we can show that $\widetilde{K}(\mathcal{G})_{11} > O$ and $\widetilde{K}(\mathcal{G})_{22} - \widetilde{K}(\mathcal{G})_{21}\,\widetilde{K}(\mathcal{G})_{11}^{-1}\,\widetilde{K}(\mathcal{G})_{12} > O$, then it follows that $\widetilde{K}(\mathcal{G}) > O$. Since $\widehat{K}(\mathcal{G}_A)_{A\setminus B,A\setminus B} > O$ and $\widehat{K}(\mathcal{G}_B)_{B\setminus A,B\setminus A} > O$ we conclude that $\widetilde{K}(\mathcal{G})_{11} > O$. Furthermore

$$\begin{aligned} \widetilde{K}(\mathcal{G})_{22} - \widetilde{K}(\mathcal{G})_{21}\,\widetilde{K}(\mathcal{G})_{11}^{-1}\,\widetilde{K}(\mathcal{G})_{12} &= \widehat{K}(\mathcal{G}_A)_{A\cap B,A\cap B} + \widehat{K}(\mathcal{G}_B)_{A\cap B,A\cap B} - s^{-1}_{A\cap B,A\cap B} \\ &\quad - \widehat{K}(\mathcal{G}_A)_{A\cap B,A\setminus B}\,\widehat{K}(\mathcal{G}_A)^{-1}_{A\setminus B,A\setminus B}\,\widehat{K}(\mathcal{G}_A)_{A\setminus B,A\cap B} \\ &\quad - \widehat{K}(\mathcal{G}_B)_{A\cap B,B\setminus A}\,\widehat{K}(\mathcal{G}_B)^{-1}_{B\setminus A,B\setminus A}\,\widehat{K}(\mathcal{G}_B)_{B\setminus A,A\cap B}. \end{aligned}$$

Considering the partition of $\widehat{K}(\mathcal{G}_A)$ and $\widehat{K}(\mathcal{G}_B)$ in (7.25) and (7.26) page 140 we get

(7.27)
$$\left(\widehat{K}(\mathcal{G}_A)^{-1}\right)^{-1}_{A\cap B,A\cap B} = \widehat{K}(\mathcal{G}_A)_{A\cap B,A\cap B} - \widehat{K}(\mathcal{G}_A)_{A\cap B,A\setminus B}\,\widehat{K}(\mathcal{G}_A)^{-1}_{A\setminus B,A\setminus B}\,\widehat{K}(\mathcal{G}_A)_{A\setminus B,A\cap B}$$

and

(7.28)
$$\left(\widehat{K}(\mathcal{G}_B)^{-1}\right)^{-1}_{A\cap B,A\cap B} = \widehat{K}(\mathcal{G}_B)_{A\cap B,A\cap B} - \widehat{K}(\mathcal{G}_B)_{A\cap B,B\setminus A}\,\widehat{K}(\mathcal{G}_B)^{-1}_{B\setminus A,B\setminus A}\,\widehat{K}(\mathcal{G}_B)_{B\setminus A,A\cap B}.$$

Thus we have

$$\widetilde{K}(\mathcal{G})_{22} - \widetilde{K}(\mathcal{G})_{21}\,\widetilde{K}(\mathcal{G})_{11}^{-1}\,\widetilde{K}(\mathcal{G})_{12} = \left(\widehat{K}(\mathcal{G}_A)^{-1}\right)^{-1}_{A\cap B,A\cap B} + \left(\widehat{K}(\mathcal{G}_B)^{-1}\right)^{-1}_{A\cap B,A\cap B} - s^{-1}_{A\cap B,A\cap B}.$$

Since $\widehat{K}(\mathcal{G}_A)$ is the unique maximum likelihood estimate of $K(\mathcal{G}_A)$ it satisfies the likelihood equations for every complete subset of A. In particular we have that $A\cap B$ is a complete subset of A, whereby it holds that $\left(\widehat{K}(\mathcal{G}_A)^{-1}\right)_{A\cap B,A\cap B} = s_{A\cap B,A\cap B}$. In a similar way we argue that $\left(\widehat{K}(\mathcal{G}_B)^{-1}\right)_{A\cap B,A\cap B} = s_{A\cap B,A\cap B}$. Thus we conclude, since $s(\mathcal{G})$ is $\mathcal{G}$-regular, that

(7.29)
$$\widetilde{K}(\mathcal{G})_{22} - \widetilde{K}(\mathcal{G})_{21}\,\widetilde{K}(\mathcal{G})_{11}^{-1}\,\widetilde{K}(\mathcal{G})_{12} = s^{-1}_{A\cap B,A\cap B} > O\,.$$

Altogether we have shown $\widetilde{K}(\mathcal{G}) \in \mathbb{C}_+(\mathcal{G})$.

Re 2:
Using the partition of $\widetilde{K}(\mathcal{G})$ in part 1 and (7.29) page 141 we obtain

$$\begin{aligned} \det\left(\widetilde{K}(\mathcal{G})\right) &= \det\left(\widetilde{K}(\mathcal{G})_{11}\right)\det\left(\widetilde{K}(\mathcal{G})_{22} - \widetilde{K}(\mathcal{G})_{21}\widetilde{K}(\mathcal{G})_{11}^{-1}\widetilde{K}(\mathcal{G})_{12}\right) \\ &= \det\left(\widehat{K}(\mathcal{G}_A)_{A\setminus B,A\setminus B}\right)\det\left(\widehat{K}(\mathcal{G}_B)_{B\setminus A,B\setminus A}\right)\det\left(s_{A\cap B,A\cap B}^{-1}\right). \end{aligned} \tag{7.30}$$

It follows from (7.25) page 140 and (7.27) page 141 that

$$\begin{aligned} \det\left(\widehat{K}(\mathcal{G}_A)\right) &= \det\left(\widehat{K}(\mathcal{G}_A)_{A\setminus B,A\setminus B}\right)\det\left(\left(\widehat{K}(\mathcal{G}_A)^{-1}\right)_{A\cap B,A\cap B}^{-1}\right) \\ &= \det\left(\widehat{K}(\mathcal{G}_A)_{A\setminus B,A\setminus B}\right)\det\left(s_{A\cap B,A\cap B}\right)^{-1}, \end{aligned}$$

where $\det\left(s_{A\cap B,A\cap B}\right) > 0$, since $s_{A\cap B,A\cap B} > O$. Similarly from (7.26) page 140 and (7.28) page 141 we have

$$\det\left(\widehat{K}(\mathcal{G}_B)\right) = \det\left(\widehat{K}(\mathcal{G}_B)_{B\setminus A,B\setminus A}\right)\det\left(s_{A\cap B,A\cap B}\right)^{-1}.$$

From (7.30) we then obtain

$$\det\left(\widetilde{K}(\mathcal{G})\right) = \det\left(\widehat{K}(\mathcal{G}_A)\right)\det\left(\widehat{K}(\mathcal{G}_B)\right)\det\left(s_{A\cap B,A\cap B}\right).$$

■

Using Lemma 7.4 page 138 and Lemma 7.5 page 140 we are able to prove that $\widetilde{K}(\mathcal{G})$ is the unique maximum likelihood estimate of $K(\mathcal{G})$.

Theorem 7.4
Let $\mathcal{G} = (V, E)$ be a simple undirected graph. Let X be an $n \times |V|$ complex random matrix with $\mathcal{L}(X) = \mathbb{C}\mathcal{N}_{n\times|V|}\left(O, I_n \otimes K(\mathcal{G})^{-1}\right)$, where $K(\mathcal{G}) \in \mathbb{C}_+(\mathcal{G})$, and let x be an observation matrix of X. Furthermore let $s = \frac{1}{n}x^*x$ and let $s(\mathcal{G})$ defined in (7.7) page 122 be $\mathcal{G}$-regular. Finally let A and B be subsets of V forming a decomposition of $\mathcal{G}$. The unique maximum likelihood estimate of $K(\mathcal{G})$ is given by

$$\widehat{K}(\mathcal{G}) = [\widehat{K}(\mathcal{G}_A)] + [\widehat{K}(\mathcal{G}_B)] - [s_{A\cap B,A\cap B}^{-1}]$$

and it holds that

$$\det\left(\widehat{K}(\mathcal{G})\right) = \det\left(\widehat{K}(\mathcal{G}_A)\right)\det\left(\widehat{K}(\mathcal{G}_B)\right)\det\left(s_{A\cap B,A\cap B}\right),$$

where $\widehat{K}(\mathcal{G}_A)$ and $\widehat{K}(\mathcal{G}_B)$ are the unique marginal maximum likelihood estimates of $K(\mathcal{G}_A) \in \mathbb{C}_+(\mathcal{G}_A)$ and $K(\mathcal{G}_B) \in \mathbb{C}_+(\mathcal{G}_B)$, respectively.

Proof:
Let the assumptions in the theorem be satisfied and let X_A and X_B be defined as on page 138.

From page 120 we know that the density function of each column of X^* w.r.t Lebesgue measure on $\mathbb{C}^{|V|}$ is positive and continuous and that the distribution of each column of X^* factorizes according to $\mathcal{G}$. Using Theorem 6.9 page 112 together with the mutual independence of the columns of $\boldsymbol{X}^*$ we get

$$f_{\boldsymbol{X}}(\boldsymbol{x}) = \frac{f_{\boldsymbol{X}_A}(\boldsymbol{x}_A)\, f_{\boldsymbol{X}_B}(\boldsymbol{x}_B)}{f_{\boldsymbol{X}_{A\cap B}}(\boldsymbol{x}_{A\cap B})} .$$

Therefore, according to the marginal distributions, the likelihood function of $\boldsymbol{K}(\mathcal{G})$ becomes

$$L(\boldsymbol{K}(\mathcal{G})|\,\boldsymbol{x}) = \frac{L\left(\left(\boldsymbol{K}(\mathcal{G})^{-1}\right)_{AA}^{-1}\Big|\,\boldsymbol{x}_A\right) L\left(\left(\boldsymbol{K}(\mathcal{G})^{-1}\right)_{BB}^{-1}\Big|\,\boldsymbol{x}_B\right)}{L\left(\left(\boldsymbol{K}(\mathcal{G})^{-1}\right)_{A\cap B,A\cap B}^{-1}\Big|\,\boldsymbol{x}_{A\cap B}\right)} .$$

Let $\widehat{\boldsymbol{K}}(\mathcal{G})$ be the unique maximum likelihood estimate of $\boldsymbol{K}(\mathcal{G}) \in \mathbb{C}_+(\mathcal{G})$, then the likelihood function of $\widehat{\boldsymbol{K}}(\mathcal{G})$ becomes

$$L\left(\widehat{\boldsymbol{K}}(\mathcal{G})\Big|\,\boldsymbol{x}\right) = \frac{L\left(\left(\widehat{\boldsymbol{K}}(\mathcal{G})^{-1}\right)_{AA}^{-1}\Big|\,\boldsymbol{x}_A\right) L\left(\left(\widehat{\boldsymbol{K}}(\mathcal{G})^{-1}\right)_{BB}^{-1}\Big|\,\boldsymbol{x}_B\right)}{L\left(\left(\widehat{\boldsymbol{K}}(\mathcal{G})^{-1}\right)_{A\cap B,A\cap B}^{-1}\Big|\,\boldsymbol{x}_{A\cap B}\right)} .$$

Furthermore let $\widehat{\boldsymbol{K}}(\mathcal{G}_A)$ and $\widehat{\boldsymbol{K}}(\mathcal{G}_B)$ denote the unique marginal maximum likelihood estimates of $\boldsymbol{K}(\mathcal{G}_A) \in \mathbb{C}_+(\mathcal{G}_A)$ and $\boldsymbol{K}(\mathcal{G}_B) \in \mathbb{C}_+(\mathcal{G}_B)$. From Lemma 7.4 page 138 we have that $\left(\widehat{\boldsymbol{K}}(\mathcal{G})^{-1}\right)_{AA}^{-1} \in \mathbb{C}_+(\mathcal{G}_A)$ and $\left(\widehat{\boldsymbol{K}}(\mathcal{G})^{-1}\right)_{BB}^{-1} \in \mathbb{C}_+(\mathcal{G}_B)$, therefore

$$L\left(\widehat{\boldsymbol{K}}(\mathcal{G}_A)\Big|\,\boldsymbol{x}_A\right) \geq L\left(\left(\widehat{\boldsymbol{K}}(\mathcal{G})^{-1}\right)_{AA}^{-1}\Big|\,\boldsymbol{x}_A\right)$$

and

$$L\left(\widehat{\boldsymbol{K}}(\mathcal{G}_B)\Big|\,\boldsymbol{x}_B\right) \geq L\left(\left(\widehat{\boldsymbol{K}}(\mathcal{G})^{-1}\right)_{BB}^{-1}\Big|\,\boldsymbol{x}_B\right) .$$

Moreover since $A \cap B$ is a complete subset of V we know that $\widehat{\boldsymbol{K}}(\mathcal{G})$ satisfies the likelihood equations for $A\cap B$, therefore $\left(\widehat{\boldsymbol{K}}(\mathcal{G})^{-1}\right)_{A\cap B,A\cap B} = \boldsymbol{s}_{A\cap B,A\cap B}$. Hereby we obtain the following inequality

$$L\left(\widehat{\boldsymbol{K}}(\mathcal{G})\Big|\,\boldsymbol{x}\right) \leq \frac{L\left(\widehat{\boldsymbol{K}}(\mathcal{G}_A)\Big|\,\boldsymbol{x}_A\right) L\left(\widehat{\boldsymbol{K}}(\mathcal{G}_B)\Big|\,\boldsymbol{x}_B\right)}{L\left(\boldsymbol{s}_{A\cap B,A\cap B}^{-1}\Big|\,\boldsymbol{x}_{A\cap B}\right)} .$$

From the density function of $\boldsymbol{X}_A$ w.r.t. Lebesgue measure on $\mathbb{C}^{n\times|A|}$, which can be found as in (7.8) page 122, and $\boldsymbol{s}_{AA} = \frac{1}{n}\boldsymbol{x}_A^*\boldsymbol{x}_A$ the likelihood function of $\boldsymbol{K}(\mathcal{G}_A)$ is given by

$$L(\boldsymbol{K}(\mathcal{G}_A)|\,\boldsymbol{x}_A) = \pi^{-n|A|}\det(\boldsymbol{K}(\mathcal{G}_A))^n \exp\left(-n\operatorname{tr}(\boldsymbol{K}(\mathcal{G}_A)\,\boldsymbol{s}_{AA})\right) .$$

Inserting $\widehat{\boldsymbol{K}}(\mathcal{G}_A)$ we find

$$L\left(\widehat{\boldsymbol{K}}(\mathcal{G}_A)\Big|\,\boldsymbol{x}_A\right) = \pi^{-n|A|}\det\left(\widehat{\boldsymbol{K}}(\mathcal{G}_A)\right)^n \exp\left(-n\operatorname{tr}\left(\widehat{\boldsymbol{K}}(\mathcal{G}_A)\,\boldsymbol{s}_{AA}\right)\right) .$$

Using that $\{\alpha, \beta\} \in E_A$ implies $\left(\widehat{K}(\mathcal{G}_A)^{-1}\right)_{\beta\alpha} = s_{\beta\alpha}$ we deduce

$$\begin{aligned}
\operatorname{tr}\left(\widehat{K}(\mathcal{G}_A)\, s_{AA}\right) &= \sum_{\alpha}\left(\widehat{K}(\mathcal{G}_A)\, s_{AA}\right)_{\alpha\alpha} \\
&= \sum_{\alpha,\beta} \widehat{K}(\mathcal{G}_A)_{\alpha\beta}\, s_{\beta\alpha} \\
&= \sum_{\alpha\sim\beta} \widehat{K}(\mathcal{G}_A)_{\alpha\beta}\, s_{\beta\alpha} + \sum_{\alpha\not\sim\beta} \widehat{K}(\mathcal{G}_A)_{\alpha\beta}\, s_{\beta\alpha} \\
&= \sum_{\alpha\sim\beta} \widehat{K}(\mathcal{G}_A)_{\alpha\beta} \left(\widehat{K}(\mathcal{G}_A)^{-1}\right)_{\beta\alpha} \\
&= \sum_{\alpha,\beta} \widehat{K}(\mathcal{G}_A)_{\alpha\beta} \left(\widehat{K}(\mathcal{G}_A)^{-1}\right)_{\beta\alpha} \\
&= \sum_{\alpha}\left(\widehat{K}(\mathcal{G}_A)\, \widehat{K}(\mathcal{G}_A)^{-1}\right)_{\alpha\alpha} \\
&= \sum_{\alpha}\left(I_{|A|}\right)_{\alpha\alpha} \\
&= |A| \,.
\end{aligned}$$

Hereby we obtain the maximum value of the likelihood function as

$$L\left(\widehat{K}(\mathcal{G}_A)\middle|\, x_A\right) = \pi^{-n|A|} \det\left(\widehat{K}(\mathcal{G}_A)\right)^n \exp\left(-n|A|\right) \,.$$

Analogous calculations on the likelihood function of $K(\mathcal{G}_B)$ give us the maximum value of this function as

$$L\left(\widehat{K}(\mathcal{G}_B)\middle|\, x_B\right) = \pi^{-n|B|} \det\left(\widehat{K}(\mathcal{G}_B)\right)^n \exp\left(-n|B|\right) \,.$$

Finally we easily obtain that

$$L\left(s_{A\cap B,A\cap B}^{-1}\middle|\, x_{A\cap B}\right) = \pi^{-n|A\cap B|} \det\left(s_{A\cap B,A\cap B}^{-1}\right)^n \exp\left(-n|A\cap B|\right) \,.$$

Combining these results leads to

$$L\left(\widehat{K}(\mathcal{G})\middle|\, x\right) \le \frac{\pi^{-n|A|} \det\left(\widehat{K}(\mathcal{G}_A)\right)^n \exp\left(-n|A|\right) \pi^{-n|B|} \det\left(\widehat{K}(\mathcal{G}_B)\right)^n \exp\left(-n|B|\right)}{\pi^{-n|A\cap B|} \det\left(s_{A\cap B,A\cap B}^{-1}\right)^n \exp\left(-n|A\cap B|\right)} .$$

Let $\widetilde{K}(\mathcal{G}) = [\![\widehat{K}(\mathcal{G}_A)]\!] + [\![\widehat{K}(\mathcal{G}_B)]\!] - [\![s_{A\cap B,A\cap B}^{-1}]\!]$, then by using Lemma 7.5 page 140 we get

$$\begin{aligned}
L\left(\widehat{K}(\mathcal{G})\middle|\, x\right) &\le \pi^{-n|V|} \det\left(\widetilde{K}(\mathcal{G})\right)^n \exp\left(-n|V|\right) \\
&= L\left(\widetilde{K}(\mathcal{G})\middle|\, x\right) .
\end{aligned}$$

Also from Lemma 7.5 we know that $\widetilde{K}(\mathcal{G}) \in \mathbb{C}_+(\mathcal{G})$, hence we conclude that $L\left(K(\mathcal{G})|\, x\right)$ is maximized uniquely, when $\widehat{K}(\mathcal{G}) = \widetilde{K}(\mathcal{G})$, i.e. when $\widehat{K}(\mathcal{G}) = [\![\widehat{K}(\mathcal{G}_A)]\!] + [\![\widehat{K}(\mathcal{G}_B)]\!] - [\![s_{A\cap B,A\cap B}^{-1}]\!]$.

The last part of the theorem follows immediately from Lemma 7.5. ■

From Theorem 7.4 we observe that the estimation problem can be reduced when $\mathcal{G}$ has a decomposition. The estimations are then performed in the A- and B-marginals and $s_{A\cap B,A\cap B}^{-1}$ can be found directly from $x_{A\cap B}$.

7.5.1 Estimation in Complex Normal Decomposable Models

A complex normal graphical model w.r.t. a simple undirected graph $\mathcal{G}$, where $\mathcal{G}$ in addition is decomposable is called a *complex normal decomposable model* w.r.t. $\mathcal{G}$. The following theorem says that, when $\mathcal{G}$ is decomposable, we can find an explicit expression of the unique maximum likelihood estimate of $\boldsymbol{K}(\mathcal{G})$.

Theorem 7.5
Let $\mathcal{G} = (V, E)$ be a simple undirected decomposable graph. Let $\boldsymbol{X}$ be an $n \times |V|$ complex random matrix with $\mathcal{L}(\boldsymbol{X}) = \mathbb{C}\mathcal{N}_{n\times|V|}\left(\boldsymbol{O}, \boldsymbol{I}_n \otimes \boldsymbol{K}(\mathcal{G})^{-1}\right)$, where $\boldsymbol{K}(\mathcal{G}) \in \mathbb{C}_+(\mathcal{G})$, and let $\boldsymbol{x}$ be an observation matrix of $\boldsymbol{X}$. Furthermore let $\boldsymbol{s} = \frac{1}{n}\boldsymbol{x}^*\boldsymbol{x}$ and let $\boldsymbol{s}(\mathcal{G})$ defined in (7.7) page 122 be $\mathcal{G}$-regular. Finally let $C_1, C_2, \ldots, C_m$ be a RIP-ordering of the cliques in $\mathcal{G}$. The unique maximum likelihood estimate of $\boldsymbol{K}(\mathcal{G})$ is given by

$$\widehat{\boldsymbol{K}}(\mathcal{G}) = \sum_{j=1}^{m} [\boldsymbol{s}_{C_jC_j}^{-1}] - \sum_{j=2}^{m} [\boldsymbol{s}_{C_j\cap D_j, C_j\cap D_j}^{-1}]$$

and it holds that

$$\det\left(\widehat{\boldsymbol{K}}(\mathcal{G})\right) = \frac{\prod_{j=2}^{m} \det\left(\boldsymbol{s}_{C_j\cap D_j, C_j\cap D_j}\right)}{\prod_{j=1}^{m} \det\left(\boldsymbol{s}_{C_jC_j}\right)},$$

where $D_j = \bigcup_{k<j} C_k$ for $j = 2, 3, \ldots, m$.

Proof:
Let the assumptions in the theorem be satisfied and let $D_j = \bigcup_{k<j} C_k$ for $j = 2, 3, \ldots, m$. Since $C_1, C_2, \ldots, C_m$ is a RIP-ordering we know that $C_j \cap D_j \subseteq C_q$ for some $q < j$, $j = 2, 3, \ldots, m$. Hereby $\mathcal{G}_{D_{j+1}} = \mathcal{G}_{C_j} \dot{\cup}\, \mathcal{G}_{D_j}$ for $j = 2, 3, \ldots, m$, where $\mathcal{G}_{D_{m+1}} = \mathcal{G}$. Thus C_j and D_j form a decomposition of $\mathcal{G}_{D_{j+1}}$ and from Theorem 7.4 page 142 we have

$$\widehat{\boldsymbol{K}}\left(\mathcal{G}_{D_{j+1}}\right) = [\widehat{\boldsymbol{K}}\left(\mathcal{G}_{D_j}\right)] + [\widehat{\boldsymbol{K}}\left(\mathcal{G}_{C_j}\right)] - [\boldsymbol{s}_{C_j\cap D_j, C_j\cap D_j}^{-1}]$$

and

$$\det\left(\widehat{\boldsymbol{K}}\left(\mathcal{G}_{D_{j+1}}\right)\right) = \det\left(\widehat{\boldsymbol{K}}\left(\mathcal{G}_{D_j}\right)\right) \det\left(\widehat{\boldsymbol{K}}\left(\mathcal{G}_{C_j}\right)\right) \det\left(\boldsymbol{s}_{C_j\cap D_j, C_j\cap D_j}\right).$$

From the remark on page 129 the unique maximum likelihood estimate of $\boldsymbol{K}\left(\mathcal{G}_{C_j}\right)$ is given by $\widehat{\boldsymbol{K}}\left(\mathcal{G}_{C_j}\right) = \boldsymbol{s}_{C_jC_j}^{-1}$, since C_j is complete. Therefore

$$\widehat{\boldsymbol{K}}\left(\mathcal{G}_{D_{j+1}}\right) = [\widehat{\boldsymbol{K}}\left(\mathcal{G}_{D_j}\right)] + [\boldsymbol{s}_{C_jC_j}^{-1}] - [\boldsymbol{s}_{C_j\cap D_j, C_j\cap D_j}^{-1}]$$

and

$$\det\left(\widehat{\boldsymbol{K}}\left(\mathcal{G}_{D_{j+1}}\right)\right) = \det\left(\widehat{\boldsymbol{K}}\left(\mathcal{G}_{D_j}\right)\right) \det\left(\boldsymbol{s}_{C_jC_j}\right)^{-1} \det\left(\boldsymbol{s}_{C_j\cap D_j, C_j\cap D_j}\right).$$

Combining these equations for $j = m, m-1, \ldots, 2$ and realizing that $\widehat{\boldsymbol{K}}(\mathcal{G}_{D_2}) = \widehat{\boldsymbol{K}}(\mathcal{G}_{C_1}) = \boldsymbol{s}_{C_1C_1}^{-1}$ completes the proof. ■

By this theorem we see that, when $\mathcal{G}$ is decomposable, we do not need the IPS-algorithm to find the maximum likelihood estimate of $\boldsymbol{K}(\mathcal{G})$, since it is expressed explicitly by submatrices of $\boldsymbol{s}$. Considering in general the estimation problem it is hereby obvious, that one decomposes the simple undirected graph as much as possible and only uses the IPS-algorithm where it is necessary.

Consider the situation having a covariance selection model with additional structure on the concentration matrix, as described in Example 7.1 page 118, where the graph is not decomposable. Possibly this model can be represented as a complex normal decomposable model, where the maximum likelihood estimate of $\boldsymbol{K}(\mathcal{G})$ can be found explicitly. This is an obvious advantage in these particular models.

Example 7.5

Let $\mathcal{G} = (V, E)$ be the graph from Example 5.10 page 93. As we have seen in this example the graph is decomposable and a RIP-ordering of the cliques in $\mathcal{G}$ is given by $C_1 = \{5,6\}$, $C_2 = \{3,4,5\}$, $C_3 = \{2,3,4\}$ and $C_4 = \{1,5\}$. Let $\boldsymbol{X}$ be an $n \times 6$ complex random matrix with $\mathcal{L}(\boldsymbol{X}) = \mathbb{C}\mathcal{N}_{n\times 6}\left(\boldsymbol{O}, \boldsymbol{I}_n \otimes \boldsymbol{K}(\mathcal{G})^{-1}\right)$, where $\boldsymbol{K}(\mathcal{G}) \in \mathbb{C}_+(\mathcal{G})$. Suppose that we have an observation matrix of $\boldsymbol{X}$, which gives an observation matrix of $\boldsymbol{S}$ and an observation matrix of $\boldsymbol{S}(\mathcal{G})$. Furthermore assume that $\boldsymbol{s}(\mathcal{G})$ is $\mathcal{G}$-regular to ensure the existence of the unique maximum likelihood estimate of $\boldsymbol{K}(\mathcal{G})$. Let $\widehat{\boldsymbol{K}}(\mathcal{G})$ denote the unique maximum likelihood estimate of $\boldsymbol{K}(\mathcal{G})$. According to Theorem 7.5 page 145 we have

$$
\begin{aligned}
\widehat{\boldsymbol{K}}(\mathcal{G}) &= \begin{pmatrix}
0 & 0 & 0 & 0 & 0 & 0 \\
0 & 0 & 0 & 0 & 0 & 0 \\
0 & 0 & 0 & 0 & 0 & 0 \\
0 & 0 & 0 & 0 & 0 & 0 \\
0 & 0 & 0 & 0 & \left(\boldsymbol{s}_{\{5,6\}\{5,6\}}^{-1}\right)_{11} & \left(\boldsymbol{s}_{\{5,6\}\{5,6\}}^{-1}\right)_{12} \\
0 & 0 & 0 & 0 & \left(\boldsymbol{s}_{\{5,6\}\{5,6\}}^{-1}\right)_{21} & \left(\boldsymbol{s}_{\{5,6\}\{5,6\}}^{-1}\right)_{22}
\end{pmatrix} \\
&+ \begin{pmatrix}
0 & 0 & 0 & 0 & 0 & 0 \\
0 & 0 & 0 & 0 & 0 & 0 \\
0 & 0 & \left(\boldsymbol{s}_{\{3,4,5\}\{3,4,5\}}^{-1}\right)_{11} & \left(\boldsymbol{s}_{\{3,4,5\}\{3,4,5\}}^{-1}\right)_{12} & \left(\boldsymbol{s}_{\{3,4,5\}\{3,4,5\}}^{-1}\right)_{13} & 0 \\
0 & 0 & \left(\boldsymbol{s}_{\{3,4,5\}\{3,4,5\}}^{-1}\right)_{21} & \left(\boldsymbol{s}_{\{3,4,5\}\{3,4,5\}}^{-1}\right)_{22} & \left(\boldsymbol{s}_{\{3,4,5\}\{3,4,5\}}^{-1}\right)_{23} & 0 \\
0 & 0 & \left(\boldsymbol{s}_{\{3,4,5\}\{3,4,5\}}^{-1}\right)_{31} & \left(\boldsymbol{s}_{\{3,4,5\}\{3,4,5\}}^{-1}\right)_{32} & \left(\boldsymbol{s}_{\{3,4,5\}\{3,4,5\}}^{-1}\right)_{33} & 0 \\
0 & 0 & 0 & 0 & 0 & 0
\end{pmatrix} \\
&+ \begin{pmatrix}
0 & 0 & 0 & 0 & 0 & 0 \\
0 & \left(\boldsymbol{s}_{\{2,3,4\}\{2,3,4\}}^{-1}\right)_{11} & \left(\boldsymbol{s}_{\{2,3,4\}\{2,3,4\}}^{-1}\right)_{12} & \left(\boldsymbol{s}_{\{2,3,4\}\{2,3,4\}}^{-1}\right)_{13} & 0 & 0 \\
0 & \left(\boldsymbol{s}_{\{2,3,4\}\{2,3,4\}}^{-1}\right)_{21} & \left(\boldsymbol{s}_{\{2,3,4\}\{2,3,4\}}^{-1}\right)_{22} & \left(\boldsymbol{s}_{\{2,3,4\}\{2,3,4\}}^{-1}\right)_{23} & 0 & 0 \\
0 & \left(\boldsymbol{s}_{\{2,3,4\}\{2,3,4\}}^{-1}\right)_{31} & \left(\boldsymbol{s}_{\{2,3,4\}\{2,3,4\}}^{-1}\right)_{32} & \left(\boldsymbol{s}_{\{2,3,4\}\{2,3,4\}}^{-1}\right)_{33} & 0 & 0 \\
0 & 0 & 0 & 0 & 0 & 0 \\
0 & 0 & 0 & 0 & 0 & 0
\end{pmatrix}
\end{aligned}
$$

$$+\begin{pmatrix} \left(s^{-1}_{\{1,5\}\{1,5\}}\right)_{11} & 0 & 0 & 0 & \left(s^{-1}_{\{1,5\}\{1,5\}}\right)_{12} & 0 \\ 0 & 0 & 0 & 0 & 0 & 0 \\ 0 & 0 & 0 & 0 & 0 & 0 \\ 0 & 0 & 0 & 0 & 0 & 0 \\ \left(s^{-1}_{\{1,5\}\{1,5\}}\right)_{21} & 0 & 0 & 0 & \left(s^{-1}_{\{1,5\}\{1,5\}}\right)_{22} & 0 \\ 0 & 0 & 0 & 0 & 0 & 0 \end{pmatrix}$$

$$-\begin{pmatrix} 0 & 0 & 0 & 0 & 0 & 0 \\ 0 & 0 & 0 & 0 & 0 & 0 \\ 0 & 0 & \left(s^{-1}_{\{3,4\}\{3,4\}}\right)_{11} & \left(s^{-1}_{\{3,4\}\{3,4\}}\right)_{12} & 0 & 0 \\ 0 & 0 & \left(s^{-1}_{\{3,4\}\{3,4\}}\right)_{21} & \left(s^{-1}_{\{3,4\}\{3,4\}}\right)_{22} & 0 & 0 \\ 0 & 0 & 0 & 0 & 2\left(s^{-1}_{55}\right) & 0 \\ 0 & 0 & 0 & 0 & 0 & 0 \end{pmatrix}$$

and

$$\det\left(\widehat{\boldsymbol{K}}(\mathcal{G})\right) = \frac{(s_{55})^2 \det\left(s_{\{3,4\}\{3,4\}}\right)}{\det\left(s_{\{5,6\}\{5,6\}}\right)\det\left(s_{\{3,4,5\}\{3,4,5\}}\right)\det\left(s_{\{2,3,4\}\{2,3,4\}}\right)\det\left(s_{\{1,5\}\{1,5\}}\right)}$$

Remark that $\left(s^{-1}_{AA}\right)_{jk}$ denotes the jk'th element of s^{-1}_{AA}. ■

7.6 Hypothesis Testing in Complex Normal Graphical Models

In this section we consider hypothesis testing in complex normal graphical models. Test in these models are concerned with the structure of the concentration matrix.

Let $\mathcal{G} = (V, E)$ be a simple undirected graph and let $\boldsymbol{X}$ be an $n \times |V|$ complex random matrix with $\mathcal{L}(\boldsymbol{X}) = \mathbb{C}\mathcal{N}_{n\times|V|}\left(\boldsymbol{O}, \boldsymbol{I}_n \otimes \boldsymbol{K}(\mathcal{G})^{-1}\right)$, where $\boldsymbol{K}(\mathcal{G}) \in \mathbb{C}_+(\mathcal{G})$. Let E' be a subset of E and consider $\mathcal{G}' = (V, E')$, which is the subgraph of $\mathcal{G}$ obtained by removing the edges in E that are not in E'. Finally let $\boldsymbol{S}(\mathcal{G})$ defined in (7.6) page 122 be $\mathcal{G}$-regular with probability one. This implies that $\boldsymbol{S}(\mathcal{G}')$ is $\mathcal{G}'$-regular with probability one. To test if removal of the edges from $\mathcal{G}$ is legal corresponds to testing the null hypothesis

$$\mathrm{H}_0 : \boldsymbol{K}(\mathcal{G}) \in \mathbb{C}_+(\mathcal{G}')$$

under the hypothesis

$$\mathrm{H} : \boldsymbol{K}(\mathcal{G}) \in \mathbb{C}_+(\mathcal{G}) \ .$$

In other words we test conditional independence of selected pairs of variables given the remaining variables in $\boldsymbol{X}$.

We perform this test by using the likelihood ratio test, which consists in rejecting H_0 if the likelihood ratio defined by

$$Q(\boldsymbol{x}) = \frac{\sup_{\boldsymbol{K}(\mathcal{G})\in\mathbb{C}_+(\mathcal{G}')} L(\boldsymbol{K}(\mathcal{G})|\,\boldsymbol{x})}{\sup_{\boldsymbol{K}(\mathcal{G})\in\mathbb{C}_+(\mathcal{G})} L(\boldsymbol{K}(\mathcal{G})|\,\boldsymbol{x})}$$

is smaller than some chosen constant providing the size of the test.

As on page 144 the maximum value of $L\left(\boldsymbol{K}(\mathcal{G})|\,\boldsymbol{x}\right)$ for $\boldsymbol{K}(\mathcal{G}) \in \mathbb{C}_+(\mathcal{G})$ is observed as

$$L\left(\widehat{\boldsymbol{K}}(\mathcal{G})\middle|\,\boldsymbol{x}\right) = \pi^{-n|V|}\det\left(\widehat{\boldsymbol{K}}(\mathcal{G})\right)^n \exp\left(-n|V|\right) ,$$

where $\widehat{\boldsymbol{K}}(\mathcal{G})$ is the unique maximum likelihood estimate of $\boldsymbol{K}(\mathcal{G}) \in \mathbb{C}_+(\mathcal{G})$ fulfilling the likelihood equations for all cliques in $\mathcal{G}$. Note that $\widehat{\boldsymbol{K}}(\mathcal{G})$ exists with probability one, since $\boldsymbol{S}(\mathcal{G})$ is $\mathcal{G}$-regular with probability one. Analogously the maximum value of $L\left(\boldsymbol{K}(\mathcal{G})|\,\boldsymbol{x}\right)$ for $\boldsymbol{K}(\mathcal{G}) \in \mathbb{C}_+(\mathcal{G}')$ is

$$L\left(\widehat{\boldsymbol{K}}(\mathcal{G}')\middle|\,\boldsymbol{x}\right) = \pi^{-n|V|}\det\left(\widehat{\boldsymbol{K}}(\mathcal{G}')\right)^n \exp\left(-n|V|\right) ,$$

where $\widehat{\boldsymbol{K}}(\mathcal{G}')$ is the unique maximum likelihood estimate of $\boldsymbol{K}(\mathcal{G}) \in \mathbb{C}_+(\mathcal{G}')$ satisfying the likelihood equations for all cliques in $\mathcal{G}'$. Equivalently $\boldsymbol{S}(\mathcal{G}')$ being $\mathcal{G}'$-regular with probability one ensures that $\widehat{\boldsymbol{K}}(\mathcal{G}')$ exists with probability one.

By these results we obtain the likelihood ratio as

$$Q\left(\boldsymbol{x}\right) = \frac{\det\left(\widehat{\boldsymbol{K}}(\mathcal{G}')\right)^n}{\det\left(\widehat{\boldsymbol{K}}(\mathcal{G})\right)^n} , \tag{7.31}$$

which is real since $\widehat{\boldsymbol{K}}(\mathcal{G}')$ and $\widehat{\boldsymbol{K}}(\mathcal{G})$ are Hermitian. Under H_0 we know that $-2\log Q\left(\boldsymbol{X}\right)$ is asymptotically chi-square distributed with $2\left(|E| - |E'|\right)$ degrees of freedom as n tends to infinity (Silvey 1975, p. 113).

Generally it is a problem to determine, when the asymptotic result can be used. The guidelines to how large we require n to be are vague. When the graph is collapsible onto a subset of the vertices the hypothesis testing problem may be reduced. The distributional results for the likelihood ratio test statistic are still asymptotic, but the number of parameters to estimate have been reduced.

Let $\mathcal{G} = (V, E)$ be a simple undirected graph and let $\boldsymbol{X}$ be an $n \times |V|$ complex random matrix with $\mathcal{L}\left(\boldsymbol{X}\right) = \mathbb{C}\mathcal{N}_{n\times|V|}\left(\boldsymbol{O}, \boldsymbol{I}_n \otimes \boldsymbol{K}(\mathcal{G})^{-1}\right)$, where $\boldsymbol{K}(\mathcal{G}) \in \mathbb{C}_+(\mathcal{G})$. Further let $\mathcal{G} = (V, E)$ be collapsible onto a subset A of V and let $\mathcal{G}_{B_j} = \left(B_j, E_{B_j}\right), j = 1, 2, \ldots, k$, be the k connectivity components of $\mathcal{G}_{V\setminus A} = \left(V \setminus A, E_{V\setminus A}\right)$, where $B_j \subseteq V \setminus A$ and $E_{B_j} = E_{V\setminus A} \cap (B_j \times B_j)$. Then $\mathrm{bd}\left(B_j\right)$ is complete for $j = 1, 2, \ldots, k$. Let $\mathcal{G}' = (V, E')$ be a subgraph of $\mathcal{G}$ satisfying that

$$E \setminus E' \subseteq (A \times A) \setminus \bigcup_{j=1}^{k} \left(\mathrm{bd}\left(B_j\right) \times \mathrm{bd}\left(B_j\right)\right) , \tag{7.32}$$

i.e. $\mathcal{G}'$ is also collapsible onto A.

To perform the test of the null hypothesis

$$\mathrm{H}_0 : \boldsymbol{K}(\mathcal{G}) \in \mathbb{C}_+(\mathcal{G}')$$

under the hypothesis

$$\mathrm{H}: \boldsymbol{K}(\mathcal{G}) \in \mathbb{C}_{+}(\mathcal{G})$$

we consider the likelihood ratio, which from (7.31) page 148 is given by

$$Q(\boldsymbol{x}) = \frac{\det\left(\widehat{\boldsymbol{K}}(\mathcal{G}')\right)^n}{\det\left(\widehat{\boldsymbol{K}}(\mathcal{G})\right)^n}.$$

To find an expression of $Q(\boldsymbol{x})$ we define the sets

$$\begin{aligned} D_0 &= A, \\ D_j &= A \cup B_1 \cup B_2 \cup \cdots \cup B_j \text{ for } j = 1, 2, \ldots, k-1, \\ D_k &= V \end{aligned}$$

and

$$F_j = B_{j+1} \cup \operatorname{bd}(B_{j+1}) \text{ for } j = 0, 1, \ldots, k-1.$$

Observe for $j = 0, 1, \ldots, k-1$ that $D_j \cap F_j = \operatorname{bd}(B_{j+1})$ is complete. Therefore we have for $j = 0, 1, \ldots, k-1$ that $\mathcal{G}_{D_{j+1}} = \mathcal{G}_{D_j} \dot{\cup} \mathcal{G}_{F_j}$, i.e. D_j and F_j form a decomposition of $\mathcal{G}_{D_{j+1}}$. Considering (7.32) page 148 we see for $j = 0, 1, \ldots, k-1$ that $E_{F_j} = E'_{F_j}$ and $\mathcal{G}'_{D_{j+1}} = \mathcal{G}'_{D_j} \dot{\cup} \mathcal{G}'_{F_j}$. Hence D_j and F_j also form a decomposition of $\mathcal{G}'_{D_{j+1}}$.

From Theorem 7.4 page 142 we then have

$$\det\left(\widehat{\boldsymbol{K}}\left(\mathcal{G}_{D_{j+1}}\right)\right) = \det\left(\widehat{\boldsymbol{K}}\left(\mathcal{G}_{D_j}\right)\right) \det\left(\widehat{\boldsymbol{K}}\left(\mathcal{G}_{F_j}\right)\right) \det\left(\boldsymbol{s}_{\operatorname{bd}(B_{j+1}),\operatorname{bd}(B_{j+1})}\right)$$

and

$$\det\left(\widehat{\boldsymbol{K}}\left(\mathcal{G}'_{D_{j+1}}\right)\right) = \det\left(\widehat{\boldsymbol{K}}\left(\mathcal{G}'_{D_j}\right)\right) \det\left(\widehat{\boldsymbol{K}}\left(\mathcal{G}'_{F_j}\right)\right) \det\left(\boldsymbol{s}_{\operatorname{bd}(B_{j+1}),\operatorname{bd}(B_{j+1})}\right).$$

Therefore by combining these equations for $j = k-1, k-2, \ldots, 0$ we find the likelihood ratio as

$$Q(\boldsymbol{x}) = \frac{\det\left(\widehat{\boldsymbol{K}}(\mathcal{G}'_A)\right)^n}{\det\left(\widehat{\boldsymbol{K}}(\mathcal{G}_A)\right)^n}. \tag{7.33}$$

Hereby we deduce that, when $\mathcal{G}$ is collapsible onto a subset A of V, testing removeability of edges belonging to $(A \times A) \setminus \bigcup_{j=1}^{k} (\operatorname{bd}(B_j) \times \operatorname{bd}(B_j))$ can be performed in the A-marginal. Again under H_0 it holds that $-2\log Q(\boldsymbol{X})$ is asymptotically chi-square distributed with $2(|E_A| - |E'_A|)$ degrees of freedom as n tends to infinity.

As the test can be performed in the A-marginal, n has become larger compared to the number of parameters in the marginal. Even though the distribution is still asymptotic our confidence to the approximation is improved. If we test removal of one regular edge from $\mathcal{G}$ we are able to determine the distribution of the likelihood ratio test statistic as a beta distribution. This is examined in Theorem 7.6 page 152, but to prove this theorem the following lemma is needed.

Lemma 7.6
Let $\mathcal{G} = (V, E)$ be a simple undirected graph and let $\boldsymbol{X}$ be an $n \times |V|$ complex random matrix with $\mathcal{L}(\boldsymbol{X}) = \mathbb{C}\mathcal{N}_{n\times|V|}\left(\boldsymbol{O}, \boldsymbol{I}_n \otimes \boldsymbol{K}(\mathcal{G})^{-1}\right)$, where $\boldsymbol{K}(\mathcal{G}) \in \mathbb{C}_+(\mathcal{G})$. Furthermore let $\boldsymbol{S} = \frac{1}{n}\boldsymbol{X}^*\boldsymbol{X}$ and let $\boldsymbol{S}(\mathcal{G})$ defined in (7.6) page 122 be $\mathcal{G}$-regular with probability one. Finally let C be a complete subset of V and let $\boldsymbol{X}$ be partitioned as $\boldsymbol{X} = \left(\boldsymbol{X}_{V\setminus C}, \boldsymbol{X}_C\right)$, where $\boldsymbol{X}_{V\setminus C}$ and $\boldsymbol{X}_C$ have dimensions $n \times |V\setminus C|$ and $n \times |C|$, respectively. For the conditional distribution of $\boldsymbol{S}$ given $\boldsymbol{X}_C$ there exists a function h, such that

$$P\left(\boldsymbol{S} \in D \middle| \boldsymbol{X}_C = \boldsymbol{x}_C\right) = h\left(D, \boldsymbol{s}_{CC}\right)$$

for all measurable sets D in the sample space of $\boldsymbol{S}$.

Proof:
Let the assumptions in the lemma be satisfied.

According to the partition of $\boldsymbol{X}$ we write

$$\begin{aligned} \boldsymbol{X}^*\boldsymbol{X} &= \begin{pmatrix} \boldsymbol{X}_{V\setminus C}^*\boldsymbol{X}_{V\setminus C} & \boldsymbol{X}_{V\setminus C}^*\boldsymbol{X}_C \\ \boldsymbol{X}_C^*\boldsymbol{X}_{V\setminus C} & \boldsymbol{X}_C^*\boldsymbol{X}_C \end{pmatrix} \\ &= \begin{pmatrix} \boldsymbol{W}_{V\setminus C,V\setminus C} & \boldsymbol{W}_{V\setminus C,C} \\ \boldsymbol{W}_{C,V\setminus C} & \boldsymbol{W}_{CC} \end{pmatrix} \\ &= \boldsymbol{W}\,. \end{aligned}$$

As on page 133 we get

$$\mathcal{L}\left(\boldsymbol{X}_{V\setminus C}\middle|\boldsymbol{X}_C\right) = \mathbb{C}\mathcal{N}_{n\times|V\setminus C|}\left(-\boldsymbol{X}_C\boldsymbol{K}(\mathcal{G})_{C,V\setminus C}\boldsymbol{K}(\mathcal{G})_{V\setminus C,V\setminus C}^{-1}, \boldsymbol{I}_n \otimes \boldsymbol{K}(\mathcal{G})_{V\setminus C,V\setminus C}^{-1}\right)\,.$$

Since $\boldsymbol{S}(\mathcal{G})$ is $\mathcal{G}$-regular with probability one and C is complete, we know that $\boldsymbol{W}_{CC} > \boldsymbol{O}$ with probability one. Hence with probability one $\boldsymbol{W}_{CC}^{-1}$ exists and furthermore with probability one there exists a $|C| \times |C|$ complex matrix $\boldsymbol{W}_{CC}^{\frac{1}{2}} > \boldsymbol{O}$, such that $\boldsymbol{W}_{CC} = \left(\boldsymbol{W}_{CC}^{\frac{1}{2}}\right)^2$. Due to Theorem 2.19 page 34 we have

$$\mathcal{L}\left(\boldsymbol{W}_{CC}^{-1}\boldsymbol{W}_{C,V\setminus C}\,|\boldsymbol{X}_C\right) = \mathbb{C}\mathcal{N}_{|C|\times|V\setminus C|}\left(-\boldsymbol{K}(\mathcal{G})_{C,V\setminus C}\boldsymbol{K}(\mathcal{G})_{V\setminus C,V\setminus C}^{-1}, \boldsymbol{W}_{CC}^{-1} \otimes \boldsymbol{K}(\mathcal{G})_{V\setminus C,V\setminus C}^{-1}\right)$$

and further

$$\mathcal{L}\left(\boldsymbol{W}_{CC}^{\frac{1}{2}}\left(\boldsymbol{W}_{CC}^{-1}\boldsymbol{W}_{C,V\setminus C}+\boldsymbol{K}(\mathcal{G})_{C,V\setminus C}\boldsymbol{K}(\mathcal{G})_{V\setminus C,V\setminus C}^{-1}\right)\middle|\boldsymbol{X}_C\right)=\mathbb{C}\mathcal{N}_{|C|\times|V\setminus C|}\left(\boldsymbol{O},\boldsymbol{I}_{|C|}\otimes\boldsymbol{K}(\mathcal{G})_{V\setminus C,V\setminus C}^{-1}\right).$$

This distribution does not depend on $\boldsymbol{X}_C$ whereby we deduce that

$$\text{(7.34)} \qquad \boldsymbol{W}_{CC}^{-\frac{1}{2}}\boldsymbol{W}_{C,V\setminus C} + \boldsymbol{W}_{CC}^{\frac{1}{2}}\boldsymbol{K}(\mathcal{G})_{C,V\setminus C}\boldsymbol{K}(\mathcal{G})_{V\setminus C,V\setminus C}^{-1} \perp\!\!\!\perp \boldsymbol{X}_C\,.$$

From Definition 3.1 page 40 we know that

$$\mathcal{L}(\boldsymbol{W}) = \mathbb{C}\mathcal{W}_{|V|}\left(\boldsymbol{K}(\mathcal{G})^{-1}, n\right)\,.$$

Thereby as in the proof of Theorem 3.6 part 2 page 45 we see that

$$(7.35) \qquad \boldsymbol{W}_{V\setminus C,V\setminus C} - \boldsymbol{W}_{V\setminus C,C}\boldsymbol{W}_{CC}^{-1}\boldsymbol{W}_{C,V\setminus C} \perp\!\!\!\perp \left(\boldsymbol{W}_{C,V\setminus C}, \boldsymbol{X}_C\right) .$$

Combining (7.34) page 150 and (7.35) we obtain that

$$\boldsymbol{W}_{V\setminus C,V\setminus C} - \boldsymbol{W}_{V\setminus C,C}\boldsymbol{W}_{CC}^{-1}\boldsymbol{W}_{C,V\setminus C}, \boldsymbol{W}_{CC}^{-\frac{1}{2}}\boldsymbol{W}_{C,V\setminus C} + \boldsymbol{W}_{CC}^{\frac{1}{2}}\boldsymbol{K}(\mathcal{G})_{C,V\setminus C}\boldsymbol{K}(\mathcal{G})_{V\setminus C,V\setminus C}^{-1} \text{ and } \boldsymbol{X}_C$$

are mutually independent. Since the density function of $\boldsymbol{X}$ w.r.t. Lebesgue measure on $\mathbb{C}^{n\times|V|}$ is continuous we get by Theorem 6.2 page 101 that there exists a function h_1, such that

(7.36)

$$P\left(\left(\boldsymbol{W}_{V\setminus C,V\setminus C} - \boldsymbol{W}_{V\setminus C,C}\boldsymbol{W}_{CC}^{-1}\boldsymbol{W}_{C,V\setminus C}, \boldsymbol{W}_{CC}^{-\frac{1}{2}}\boldsymbol{W}_{C,V\setminus C} + \boldsymbol{W}_{CC}^{\frac{1}{2}}\boldsymbol{K}(\mathcal{G})_{C,V\setminus C}\boldsymbol{K}(\mathcal{G})_{V\setminus C,V\setminus C}^{-1}\right) \in F_1 \middle| \boldsymbol{X}_C = \boldsymbol{x}_C\right)$$
$$= h_1(F_1)$$

for all measurable sets F_1 in the sample space of the random matrix above.

Now consider $P(\boldsymbol{W} \in F|\, \boldsymbol{X}_C = \boldsymbol{x}_C)$ for all measurable sets F in the sample space of $\boldsymbol{W}$. Given $\boldsymbol{X}_C$, and hereby $\boldsymbol{W}_{CC}$, there exists a one-to-one correspondence between $\boldsymbol{W}$ and $\left(\boldsymbol{W}_{V\setminus C,V\setminus C}, \boldsymbol{W}_{C,V\setminus C}\right)$, which can be established by a bijective function $g_{\boldsymbol{W}_{CC}}$. This means that

$$\begin{aligned} P(\boldsymbol{W} \in F|\, \boldsymbol{X}_C = \boldsymbol{x}_C) &= P(g_{\boldsymbol{W}_{CC}}(\boldsymbol{W}) \in g_{\boldsymbol{W}_{CC}}(F)|\, \boldsymbol{X}_C = \boldsymbol{x}_C) \\ &= P\left(\left(\boldsymbol{W}_{V\setminus C,V\setminus C}, \boldsymbol{W}_{C,V\setminus C}\right) \in F_2 \middle| \boldsymbol{X}_C = \boldsymbol{x}_C\right) , \end{aligned}$$

where F_2 is the image of F under $g_{\boldsymbol{W}_{CC}}$.

Also given $\boldsymbol{W}_{CC}$ there is a one-to-one correspondence between $\left(\boldsymbol{W}_{V\setminus C,V\setminus C}, \boldsymbol{W}_{C,V\setminus C}\right)$ and $\left(\boldsymbol{W}_{V\setminus C,V\setminus C} - \boldsymbol{W}_{V\setminus C,C}\boldsymbol{W}_{CC}^{-1}\boldsymbol{W}_{C,V\setminus C}, \boldsymbol{W}_{CC}^{-\frac{1}{2}}\boldsymbol{W}_{C,V\setminus C} + \boldsymbol{W}_{CC}^{\frac{1}{2}}\boldsymbol{K}(\mathcal{G})_{C,V\setminus C}\boldsymbol{K}(\mathcal{G})_{V\setminus C,V\setminus C}^{-1}\right)$ given by a bijective function $\tilde{g}_{\boldsymbol{W}_{CC}}$. Hence the above and (7.36) lead to

$$\begin{aligned} P(\boldsymbol{W} \in F|\, \boldsymbol{X}_C = \boldsymbol{x}_C) = {} & P\left(\left(\boldsymbol{W}_{V\setminus C,V\setminus C} - \boldsymbol{W}_{V\setminus C,C}\boldsymbol{W}_{CC}^{-1}\boldsymbol{W}_{C,V\setminus C},\right.\right. \\ & \left.\left. \boldsymbol{W}_{CC}^{-\frac{1}{2}}\boldsymbol{W}_{C,V\setminus C} + \boldsymbol{W}_{CC}^{\frac{1}{2}}\boldsymbol{K}(\mathcal{G})_{C,V\setminus C}\boldsymbol{K}(\mathcal{G})_{V\setminus C,V\setminus C}^{-1}\right) \in F_1 \middle| \boldsymbol{X}_C = \boldsymbol{x}_C\right) \\ = {} & h_1(F_1) , \end{aligned}$$

where F_1 is the image of F under the composite function $\tilde{g}_{\boldsymbol{W}_{CC}} \circ g_{\boldsymbol{W}_{CC}}$. Since F_1 only depends on F and the given value of $\boldsymbol{W}_{CC}$ we conclude that

$$P(\boldsymbol{W} \in F\,|\boldsymbol{X}_C = \boldsymbol{x}_C) = h_2(F, \boldsymbol{w}_{CC})$$

and hereby, since $\boldsymbol{W} = n\boldsymbol{S}$, we get

$$P(\boldsymbol{S} \in D|\, \boldsymbol{X}_C = \boldsymbol{x}_C) = h(D, \boldsymbol{s}_{CC}) .$$

∎

We are now able to prove the theorem saying that when removal of a regular edge is tested, then the likelihood ratio test statistic has a beta distribution.

Theorem 7.6
Let $\mathcal{G} = (V, E)$ be a simple undirected graph and let $\{\alpha, \beta\} \in E$ be a regular edge. Furthermore let $\mathcal{G}' = (V, E \setminus \{\alpha, \beta\})$ and let $\boldsymbol{X}$ be an $n \times |V|$ complex random matrix with $\mathcal{L}(\boldsymbol{X}) = \mathbb{C}\mathcal{N}_{n\times|V|}\left(\boldsymbol{O}, \boldsymbol{I}_n \otimes \boldsymbol{K}(\mathcal{G})^{-1}\right)$, where $\boldsymbol{K}(\mathcal{G}) \in \mathbb{C}_+(\mathcal{G})$. Moreover let $\boldsymbol{S} = \frac{1}{n}\boldsymbol{X}^*\boldsymbol{X}$ and let $\boldsymbol{S}(\mathcal{G})$ defined in (7.6) page 122 be $\mathcal{G}$-regular with probability one. The likelihood ratio test of the null hypothesis

$$\mathrm{H}_0 : \boldsymbol{K}(\mathcal{G}) \in \mathbb{C}_+(\mathcal{G}')$$

under the hypothesis

$$\mathrm{H} : \boldsymbol{K}(\mathcal{G}) \in \mathbb{C}_+(\mathcal{G})$$

consists in rejecting H_0 *if*

$$Q^{\frac{1}{n}}(\boldsymbol{x}) \leq q\,.$$

The likelihood ratio test statistic is given by

$$Q^{\frac{1}{n}}(\boldsymbol{X}) = \frac{\det\left(\boldsymbol{S}_{CC}\right)\det\left(\boldsymbol{S}_{C\cup\{\alpha,\beta\},C\cup\{\alpha,\beta\}}\right)}{\det\left(\boldsymbol{S}_{C\cup\alpha,C\cup\alpha}\right)\det\left(\boldsymbol{S}_{C\cup\beta,C\cup\beta}\right)}\,,$$

where $C = \mathrm{bd}(\alpha) \cap \mathrm{bd}(\beta)$. Furthermore under H_0 *the distribution of $Q^{\frac{1}{n}}(\boldsymbol{X})$ is*

$$\mathcal{L}\left(Q^{\frac{1}{n}}(\boldsymbol{X})\right) = \mathcal{B}(n - |C| - 1, 1)\,,$$

and $Q^{\frac{1}{n}}(\boldsymbol{X})$ and $\widehat{\boldsymbol{K}}(\mathcal{G}')$ are independent, where $\widehat{\boldsymbol{K}}(\mathcal{G}')$ is the maximum likelihood estimate of $\boldsymbol{K}(\mathcal{G}) \in \mathbb{C}_+(\mathcal{G}')$.

The constant q is chosen to provide a size γ test, i.e. q must fulfill $P\left(Q^{\frac{1}{n}}(\boldsymbol{X}) \leq q \,\middle|\, H_0\right) = \gamma$.

Proof:
Let the assumptions in the theorem be satisfied and let $C = \mathrm{bd}(\alpha) \cap \mathrm{bd}(\beta)$.

Since $\{\alpha, \beta\} \in E$ is a regular edge and C is defined as above we deduce from Theorem 5.4 page 95 that $C \cup \{\alpha, \beta\}$ is a clique in $\mathcal{G}$. We know that a simple undirected graph always is collapsible onto a clique, hence $\mathcal{G}$ is collapsible onto $C \cup \{\alpha, \beta\}$. Let $\mathcal{G}_{B_j} = \left(B_j, E_{B_j}\right), j = 1, 2, \ldots, k$, be the k connectivity components of $\mathcal{G}_{V\setminus(C\cup\{\alpha,\beta\})} = \left(V \setminus (C \cup \{\alpha, \beta\}), E_{V\setminus(C\cup\{\alpha,\beta\})}\right)$, where $B_j \subseteq V \setminus (C \cup \{\alpha, \beta\})$ and $E_{B_j} = E_{V\setminus(C\cup\{\alpha,\beta\})} \cap (B_j \times B_j)$. Then we observe that $\{\alpha, \beta\} \in ((C \cup \{\alpha, \beta\}) \times (C \cup \{\alpha, \beta\})) \setminus \cup_{j=1}^{k} (\mathrm{bd}(B_j) \times \mathrm{bd}(B_j))$. Hereby the test of the null hypothesis

$$\mathrm{H}_0 : \boldsymbol{K}(\mathcal{G}) \in \mathbb{C}_+(\mathcal{G}')$$

under the hypothesis

$$\mathrm{H} : \boldsymbol{K}(\mathcal{G}) \in \mathbb{C}_+(\mathcal{G})$$

can be carried out in the $(C \cup \{\alpha, \beta\})$-marginal. From (7.33) page 149 the likelihood ratio becomes

$$Q^{\frac{1}{n}}(\boldsymbol{x}) = \frac{\det\left(\widehat{\boldsymbol{K}}\left(\mathcal{G}'_{C\cup\{\alpha,\beta\}}\right)\right)}{\det\left(\widehat{\boldsymbol{K}}\left(\mathcal{G}_{C\cup\{\alpha,\beta\}}\right)\right)} .$$

Since $C \cup \{\alpha, \beta\}$ is a clique, the graph $\mathcal{G}_{C\cup\{\alpha,\beta\}}$ is decomposable. Furthermore $\mathcal{G}'_{C\cup\{\alpha,\beta\}}$ is decomposable, because $C \cup \alpha$ and $C \cup \beta$ decompose $\mathcal{G}'_{C\cup\{\alpha,\beta\}}$ into the components $\mathcal{G}'_{C\cup\alpha}$ and $\mathcal{G}'_{C\cup\beta}$ and $C \cup \alpha$ and $C \cup \beta$ are the cliques in $\mathcal{G}'_{C\cup\{\alpha,\beta\}}$. Using Theorem 7.5 page 145 the likelihood ratio test statistic becomes

$$Q^{\frac{1}{n}}(\boldsymbol{X}) = \frac{\det(\boldsymbol{S}_{CC})\det\left(\boldsymbol{S}_{C\cup\{\alpha,\beta\},C\cup\{\alpha,\beta\}}\right)}{\det(\boldsymbol{S}_{C\cup\alpha,C\cup\alpha})\det(\boldsymbol{S}_{C\cup\beta,C\cup\beta})} . \tag{7.37}$$

Remark that all the determinants are positive with probability one, as $\boldsymbol{S}(\mathcal{G})$ being $\mathcal{G}$-regular with probability one implies $\boldsymbol{S}_{C\cup\{\alpha,\beta\},C\cup\{\alpha,\beta\}} > \boldsymbol{O}$ with probability one.

To determine the distribution of $Q^{\frac{1}{n}}(\boldsymbol{X})$ under H_0 we initially consider the distribution of $\boldsymbol{S}_{C\cup\{\alpha,\beta\},C\cup\{\alpha,\beta\}}$. Let $\boldsymbol{X} = \left(\boldsymbol{X}_{C\cup\{\alpha,\beta\}}, \boldsymbol{X}_{V\setminus(C\cup\{\alpha,\beta\})}\right)$, where $\boldsymbol{X}_{C\cup\{\alpha,\beta\}}$ and $\boldsymbol{X}_{V\setminus(C\cup\{\alpha,\beta\})}$ have dimensions $n \times |C \cup \{\alpha, \beta\}|$ and $n \times |V \setminus (C \cup \{\alpha, \beta\})|$, respectively. Then from Theorem 2.23 page 36 we have

$$\mathcal{L}\left(\boldsymbol{X}_{C\cup\{\alpha,\beta\}}\right) = \mathbb{C}\mathcal{N}_{n\times|C\cup\{\alpha,\beta\}|}\left(\boldsymbol{O}, \boldsymbol{I}_n \otimes \left(\boldsymbol{K}(\mathcal{G})^{-1}\right)_{C\cup\{\alpha,\beta\},C\cup\{\alpha,\beta\}}\right),$$

where $\boldsymbol{K}(\mathcal{G}) \in \mathbb{C}_+(\mathcal{G}')$. Hereby from Definition 3.1 page 40 we observe

$$\mathcal{L}\left(\boldsymbol{S}_{C\cup\{\alpha,\beta\},C\cup\{\alpha,\beta\}}\right) = \frac{1}{n}\mathbb{C}\mathcal{W}_{|C\cup\{\alpha,\beta\}|}\left(\left(\boldsymbol{K}(\mathcal{G})^{-1}\right)_{C\cup\{\alpha,\beta\},C\cup\{\alpha,\beta\}}, n\right).$$

Let $\boldsymbol{S}_{C\cup\{\alpha,\beta\},C\cup\{\alpha,\beta\}}$ be partitioned as

$$\boldsymbol{S}_{C\cup\{\alpha,\beta\},C\cup\{\alpha,\beta\}} = \begin{pmatrix} \boldsymbol{S}_{\alpha\alpha} & \boldsymbol{S}_{\alpha\beta} & \boldsymbol{S}_{\alpha C} \\ \boldsymbol{S}_{\beta\alpha} & \boldsymbol{S}_{\beta\beta} & \boldsymbol{S}_{\beta C} \\ \boldsymbol{S}_{C\alpha} & \boldsymbol{S}_{C\beta} & \boldsymbol{S}_{CC} \end{pmatrix} .$$

We see that $\boldsymbol{S}_{\{\alpha,\beta\}\{\alpha,\beta\}} - \boldsymbol{S}_{\{\alpha,\beta\}C}\boldsymbol{S}_{CC}^{-1}\boldsymbol{S}_{C\{\alpha,\beta\}}$ can be written as

(7.38)

$$\boldsymbol{S}_{\{\alpha,\beta\}\{\alpha,\beta\}} - \boldsymbol{S}_{\{\alpha,\beta\}C}\boldsymbol{S}_{CC}^{-1}\boldsymbol{S}_{C\{\alpha,\beta\}} = \begin{pmatrix} \boldsymbol{S}_{\alpha\alpha} - \boldsymbol{S}_{\alpha C}\boldsymbol{S}_{CC}^{-1}\boldsymbol{S}_{C\alpha} & \boldsymbol{S}_{\alpha\beta} - \boldsymbol{S}_{\alpha C}\boldsymbol{S}_{CC}^{-1}\boldsymbol{S}_{C\beta} \\ \boldsymbol{S}_{\beta\alpha} - \boldsymbol{S}_{\beta C}\boldsymbol{S}_{CC}^{-1}\boldsymbol{S}_{C\alpha} & \boldsymbol{S}_{\beta\beta} - \boldsymbol{S}_{\beta C}\boldsymbol{S}_{CC}^{-1}\boldsymbol{S}_{C\beta} \end{pmatrix}.$$

Observe that the following equations hold

$$\begin{aligned} \det\left(\boldsymbol{S}_{C\cup\{\alpha,\beta\},C\cup\{\alpha,\beta\}}\right) &= \det(\boldsymbol{S}_{CC})\det\left(\boldsymbol{S}_{\{\alpha,\beta\}\{\alpha,\beta\}} - \boldsymbol{S}_{\{\alpha,\beta\}C}\boldsymbol{S}_{CC}^{-1}\boldsymbol{S}_{C\{\alpha,\beta\}}\right) \\ \det(\boldsymbol{S}_{C\cup\alpha,C\cup\alpha}) &= \det(\boldsymbol{S}_{CC})\left(\boldsymbol{S}_{\alpha\alpha} - \boldsymbol{S}_{\alpha C}\boldsymbol{S}_{CC}^{-1}\boldsymbol{S}_{C\alpha}\right) \\ \det(\boldsymbol{S}_{C\cup\beta,C\cup\beta}) &= \det(\boldsymbol{S}_{CC})\left(\boldsymbol{S}_{\beta\beta} - \boldsymbol{S}_{\beta C}\boldsymbol{S}_{CC}^{-1}\boldsymbol{S}_{C\beta}\right) . \end{aligned}$$

Inserting these results in (7.37) page 153 the likelihood ratio test statistic becomes

$$Q^{\frac{1}{n}}(\boldsymbol{X}) = \frac{\det\left(\boldsymbol{S}_{\{\alpha,\beta\}\{\alpha,\beta\}} - \boldsymbol{S}_{\{\alpha,\beta\}C}\boldsymbol{S}_{CC}^{-1}\boldsymbol{S}_{C\{\alpha,\beta\}}\right)}{\left(\boldsymbol{S}_{\alpha\alpha} - \boldsymbol{S}_{\alpha C}\boldsymbol{S}_{CC}^{-1}\boldsymbol{S}_{C\alpha}\right)\left(\boldsymbol{S}_{\beta\beta} - \boldsymbol{S}_{\beta C}\boldsymbol{S}_{CC}^{-1}\boldsymbol{S}_{C\beta}\right)} .$$

Using (7.38) page 153 we find that

$$\begin{aligned}
&\det\left(\boldsymbol{S}_{\{\alpha,\beta\}\{\alpha,\beta\}} - \boldsymbol{S}_{\{\alpha,\beta\}C}\boldsymbol{S}_{CC}^{-1}\boldsymbol{S}_{C\{\alpha,\beta\}}\right)\\
&= \left(\boldsymbol{S}_{\alpha\alpha} - \boldsymbol{S}_{\alpha C}\boldsymbol{S}_{CC}^{-1}\boldsymbol{S}_{C\alpha}\right)\left(\boldsymbol{S}_{_eta\beta} - \boldsymbol{S}_{\beta C}\boldsymbol{S}_{CC}^{-1}\boldsymbol{S}_{C\beta}\right) - \left(\boldsymbol{S}_{\alpha\beta} - \boldsymbol{S}_{\alpha C}\boldsymbol{S}_{CC}^{-1}\boldsymbol{S}_{C\beta}\right)\left(\boldsymbol{S}_{\beta\alpha} - \boldsymbol{S}_{\beta C}\boldsymbol{S}_{CC}^{-1}\boldsymbol{S}_{C\alpha}\right).
\end{aligned}$$

For reasons of simplicity we introduce the complex random variables

$$\begin{aligned}
U_1 &= \left(\boldsymbol{S}_{\alpha\alpha} - \boldsymbol{S}_{\alpha C}\boldsymbol{S}_{CC}^{-1}\boldsymbol{S}_{C\alpha}\right) - \left(\boldsymbol{S}_{\alpha\beta} - \boldsymbol{S}_{\alpha C}\boldsymbol{S}_{CC}^{-1}\boldsymbol{S}_{C\beta}\right)\left(\boldsymbol{S}_{\beta\beta} - \boldsymbol{S}_{\beta C}\boldsymbol{S}_{CC}^{-1}\boldsymbol{S}_{C\beta}\right)^{-1}\left(\boldsymbol{S}_{\beta\alpha} - \boldsymbol{S}_{\beta C}\boldsymbol{S}_{CC}^{-1}\boldsymbol{S}_{C\alpha}\right)\\
U_2 &= \left(\boldsymbol{S}_{\alpha\beta} - \boldsymbol{S}_{\alpha C}\boldsymbol{S}_{CC}^{-1}\boldsymbol{S}_{C\beta}\right)\left(\boldsymbol{S}_{\beta\beta} - \boldsymbol{S}_{\beta C}\boldsymbol{S}_{CC}^{-1}\boldsymbol{S}_{C\beta}\right)^{-1}\left(\boldsymbol{S}_{\beta\alpha} - \boldsymbol{S}_{\beta C}\boldsymbol{S}_{CC}^{-1}\boldsymbol{S}_{C\alpha}\right).
\end{aligned}$$

Hence the likelihood ratio test statistic is

$$Q^{\frac{1}{n}}(\boldsymbol{X}) = \frac{U_1}{U_1 + U_2} .$$

From Theorem 3.6 page 44 the distribution of $\boldsymbol{S}_{\{\alpha,\beta\}\{\alpha,\beta\}} - \boldsymbol{S}_{\{\alpha,\beta\}C}\boldsymbol{S}_{CC}^{-1}\boldsymbol{S}_{C\{\alpha,\beta\}}$ is determined as

$$\begin{aligned}
&\mathcal{L}\left(\boldsymbol{S}_{\{\alpha,\beta\}\{\alpha,\beta\}} - \boldsymbol{S}_{\{\alpha,\beta\}C}\boldsymbol{S}_{CC}^{-1}\boldsymbol{S}_{C\{\alpha,\beta\}}\right)\\
&= \frac{1}{n}\mathbb{C}\mathcal{W}_2\left(\left(\boldsymbol{K}(\mathcal{G})^{-1}\right)_{\{\alpha,\beta\}\{\alpha,\beta\}} - \left(\boldsymbol{K}(\mathcal{G})^{-1}\right)_{\{\alpha,\beta\}C}\left(\boldsymbol{K}(\mathcal{G})^{-1}\right)_{CC}^{-1}\left(\boldsymbol{K}(\mathcal{G})^{-1}\right)_{C\{\alpha,\beta\}}, n-|C|\right)\\
&= \frac{1}{n}\mathbb{C}\mathcal{W}_2\left(\boldsymbol{K}(\mathcal{G})_{\{\alpha,\beta\}\{\alpha,\beta\}}^{-1}, n-|C|\right)
\end{aligned}$$

and

$$\boldsymbol{S}_{\{\alpha,\beta\}\{\alpha,\beta\}} - \boldsymbol{S}_{\{\alpha,\beta\}C}\boldsymbol{S}_{CC}^{-1}\boldsymbol{S}_{C\{\alpha,\beta\}} \perp\!\!\!\perp \left(\boldsymbol{S}_{\{\alpha,\beta\}C}, \boldsymbol{S}_{CC}\right) . \tag{7.39}$$

Note that it is not necessary to assume $n \geq |C|$ to apply the theorem, since $\boldsymbol{S}(\mathcal{G})$ being $\mathcal{G}$-regular with probability one implies that $\boldsymbol{S}_{CC}^{-1}$ exists with probability one.

Since $\boldsymbol{K}(\mathcal{G}) \in \mathbb{C}_+(\mathcal{G}')$ we have $\boldsymbol{K}(\mathcal{G})_{\alpha\beta} = \boldsymbol{K}(\mathcal{G})_{\beta\alpha} = 0$. Hereby

$$\boldsymbol{K}(\mathcal{G})_{\{\alpha,\beta\}\{\alpha,\beta\}}^{-1} = \begin{pmatrix} \boldsymbol{K}(\mathcal{G})_{\alpha\alpha}^{-1} & 0 \\ 0 & \boldsymbol{K}(\mathcal{G})_{\beta\beta}^{-1} \end{pmatrix}$$

and again according to Theorem 3.6 the distributions of U_1 and U_2 are

$$\begin{aligned}
\mathcal{L}(U_1) &= \frac{1}{n}\mathbb{C}\mathcal{W}_1\left(\boldsymbol{K}(\mathcal{G})_{\alpha\alpha}^{-1}, n-|C|-1\right)\\
&= \frac{1}{2n}\boldsymbol{K}(\mathcal{G})_{\alpha\alpha}^{-1}\chi^2_{2(n-|C|-1)}
\end{aligned}$$

and

$$\begin{aligned}\mathcal{L}(U_2) &= \frac{1}{n}\mathbb{C}\mathcal{W}_1\left(\boldsymbol{K}(\mathcal{G})^{-1}_{\alpha\alpha}, 1\right) \\ &= \frac{1}{2n}\boldsymbol{K}(\mathcal{G})^{-1}_{\alpha\alpha}\chi^2_2 .\end{aligned}$$

Furthermore U_1, U_2 and $\boldsymbol{S}_{\beta\beta} - \boldsymbol{S}_{\beta C}\boldsymbol{S}^{-1}_{CC}\boldsymbol{S}_{C\beta}$ are mutually independent. From the above we know under H_0 that

$$\mathcal{L}\left(Q^{\frac{1}{n}}(\boldsymbol{X})\right) = \mathcal{B}(n - |C| - 1, 1)$$

and

$$Q^{\frac{1}{n}}(\boldsymbol{X}) \perp\!\!\!\perp U_1 + U_2 .$$

Knowing the distribution of $Q^{\frac{1}{n}}(\boldsymbol{X})$ under H_0 we are able to determine the critical region for the likelihood ratio test of size γ of H_0 under H.

We turn to the problem of showing under H_0 that $Q^{\frac{1}{n}}(\boldsymbol{X})$ and $\widehat{\boldsymbol{K}}(\mathcal{G}')$ are independent, where $\widehat{\boldsymbol{K}}(\mathcal{G}')$ is the maximum likelihood estimate of $\boldsymbol{K}(\mathcal{G}) \in \mathbb{C}_+(\mathcal{G}')$. First we show that

i. $Q^{\frac{1}{n}}(\boldsymbol{X})$ and $\left(\boldsymbol{S}_{\alpha\alpha} - \boldsymbol{S}_{\alpha C}\boldsymbol{S}^{-1}_{CC}\boldsymbol{S}_{C\alpha}, \boldsymbol{S}_{\beta\beta} - \boldsymbol{S}_{\beta C}\boldsymbol{S}^{-1}_{CC}\boldsymbol{S}_{C\beta}, \boldsymbol{S}_{\{\alpha,\beta\}C}, \boldsymbol{S}_{CC}\right)$ are independent under H_0.

From (7.39) page 154 we know that

$$\boldsymbol{S}_{\{\alpha,\beta\}\{\alpha,\beta\}} - \boldsymbol{S}_{\{\alpha,\beta\}C}\boldsymbol{S}^{-1}_{CC}\boldsymbol{S}_{C\{\alpha,\beta\}} \perp\!\!\!\perp \left(\boldsymbol{S}_{\{\alpha,\beta\}C}, \boldsymbol{S}_{CC}\right) .$$

Thus it holds that

$$\left(U_1, U_2, \boldsymbol{S}_{\beta\beta} - \boldsymbol{S}_{\beta C}\boldsymbol{S}^{-1}_{CC}\boldsymbol{S}_{C\beta}\right) \perp\!\!\!\perp \left(\boldsymbol{S}_{\{\alpha,\beta\}C}, \boldsymbol{S}_{CC}\right) ,$$

which implies

$$\left(Q^{\frac{1}{n}}(\boldsymbol{X}), U_1 + U_2, \boldsymbol{S}_{\beta\beta} - \boldsymbol{S}_{\beta C}\boldsymbol{S}^{-1}_{CC}\boldsymbol{S}_{C\beta}\right) \perp\!\!\!\perp \left(\boldsymbol{S}_{\{\alpha,\beta\}C}, \boldsymbol{S}_{CC}\right) . \tag{7.40}$$

Also we have from above that U_1, U_2 and $\boldsymbol{S}_{\beta\beta} - \boldsymbol{S}_{\beta C}\boldsymbol{S}^{-1}_{CC}\boldsymbol{S}_{C\beta}$ are mutually independent, whereby

$$(U_1, U_2) \perp\!\!\!\perp \boldsymbol{S}_{\beta\beta} - \boldsymbol{S}_{\beta C}\boldsymbol{S}^{-1}_{CC}\boldsymbol{S}_{C\beta}$$

implying that

$$\left(Q^{\frac{1}{n}}(\boldsymbol{X}), U_1 + U_2\right) \perp\!\!\!\perp \boldsymbol{S}_{\beta\beta} - \boldsymbol{S}_{\beta C}\boldsymbol{S}^{-1}_{CC}\boldsymbol{S}_{C\beta} .$$

Moreover we also know that

$$Q^{\frac{1}{n}}(\boldsymbol{X}) \perp\!\!\!\perp U_1 + U_2 ,$$

whereby it holds that

$$Q^{\frac{1}{n}}(\boldsymbol{X}),\ U_1+U_2 \text{ and } \boldsymbol{S}_{\beta\beta}-\boldsymbol{S}_{\beta C}\boldsymbol{S}_{CC}^{-1}\boldsymbol{S}_{C\beta}$$

are mutually independent. Combining this with (7.40) page 155 we observe that

$$Q^{\frac{1}{n}}(\boldsymbol{X}),\ U_1+U_2,\ \boldsymbol{S}_{\beta\beta}-\boldsymbol{S}_{\beta C}\boldsymbol{S}_{CC}^{-1}\boldsymbol{S}_{C\beta} \text{ and } \left(\boldsymbol{S}_{\{\alpha,\beta\}C},\boldsymbol{S}_{CC}\right)$$

are mutually independent. Thus we finally conclude under H_0 that

$$Q^{\frac{1}{n}}(\boldsymbol{X}) \perp\!\!\!\perp \left(\boldsymbol{S}_{\alpha\alpha}-\boldsymbol{S}_{\alpha C}\boldsymbol{S}_{CC}^{-1}\boldsymbol{S}_{C\alpha},\boldsymbol{S}_{\beta\beta}-\boldsymbol{S}_{\beta C}\boldsymbol{S}_{CC}^{-1}\boldsymbol{S}_{C\beta},\boldsymbol{S}_{\{\alpha,\beta\}C},\boldsymbol{S}_{CC}\right).$$

We introduce

$$\tilde{\boldsymbol{S}}=\left(\boldsymbol{S}_{C\cup\alpha,C\cup\alpha},\boldsymbol{S}_{C\cup\beta,C\cup\beta}\right)$$

and by using i. we show that

ii. $Q^{\frac{1}{n}}(\boldsymbol{X})$ and $\tilde{\boldsymbol{S}}$ are independent under H_0.

We observe from (7.38) page 153 that

$$\begin{aligned}\boldsymbol{S}_{\alpha\alpha} &= \boldsymbol{S}_{\alpha\alpha}-\boldsymbol{S}_{\alpha C}\boldsymbol{S}_{CC}^{-1}\boldsymbol{S}_{C\alpha}+\left(\boldsymbol{S}_{\{\alpha,\beta\}C}\boldsymbol{S}_{CC}^{-1}\boldsymbol{S}_{C\{\alpha,\beta\}}\right)_{\alpha\alpha}\\ \boldsymbol{S}_{\beta\beta} &= \boldsymbol{S}_{\beta\beta}-\boldsymbol{S}_{\beta C}\boldsymbol{S}_{CC}^{-1}\boldsymbol{S}_{C\beta}+\left(\boldsymbol{S}_{\{\alpha,\beta\}C}\boldsymbol{S}_{CC}^{-1}\boldsymbol{S}_{C\{\alpha,\beta\}}\right)_{\beta\beta}.\end{aligned}$$

Hereby from i. we deduce under H_0 that

$$Q^{\frac{1}{n}}(\boldsymbol{X}) \perp\!\!\!\perp \left(\boldsymbol{S}_{\alpha\alpha},\boldsymbol{S}_{\beta\beta},\boldsymbol{S}_{\{\alpha,\beta\}C},\boldsymbol{S}_{CC}\right),$$

which implies ii.

To continue the proof we define the sets A and B given by

$$A=\{\gamma\in V \mid \gamma \text{ and } \beta \text{ are separated by } C\cup\alpha\} \text{ and } B=(V\setminus A)\cup C.$$

We seek to establish

iii. $Q^{\frac{1}{n}}(\boldsymbol{X})$ and $(\boldsymbol{S}_{AA},\boldsymbol{S}_{BB})$ are independent under H_0.

Let D_1, D_2 and D_3 be measurable sets in the sample spaces of $Q^{\frac{1}{n}}(\boldsymbol{X})$, $\boldsymbol{S}_{AA}$ and $\boldsymbol{S}_{BB}$, respectively. By the law of total probability (Theorem 6.1 page 100) we get

$$\begin{aligned}&P\left(Q^{\frac{1}{n}}(\boldsymbol{X})\in D_1,\boldsymbol{S}_{AA}\in D_2,\boldsymbol{S}_{BB}\in D_3\right)\\ &=\int P\left(Q^{\frac{1}{n}}(\boldsymbol{X})\in D_1,\boldsymbol{S}_{AA}\in D_2,\boldsymbol{S}_{BB}\in D_3 \,\middle|\, \boldsymbol{X}_{C\cup\{\alpha,\beta\}}=\boldsymbol{x}_{C\cup\{\alpha,\beta\}}\right) f_{\boldsymbol{X}_{C\cup\{\alpha,\beta\}}}\left(\boldsymbol{x}_{C\cup\{\alpha,\beta\}}\right)d\boldsymbol{x}_{C\cup\{\alpha,\beta\}}.\end{aligned}$$

Define for an arbitrary measurable set D in the sample space of a complex random matrix $\boldsymbol{X}$ the indicator function given by

$$\mathbf{1}_D(\boldsymbol{x}) = \begin{cases} 1 & \boldsymbol{x} \in D \\ 0 & \boldsymbol{x} \notin D . \end{cases}$$

As $Q^{\frac{1}{n}}(\boldsymbol{X})$ only depends on $\boldsymbol{X}_{C\cup\{\alpha,\beta\}}$ we can write

$$\begin{aligned} &P\left(Q^{\frac{1}{n}}(\boldsymbol{X}) \in D_1, \boldsymbol{S}_{AA} \in D_2, \boldsymbol{S}_{BB} \in D_3\right) \\ &= \int \mathbf{1}_{D_1}\left(Q^{\frac{1}{n}}(\boldsymbol{x})\right) P\left(\boldsymbol{S}_{AA} \in D_2, \boldsymbol{S}_{BB} \in D_3 \middle| \boldsymbol{X}_{C\cup\{\alpha,\beta\}} = \boldsymbol{x}_{C\cup\{\alpha,\beta\}}\right) f_{\boldsymbol{X}_{C\cup\{\alpha,\beta\}}}\left(\boldsymbol{x}_{C\cup\{\alpha,\beta\}}\right) d\boldsymbol{x}_{C\cup\{\alpha,\beta\}}. \end{aligned}$$

From the definition of A and B we deduce that $A \cup \beta$ and $B \cup \alpha$ form a decomposition of $\mathcal{G}$ and that $(A \cup \beta) \cap (B \cup \alpha) = C \cup \{\alpha, \beta\}$. As stated on page 120 the density function of each column of $\boldsymbol{X}^*$ w.r.t. Lebesgue measure on $\mathbb{C}^{|V|}$ is positive and continuous and the distribution of each column of $\boldsymbol{X}^*$ factorizes according to $\mathcal{G}$. Using Theorem 6.9 page 112 together with the mutual independence of the columns of $\boldsymbol{X}^*$ we deduce that

$$f_{\boldsymbol{X}}(\boldsymbol{x}) = \frac{f_{\boldsymbol{X}_{A\cup\beta}}(\boldsymbol{x}_{A\cup\beta}) f_{\boldsymbol{X}_{B\cup\alpha}}(\boldsymbol{x}_{B\cup\alpha})}{f_{\boldsymbol{X}_{C\cup\{\alpha,\beta\}}}\left(\boldsymbol{x}_{C\cup\{\alpha,\beta\}}\right)},$$

whereby Theorem 6.4 page 103 tells us that

$$\boldsymbol{X}_{(A\cup\beta)\setminus(C\cup\{\alpha,\beta\})} \perp\!\!\!\perp \boldsymbol{X}_{(B\cup\alpha)\setminus(C\cup\{\alpha,\beta\})} | \boldsymbol{X}_{C\cup\{\alpha,\beta\}} .$$

By use of Theorem 6.5 page 104 it follows that

$$\boldsymbol{X}_A \perp\!\!\!\perp \boldsymbol{X}_B \mid \boldsymbol{X}_{C\cup\{\alpha,\beta\}} ,$$

which implies

$$\boldsymbol{S}_{AA} \perp\!\!\!\perp \boldsymbol{S}_{BB} \mid \boldsymbol{X}_{C\cup\{\alpha,\beta\}} .$$

Using this we obtain

$$\begin{aligned} &P\left(Q^{\frac{1}{n}}(\boldsymbol{X}) \in D_1, \boldsymbol{S}_{AA} \in D_2, \boldsymbol{S}_{BB} \in D_3\right) \\ &= \int \mathbf{1}_{D_1}\left(Q^{\frac{1}{n}}(\boldsymbol{x})\right) P\left(\boldsymbol{S}_{AA} \in D_2 \middle| \boldsymbol{X}_{C\cup\{\alpha,\beta\}} = \boldsymbol{x}_{C\cup\{\alpha,\beta\}}\right) P\left(\boldsymbol{S}_{BB} \in D_3 \middle| \boldsymbol{X}_{C\cup\{\alpha,\beta\}} = \boldsymbol{x}_{C\cup\{\alpha,\beta\}}\right) \\ &\quad f_{\boldsymbol{X}_{C\cup\{\alpha,\beta\}}}\left(\boldsymbol{x}_{C\cup\{\alpha,\beta\}}\right) d\boldsymbol{x}_{C\cup\{\alpha,\beta\}}. \end{aligned}$$

According to Theorem 5.7 page 96 we have that A and $C \cup \{\alpha, \beta\}$ form a decomposition of $\mathcal{G}_{A\cup\beta}$ and that $A \cap (C \cup \{\alpha, \beta\}) = C \cup \alpha$. As above we obtain that

$$\boldsymbol{X}_{A\setminus(C\cup\alpha)} \perp\!\!\!\perp \boldsymbol{X}_{(C\cup\{\alpha,\beta\})\setminus(C\cup\alpha)} \Big| \boldsymbol{X}_{C\cup\alpha} ,$$

which yields that

$$\boldsymbol{X}_A \perp\!\!\!\perp \boldsymbol{X}_\beta \mid \boldsymbol{X}_{C\cup\alpha}$$

and further

$$S_{AA} \perp\!\!\!\perp X_\beta \mid X_{C\cup\alpha} ,$$

This means according to Theorem 6.2 page 101 that

$$P\left(S_{AA} \in D_2 \mid X_{C\cup\{\alpha,\beta\}} = x_{C\cup\{\alpha,\beta\}}\right) = P\left(S_{AA} \in D_2 \mid X_{C\cup\alpha} = x_{C\cup\alpha}\right) .$$

For further calculations the marginal distribution of X_A is a necessity. According to Theorem 2.23 page 36 we have

$$\mathcal{L}(X_A) = \mathbb{C}\mathcal{N}_{n\times|A|}\left(O, I_n \otimes \left(K(\mathcal{G})^{-1}\right)_{AA}\right) ,$$

where $\left(K(\mathcal{G})^{-1}\right)_{AA}^{-1} \in \mathbb{C}_+(\mathcal{G}_A)$ according to Lemma 7.4 page 138 and the fact that A and B form a decomposition of $\mathcal{G}'$.

Using Lemma 7.6 page 150 we obtain

$$P\left(S_{AA} \in D_2 \mid X_{C\cup\alpha} = x_{C\cup\alpha}\right) = h_1\left(D_2, s_{C\cup\alpha,C\cup\alpha}\right) .$$

Similarly from Theorem 5.7 we obtain that B and $C \cup \{\alpha,\beta\}$ form a decomposition of $\mathcal{G}_{B\cup\alpha}$ and that $B \cap (C \cup \{\alpha,\beta\}) = C \cup \beta$. Hence as above

$$\begin{aligned} P\left(S_{BB} \in D_3 \mid X_{C\cup\{\alpha,\beta\}} = x_{C\cup\{\alpha,\beta\}}\right) &= P\left(S_{BB} \in D_3 \mid X_{C\cup\beta} = x_{C\cup\beta}\right) \\ &= h_2\left(D_3, s_{C\cup\beta,C\cup\beta}\right) . \end{aligned}$$

From these two results we deduce that

$$P\left(Q^{\frac{1}{n}}(X) \in D_1, S_{AA} \in D_2, S_{BB} \in D_3\right) = \int \mathbf{1}_{D_1}\left(Q^{\frac{1}{n}}(x)\right) h(\tilde{s}, D_2, D_3) f_{X_{C\cup\{\alpha,\beta\}}}\left(x_{C\cup\{\alpha,\beta\}}\right) dx_{C\cup\{\alpha,\beta\}},$$

where $h(\tilde{s}, D_2, D_3) = h_1\left(D_2, s_{C\cup\alpha,C\cup\alpha}\right) h_2\left(D_3, s_{C\cup\beta,C\cup\beta}\right)$ and $\tilde{s}$ represents an observation matrix of $\tilde{S}$. Hereby it holds that

$$P\left(Q^{\frac{1}{n}}(X) \in D_1, S_{AA} \in D_2, S_{BB} \in D_3\right) = \mathbb{E}\left(\mathbf{1}_{D_1}\left(Q^{\frac{1}{n}}(X)\right) h\left(\tilde{S}, D_2, D_3\right)\right) .$$

Therefore using ii. and Theorem 1.10 page 13 we find under H_0 that

$$\begin{aligned} P\left(Q^{\frac{1}{n}}(X) \in D_1, S_{AA} \in D_2, S_{BB} \in D_3\right) &= \mathbb{E}\left(\mathbf{1}_{D_1}\left(Q^{\frac{1}{n}}(X)\right)\right) \mathbb{E}\left(h\left(\tilde{S}, D_2, D_3\right)\right) \\ &= P\left(Q^{\frac{1}{n}}(X) \in D_1\right) k(D_2, D_3) , \end{aligned}$$

where $k(D_2, D_3) = \mathbb{E}\left(h\left(\tilde{S}, D_2, D_3\right)\right)$. According to Theorem 6.2 we get

$$Q^{\frac{1}{n}}(X) \perp\!\!\!\perp (S_{AA}, S_{BB}) .$$

As the maximum likelihood estimate of $K(\mathcal{G}) \in \mathbb{C}_+(\mathcal{G}')$ denoted by $\widehat{K}(\mathcal{G}')$ can be found from (S_{AA}, S_{BB}) we finally conclude from iii. that

$$\widehat{K}(\mathcal{G}') \perp\!\!\!\perp Q^{\frac{1}{n}}(X) .$$

Hereby the proof is completed. ■

In addition to the theorem above it should be noted that test of removal of a regular edge $\{\alpha,\beta\}$ can be performed, if only $n \geq |\mathrm{bd}(\alpha) \cap \mathrm{bd}(\beta)| + 2$ is fulfilled. Hereby the assumption of $S(\mathcal{G})$ being $\mathcal{G}$-regular with probability one is not needed.

7.6.1 Hypothesis Testing in Complex Normal Decomposable Models

When performing test in complex normal decomposable models, the distribution of the likelihood ratio test statistic can be found as the distribution of a product of mutually independent beta distributed random variables. This is studied in the following.

Theorem 7.7
Let $\mathcal{G} = (V, E)$ be a simple undirected decomposable graph and let $\mathcal{G}' = (V, E')$ be a decomposable subgraph of $\mathcal{G}$ such that $|E| - |E'| = k$. Furthermore let $\boldsymbol{X}$ be an $n \times |V|$ complex random matrix with $\mathcal{L}(\boldsymbol{X}) = \mathbb{C}\mathcal{N}_{n\times|V|}\left(\boldsymbol{O}, \boldsymbol{I}_n \otimes \boldsymbol{K}(\mathcal{G})^{-1}\right)$, where $\boldsymbol{K}(\mathcal{G}) \in \mathbb{C}_+(\mathcal{G})$. Moreover let $\boldsymbol{S} = \frac{1}{n}\boldsymbol{X}^*\boldsymbol{X}$ and let $\boldsymbol{S}(\mathcal{G})$ defined in (7.6) page 122 be $\mathcal{G}$-regular with probability one. Finally let $C_1, C_2, \ldots, C_m$ and $C_1', C_2', \ldots, C_q'$ be RIP-orderings of the cliques in $\mathcal{G}$ and $\mathcal{G}'$, respectively. The likelihood ratio test statistic from the test of the null hypothesis

$$\mathrm{H}_0 : \boldsymbol{K}(\mathcal{G}) \in \mathbb{C}_+(\mathcal{G}')$$

under the hypothesis

$$\mathrm{H} : \boldsymbol{K}(\mathcal{G}) \in \mathbb{C}_+(\mathcal{G})$$

is given by

$$Q^{\frac{1}{n}}(\boldsymbol{X}) = \frac{\prod_{j=2}^{q} \det\left(\boldsymbol{S}_{D_j'\cap C_j', D_j'\cap C_j'}\right) \prod_{j=1}^{m} \det\left(\boldsymbol{S}_{C_j C_j}\right)}{\prod_{j=2}^{m} \det\left(\boldsymbol{S}_{D_j\cap C_j, D_j\cap C_j}\right) \prod_{j=1}^{q} \det\left(\boldsymbol{S}_{C_j' C_j'}\right)},$$

where $D_j = \bigcup_{k<j} C_k$, $j = 2, 3, \ldots, m$, and $D_j' = \bigcup_{k<j} C_k'$, $j = 2, 3, \ldots, q$. Furthermore under H_0 the distribution of $Q^{\frac{1}{n}}(\boldsymbol{X})$ is given as the distribution of a product of k mutually independent beta distributed random variables and

$$Q^{\frac{1}{n}}(\boldsymbol{X}) \perp\!\!\!\perp \widehat{\boldsymbol{K}}(\mathcal{G}'),$$

where $\widehat{\boldsymbol{K}}(\mathcal{G}')$ is the maximum likelihood estimate of $\boldsymbol{K}(\mathcal{G}) \in \mathbb{C}_+(\mathcal{G}')$.

Proof:
Let the assumptions in the theorem be satisfied.

The likelihood ratio of the test of the null hypothesis

$$\mathrm{H}_0 : \boldsymbol{K}(\mathcal{G}) \in \mathbb{C}_+(\mathcal{G}')$$

under the hypothesis

$$\mathrm{H} : \boldsymbol{K}(\mathcal{G}) \in \mathbb{C}_+(\mathcal{G})$$

is from (7.31) page 148 given by

$$Q^{\frac{1}{n}}(\boldsymbol{x}) = \frac{\det\left(\widehat{\boldsymbol{K}}(\mathcal{G}')\right)}{\det\left(\widehat{\boldsymbol{K}}(\mathcal{G})\right)},$$

where $\widehat{K}(\mathcal{G}')$ and $\widehat{K}(\mathcal{G})$ are the maximum likelihood estimates of $K(\mathcal{G})$ under H_0 and H, respectively. According to Theorem 7.5 page 145 the likelihood ratio test statistic becomes

$$Q^{\frac{1}{n}}(X) = \frac{\prod_{j=2}^{q} \det\left(S_{D'_j \cap C'_j, D'_j \cap C'_j}\right) \prod_{j=1}^{m} \det\left(S_{C_j C_j}\right)}{\prod_{j=2}^{m} \det\left(S_{D_j \cap C_j, D_j \cap C_j}\right) \prod_{j=1}^{q} \det\left(S_{C'_j C'_j}\right)} ,$$

where $D_j = \bigcup_{k<j} C_k$, $j = 2, 3, \ldots, m$, and $D'_j = \bigcup_{k<j} C'_k$, $j = 2, 3, \ldots, q$.

By Theorem 5.6 page 96 we know that there exists a sequence $\mathcal{G}' = \mathcal{G}_0 \subset \mathcal{G}_1 \subset \cdots \subset \mathcal{G}_k = \mathcal{G}$ of decomposable graphs which differ by exactly one regular edge. Hereby the likelihood ratio can be factorized as

$$Q^{\frac{1}{n}}(x) = \frac{\det\left(\widehat{K}(\mathcal{G}_0)\right)}{\det\left(\widehat{K}(\mathcal{G}_1)\right)} \frac{\det\left(\widehat{K}(\mathcal{G}_1)\right)}{\det\left(\widehat{K}(\mathcal{G}_2)\right)} \cdots \frac{\det\left(\widehat{K}(\mathcal{G}_{k-1})\right)}{\det\left(\widehat{K}(\mathcal{G}_k)\right)} ,$$

where $\widehat{K}(\mathcal{G}_j)$ is maximum likelihood estimate of $K(\mathcal{G}) \in \mathbb{C}_+(\mathcal{G}_j)$ for $j = 0, 1, \ldots, k$. Note for $j = 1, 2, \ldots, k$ that $Q_j^{\frac{1}{n}}(x) = \frac{\det\left(\widehat{K}(\mathcal{G}_{j-1})\right)}{\det\left(\widehat{K}(\mathcal{G}_j)\right)}$ is the likelihood ratio from the test of removal of the regular edge which $\mathcal{G}_j$ and $\mathcal{G}_{j-1}$ differ by. We denote this particular regular edge by $\{\alpha_j, \beta_j\}$.

It appears from the above that

$$Q^{\frac{1}{n}}(X) = \prod_{j=1}^{k} Q_j^{\frac{1}{n}}(X) ,$$

and for $j = 1, 2, \ldots, k$ we conclude under H_0 by Theorem 7.6 page 152 that

$$\mathcal{L}\left(Q_j^{\frac{1}{n}}(X)\right) = \mathcal{B}(n - |\mathrm{bd}\,(\alpha_j) \cap \mathrm{bd}\,(\beta_j)| - 1, 1) ,$$

and that

$$Q_j^{\frac{1}{n}}(X) \perp\!\!\!\perp \widehat{K}(\mathcal{G}_{j-1}) . \tag{7.41}$$

This means that

$$Q_k^{\frac{1}{n}}(X) \perp\!\!\!\perp \widehat{K}(\mathcal{G}_{k-1}) .$$

Since $Q_{k-1}^{\frac{1}{n}}(X)$ and $\widehat{K}(\mathcal{G}_{k-2})$ are functions of $\widehat{K}(\mathcal{G}_{k-1})$, we hereby get

$$Q_k^{\frac{1}{n}}(X) \perp\!\!\!\perp \left(Q_{k-1}^{\frac{1}{n}}(X), \widehat{K}(\mathcal{G}_{k-2})\right) .$$

According to (7.41) it follows that

$$Q_k^{\frac{1}{n}}(X), Q_{k-1}^{\frac{1}{n}}(X) \text{ and } \widehat{K}(\mathcal{G}_{k-2}) \tag{7.42}$$

are mutually independent. Hence we know

$$\left(Q_k^{\frac{1}{n}}(X), Q_{k-1}^{\frac{1}{n}}(X)\right) \perp\!\!\!\perp \widehat{K}(\mathcal{G}_{k-2}) ,$$

which implies

$$\left(Q_k^{\frac{1}{n}}(\boldsymbol{X}), Q_{k-1}^{\frac{1}{n}}(\boldsymbol{X})\right) \perp\!\!\!\perp \left(Q_{k-2}^{\frac{1}{n}}(\boldsymbol{X}), \widehat{\boldsymbol{K}}(\mathcal{G}_{k-3})\right) .$$

Applying (7.42) and (7.41) page 160 tells us that

$$Q_k^{\frac{1}{n}}(\boldsymbol{X}), Q_{k-1}^{\frac{1}{n}}(\boldsymbol{X}), Q_{k-2}^{\frac{1}{n}}(\boldsymbol{X}) \text{ and } \widehat{\boldsymbol{K}}(\mathcal{G}_{k-3})$$

are mutually independent. Using this result successively we deduce that

$$Q_k^{\frac{1}{n}}(\boldsymbol{X}), Q_{k-1}^{\frac{1}{n}}(\boldsymbol{X}), \ldots, Q_1^{\frac{1}{n}}(\boldsymbol{X}) \text{ and } \widehat{\boldsymbol{K}}(\mathcal{G}_0)$$

are mutually independent. Hereby the $Q_j^{\frac{1}{n}}(\boldsymbol{X})$'s for $j = 1, 2, \ldots, k$ are mutually independent and moreover $Q^{\frac{1}{n}}(\boldsymbol{X})$ and $\widehat{\boldsymbol{K}}(\mathcal{G}')$ are independent under H_0. ■

Bibliography

Andersson, S. A., Brøns, H. K. & Jensen, S. T. (1983), 'Distribution of eigenvalues in multivariate statistical analysis', *The Annals of Statistics* **11**, 392–415.

Bain, L. J. & Engelhardt, M. (1989), *Introduction to Probability and Mathematical Statistics*, PWS-KENT Publishing Company.

Brillinger, D. R. (1975), *Time Series, Data Analysis and Theory*, Holt, Rinehart and Winston.

Conway, J. B. (1978), *Functions of One Complex Variable*, Graduate Texts in Mathematics 11, 2nd edn, Springer Verlag.

Cramér, H. (1945), *Mathematical Methods of Statistics*, Almqvist & Wiksells.

Darroch, J. N., Lauritzen, S. L. & Speed, T. P. (1980), 'Markov fields and log-linear interaction models for contingency tables', *The Annals of Statistics* **8**, 522–539.

Dawid, A. P. (1979), 'Conditional independence in statistical theory (with discussion)', *Journal of the Royal Statistical Society* **41**, 1–31.

Dempster, A. P. (1972), 'Covariance selection', *Biometrics* **28**, 157–175.

Eaton, M. L. (1983), *Multivariate Statistics, A Vector Space Approach*, John Wiley & Sons.

Edwards, C. H. & Penney, D. E. (1988), *Elementary Linear Algebra*, Prentice Hall International.

Eriksen, P. S. (1992), 'Covariance selection and graphical chain models, draft version', Department of Mathematics and Computer Science, Aalborg University. Lecture Notes.

Frederiksen, E. & Ragnarsson, R. (1986), En stokastisk model for mobil radio kommunikation, Master's thesis, Department of Mathematics and Computer Science, Aalborg University. In Danish.

Giri, N. (1965), 'On the complex analogues of t^2- and r^2-tests', *The Annals of Mathematical Statistics* **36**, 664–670.

Goodman, N. R. (1963), 'Statistical analysis based on a certain multivariate complex Gaussian distribution (an introduction)', *The Annals of Mathematical Statistics* **34**, 152 – 177.

Gupta, A. K. (1971), 'Distribution of Wilks' likelihood-ratio criterion in the complex case', *Annals of the Institute of Statistical Mathematics* **23**, 77–78.

Halmos, P. R. (1974), *Finite-Dimensional Vector Spaces*, Undergraduate Texts in Mathematics, Springer Verlag. Reprint of the 2nd edition published in 1958 by D. Van Nostrand Company, Princeton, N. J. ., in series: The University Series in Undergraduate Mathematics.

Jensen, J. L. (1991), 'A large deviation-type approximation for the "Box class" of the likelihood ratio criteria', *Journal of the American Statistical Association* **86**, 437–440.

Khatri, C. G. (1965*a*), 'Classical statistical analysis based on a certain multivariate complex distribution', *The Annals of Mathematical Statistics* **36**, 98–114.

Khatri, C. G. (1965*b*), 'A test for reality of a covariance matrix in a certain complex Gaussian distribution', *The Annals of Mathematical Statistics* **36**, 115–119.

Krishnaiah, P. R. (1976), 'Some recent developments on complex multivariate distributions', *Journal of Multivariate Analysis* **6**, 1–30.

Lauritzen, S. L. (1985), *Lectures on Multivariate Analysis*, 2nd edn, Institute of Mathematical Statistics, University of Copenhagen.

Lauritzen, S. L. (1989), *Lectures on Contingency Tables*, 3rd edn, Department of Mathematics and Computer Science, Aalborg University.

Lauritzen, S. L. (1993), 'Graphical association models, draft version', Department of Mathematics and Computer Science, Aalborg University.

Lauritzen, S. L. & Frydenberg, M. (1989), 'Decomposition of maximum likelihood in mixed graphical interaction models', *Biometrika* **76**, 539–555.

Leimer, H.-G. (1989), 'Triangulated graphs with marked vertices', *Annals of Discrete Mathematics* **41**, 311–324.

MacDuffee, C. C. (1956), *The Theory of Matrices*, Chelsea Publishing Company, New York.

Muirhead, R. J. (1982), *Aspects of Multivariate Statistical Theory*, John Wiley & Sons.

Pearl, J. (1988), *Probabilistic Reasoning in Intelligent Systems: Networks of Plausible Inference*, Morgan Kaufmann, San Mateo.

Pearl, J. & Paz, A. (1986), Graphoids: A graph based logic for reasoning about relevancy relations, *in* 'Proceedings of the European Conference on Artificial Intellligence, Brighton, United Kingdom'.

Pedersen, J. G. (1985), 'Den flerdimensionale normale fordeling', Department of Theoretical Statistics, University of Aarhus. Lecture Notes in Danish.

Rockafellar, R. T. (1970), *Convex Analysis*, Princeton University Press.

Rudin, W. (1987), *Principles of Mathematical Analysis*, McGraw-hill.

Saxena, A. K. (1978), 'Complex multivariate statistical analysis: an annotated bibliography', *International Statistical Review* **46**, 209–214.

Seber, G. A. F. (1984), *Multivariate Observations*, John Wiley & Sons.

Silvey, S. D. (1975), *Statistical Inference*, Monographs on Statistics and Applied Probability 7, Chapman & Hall.

Speed, T. P. (1979), 'A note on nearest-neighbour gibbs and markov probablilties', *Sankhyā: The Indian Journal of Statiatics* **41**, 184–197.

Speed, T. P. & Kiiveri, H. T. (1986), 'Gaussian Markov distributions over finite graphs', *The Annals of Statistics* **14**, 138–150.

Tarjan, R. E. & Yannakakis, M. (1984), 'Simple linear-time algorithms to test chordality of graphs, test acyclicity of hypergraphs, and selectively reduce acyclic hypergraphs', *SIAM Journal on Computing* **13**, 566–579.

Wermuth, N. (1976), 'Analogies between multiplicative models in contingency tables and covariance selection', *Biometrics* **32**, 95–108.

Whittaker, J. (1990), *Graphical Models in Applied Multivariate Statistics*, John Wiley & Sons.

Wooding, R. A. (1956), 'The multivariate distribution of complex normal variables', *Biometrika* **43**, 112–115.

A

Complex Matrices

This appendix contains results on complex matrices. It is rather elementary and it is only meant as help for readers who are not familiar with complex matrix algebra. The proofs are omitted but can be found in Eaton (1983), Halmos (1974), MacDuffee (1956), Giri (1965) or Goodman (1963).

A.1 Complex Vector Space

Definition A.1 Vector Space
Let F be a field. A set V is called a vector space over F if the following axioms hold.

1. *For all elements $\mathbf{c}, \mathbf{d} \in V$ there corresponds an element $\mathbf{c} + \mathbf{d} \in V$, called the sum, such that*

   i. *addition is commutative, $\mathbf{c} + \mathbf{d} = \mathbf{d} + \mathbf{c}$.*

   ii. *addition is associative, $(\mathbf{c} + \mathbf{d}) + \boldsymbol{f} = \mathbf{c} + (\mathbf{d} + \boldsymbol{f})$ for $\boldsymbol{f} \in V$.*

   iii. *there exists a unique element $\mathbf{0} \in V$ such that $\mathbf{c} + \mathbf{0} = \mathbf{c}$ for all $\mathbf{c} \in V$. The element $\mathbf{0}$ is called the zero element.*

   iv. *for all $\mathbf{c} \in V$ there corresponds a unique element $-\mathbf{c}$ such that $\mathbf{c} + (-\mathbf{c}) = \mathbf{0}$. The element $-\mathbf{c}$ is called the additive inverse.*

2. *For all $\alpha \in F$ and all $\mathbf{c} \in V$ there corresponds an element $\alpha\mathbf{c} \in V$, called the product of α and $\mathbf{c}$, such that*

   i. *multiplication by scalars is associative, $\alpha(\beta\mathbf{c}) = (\alpha\beta)\mathbf{c}$ for $\beta \in F$.*

   ii. *there exists a unique element $1 \in F$, such that $1\mathbf{c} = \mathbf{c}$ for all $\mathbf{c} \in V$. The element 1 is called the multiplicative identity.*

   iii. *multiplication by scalars is distributive w.r.t. addition of elements in F, $(\alpha + \beta)\mathbf{c} = \alpha\mathbf{c} + \beta\mathbf{c}$ for $\beta \in F$.*

   iv. *multiplication by scalars is distributive w.r.t. addition of elements in V, $\alpha(\mathbf{c} + \mathbf{d}) = \alpha\mathbf{c} + \alpha\mathbf{d}$ for $\mathbf{d} \in V$.*

If F in Definition A.1 is the field of real numbers, $\mathbb{R}$, the set V is called a *real vector space* and if F is the field of complex numbers, $\mathbb{C}$, we call V a *complex vector space*.

Definition A.2 Complex vector
A vector $\mathbf{c} = (c_k)$, where $c_k \in \mathbb{C}$ for $k = 1, 2, \ldots, p$, is called a complex vector.

The set of p-dimensional complex vectors, denoted by $\mathbb{C}^p$, is a p-dimensional complex vector space since the axioms for a vector space over $\mathbb{C}$ are fulfilled. Note that $\mathbb{C}^p$ also can be considered as a real vector space. The standard basis for $\mathbb{C}^p$ considered as a complex vector space is $(\mathbf{e}_1, \mathbf{e}_2, \ldots, \mathbf{e}_p)$, where $\mathbf{e}_j$ has the j'th element equal to one and the remaining elements equal to zero.

A.2 Basic Operations of Complex Matrices

Definition A.3 Complex matrix
An $n \times p$ array $\boldsymbol{C} = (c_{jk})$, where $c_{jk} \in \mathbb{C}$ for $j = 1, 2, \ldots, n$ and $k = 1, 2, \ldots, p$, is called an $n \times p$ complex matrix.

The set of all $n \times p$ complex matrices is denoted by $\mathbb{C}^{n \times p}$.

Definition A.4 Matrix addition
Let $\boldsymbol{C} = (c_{jk})$ and $\boldsymbol{D} = (d_{jk})$ be $n \times p$ complex matrices, then

$$\boldsymbol{C} + \boldsymbol{D} = (c_{jk} + d_{jk})$$

is the sum of $\boldsymbol{C}$ and $\boldsymbol{D}$, where $\boldsymbol{C} + \boldsymbol{D} \in \mathbb{C}^{n \times p}$.

Definition A.5 Matrix multiplication
Let $\boldsymbol{C} = (c_{jk}) \in \mathbb{C}^{n \times p}$ and $\boldsymbol{D} \in (d_{kl}) \in \mathbb{C}^{p \times m}$, then

$$\boldsymbol{C}\boldsymbol{D} = \left(\left(\sum_{k=1}^{p} c_{jk} d_{kl} \right)_{jl} \right)$$

is the product of $\boldsymbol{C}$ and $\boldsymbol{D}$, where $\boldsymbol{C}\boldsymbol{D} \in \mathbb{C}^{n \times m}$.

Notice that $\boldsymbol{D}\boldsymbol{C}$ is not defined.

Theorem A.1 Rules for complex matrix algebra
For suitable complex matrices and $\alpha, \beta \in \mathbb{C}$ the following properties hold.

1. $\boldsymbol{C} + \boldsymbol{D} = \boldsymbol{D} + \boldsymbol{C}$.

2. $(\boldsymbol{C}+\boldsymbol{D})+\boldsymbol{F}=\boldsymbol{C}+(\boldsymbol{D}+\boldsymbol{F})$.
3. $\boldsymbol{C}(\boldsymbol{D}+\boldsymbol{F})=\boldsymbol{C}\boldsymbol{D}+\boldsymbol{C}\boldsymbol{F}$.
4. $(\boldsymbol{C}+\boldsymbol{D})\boldsymbol{F}=\boldsymbol{C}\boldsymbol{F}+\boldsymbol{D}\boldsymbol{F}$.
5. $\boldsymbol{C}(\boldsymbol{D}\boldsymbol{F})=(\boldsymbol{C}\boldsymbol{D})\boldsymbol{F}$.
6. $\alpha(\beta\boldsymbol{C})=(\alpha\beta)\boldsymbol{C}=(\beta\alpha)\boldsymbol{C}=\beta(\alpha\boldsymbol{C})$.
7. $(\alpha+\beta)\boldsymbol{C}=\alpha\boldsymbol{C}+\beta\boldsymbol{C}$.
8. $\alpha(\boldsymbol{C}+\boldsymbol{D})=\alpha\boldsymbol{C}+\alpha\boldsymbol{D}$.
9. $\alpha(\boldsymbol{C}\boldsymbol{D})=\boldsymbol{C}(\alpha\boldsymbol{D})=(\alpha\boldsymbol{C})\boldsymbol{D}$.

With the definition of matrix addition and some of the rules of Theorem A.1 we see that the elements of $\mathbb{C}^{n\times p}$ fulfill the axioms for a vector space over $\mathbb{C}$. Hence $\mathbb{C}^{n\times p}$ is a complex vector space. Note that $\mathbb{C}^{n\times p}$ also can be considered as a real vector space.

A.3 Inverse Matrix

Definition A.6 Inverse Matrix
Let $\boldsymbol{C}\in\mathbb{C}^{p\times p}$. If there exists a $\boldsymbol{D}\in\mathbb{C}^{p\times p}$ such that $\boldsymbol{C}\boldsymbol{D}=\boldsymbol{D}\boldsymbol{C}=\boldsymbol{I}$, then $\boldsymbol{D}$ is called the inverse of $\boldsymbol{C}$ and is denoted by $\boldsymbol{C}^{-1}$.

Theorem A.2 Rules for inverse matrix
For suitable $\boldsymbol{C},\boldsymbol{D}\in\mathbb{C}^{p\times p}$ the following properties hold.

1. $\left(\boldsymbol{C}^{-1}\right)^{-1}=\boldsymbol{C}$.
2. $(\boldsymbol{C}\boldsymbol{D})^{-1}=\boldsymbol{D}^{-1}\boldsymbol{C}^{-1}$.

A.4 Determinant and Eigenvalues

Let $(j_1,j_2,\ldots,j_p)$ be a permutation of the first p positive integers. Let α_k be the number of integers following j_k that are smaller than j_k. Note that α_p is always zero. The sum $\sum_{k=1}^{p}\alpha_k$ is called the number of inversions in the permutation $(j_1,j_2,\ldots,j_p)$. A permutation is said to have even or odd parity according to whether its number of inversions is even or odd. We define

$$\delta(j_1,j_2,\ldots,j_p)=\begin{cases}1 & \text{if the parity of } (j_1,j_2,\ldots,j_p) \text{ is even}\\ -1 & \text{if the parity of } (j_1,j_2,\ldots,j_p) \text{ is odd}\end{cases},$$

which enables us to define the determinant of a square complex matrix.

Definition A.7 Determinant
The transformation det : $\mathbb{C}^{p\times p} \mapsto \mathbb{C}$ *is called the determinant and for a* $p \times p$ *complex matrix* $\boldsymbol{C} = (c_{jk})$ *it is given by*

$$\det(\boldsymbol{C}) = \sum \delta(j_1, j_2, \ldots, j_p)\, c_{1j_1} c_{2j_2} \cdots c_{pj_p} ,$$

where the sum is over all possible permutations $(j_1, j_2, \ldots, j_p)$ *of* $\{1, 2, \ldots, p\}$.

Definition A.8 Eigenvalue
Let $\boldsymbol{C} \in \mathbb{C}^{p\times p}$. *A number* $\lambda \in \mathbb{C}$ *is called an eigenvalue of* $\boldsymbol{C}$ *if*

(A.1) $$\det(\lambda \boldsymbol{I}_p - \boldsymbol{C}) = 0 .$$

Equation (A.1) is called the characteristic equation of $\boldsymbol{C}$. It is a p'th degree polynomial in λ. Hence a $p \times p$ complex matrix has p eigenvalues.

Theorem A.3 Rules of determinant
For $\boldsymbol{C}, \boldsymbol{D} \in \mathbb{C}^{p\times p}$ *the following properties hold.*

1. $\det(\boldsymbol{CD}) = \det(\boldsymbol{C})\det(\boldsymbol{D})$.
2. $\det(\boldsymbol{C}) \neq 0$ *iff the columns of* $\boldsymbol{C}$ *are linear independent vectors in* $\mathbb{C}^p$.
3. *If* $\lambda_1, \lambda_2, \ldots, \lambda_p$ *are the eigenvalues of* $\boldsymbol{C}$, *then* $\det(\boldsymbol{C}) = \prod_{j=1}^{p} \lambda_j$.
4. *If* $\lambda_1, \lambda_2, \ldots, \lambda_p$ *are the eigenvalues of* $\boldsymbol{C}$, *then* $\det(\boldsymbol{I}_p + \boldsymbol{C}) = \prod_{j=1}^{p} (1 + \lambda_j)$.
5. *The eigenvalues of* $\boldsymbol{CD}$ *are the same as the eigenvalues of* $\boldsymbol{DC}$.

Definition A.9 Nonsingular
Let $\boldsymbol{C} \in \mathbb{C}^{p\times p}$. *If* $\det(\boldsymbol{C}) \neq 0$, *then* $\boldsymbol{C}$ *is nonsingular.*

Notice that $\boldsymbol{C}$ is nonsingular iff $\boldsymbol{C}^{-1}$ exists.

A.5 Trace and Rank

Definition A.10 Trace operator
Let $\boldsymbol{C} = (c_{jk}) \in \mathbb{C}^{p\times p}$. *The trace operator,* tr : $\mathbb{C}^{p\times p} \mapsto \mathbb{C}$, *of* $\boldsymbol{C}$ *is given by*

$$\operatorname{tr}(\boldsymbol{C}) = \sum_{j=1}^{p} c_{jj} .$$

The value of $\operatorname{tr}(\boldsymbol{C})$ is called the trace of $\boldsymbol{C}$.

Theorem A.4 Rules for trace
For $\boldsymbol{C}, \boldsymbol{D} \in \mathbb{C}^{p \times p}$ the following properties hold.

1. $\operatorname{tr}(\boldsymbol{C}+\boldsymbol{D}) = \operatorname{tr}(\boldsymbol{C}) + \operatorname{tr}(\boldsymbol{D})$.
2. $\operatorname{tr}(\boldsymbol{CD}) = \operatorname{tr}(\boldsymbol{DC})$.
3. *If $\lambda_1, \lambda_2, \dots, \lambda_p$ are the eigenvalues of $\boldsymbol{C}$, then* $\operatorname{tr}(\boldsymbol{C}) = \sum_{j=1}^{p} \lambda_j$.

Definition A.11 Rank
The maximum number of linear independent columns of a complex matrix $\boldsymbol{C}$ is called the rank of the matrix and is denoted by $\operatorname{rank}(\boldsymbol{C})$.

A.6 Conjugate Transpose Matrix

Definition A.12 Conjugate transpose matrix
The conjugate transpose of an $n \times p$ complex matrix $\boldsymbol{C} = (c_{jk})$ is the $p \times n$ complex matrix given by $\boldsymbol{C}^* = (\bar{c}_{kj})$ for $k = 1, 2, \dots, p$ and $j = 1, 2, \dots, n$.

Rules for the conjugate transpose operation are equivalent to the rules for the transpose operation for real matrices.

Theorem A.5 Rules for the conjugate transpose operation
For suitable complex matrices and $c \in \mathbb{C}$ the following properties hold.

1. $(\boldsymbol{C}^*)^* = \boldsymbol{C}$.
2. $(c\boldsymbol{C})^* = \bar{c}\boldsymbol{C}^*$.
3. $(\boldsymbol{C}+\boldsymbol{D})^* = \boldsymbol{C}^* + \boldsymbol{D}^*$.
4. $(\boldsymbol{CD})^* = \boldsymbol{D}^*\boldsymbol{C}^*$.
5. $\det(\boldsymbol{C}^*) = \overline{\det(\boldsymbol{C})}$.
6. $\det(\{\boldsymbol{C}\}) = \det(\boldsymbol{C})\det(\boldsymbol{C}^*)$.
7. $\operatorname{tr}(\boldsymbol{C}^*) = \overline{\operatorname{tr}(\boldsymbol{C})}$.
8. $\left(\boldsymbol{C}^{-1}\right)^* = (\boldsymbol{C}^*)^{-1}$.
9. $\operatorname{rank}(\boldsymbol{C}) = \operatorname{rank}(\boldsymbol{C}^*) = \operatorname{rank}(\boldsymbol{CC}^*) = \operatorname{rank}(\boldsymbol{C}^*\boldsymbol{C})$.

A.7 Hermitian Matrix

An important property of a complex matrix is that it can be Hermitian. A complex matrix being Hermitian corresponds to a real matrix being symmetric.

Definition A.13 Hermitian
Let $C \in \mathbb{C}^{p \times p}$. If $C = C^*$, then C is Hermitian.

The set of all $p \times p$ Hermitian matrices is denoted by $\mathbb{C}_H^{p \times p}$. Notice that this set satisfies the axioms of a vector space over $\mathbb{C}$, hence $\mathbb{C}_H^{p \times p}$ is a complex vector space.

Definition A.14 Skew Hermitian
Let $C \in \mathbb{C}^{p \times p}$. If $C = -C^*$, then C is skew Hermitian.

Theorem A.6 Rules for Hermitian matrices
For $C \in \mathbb{C}^{p \times p}$ the following properties hold.

1. *Let $C = A + iB$, where $A, B \in \mathbb{R}^{p \times p}$. If C is Hermitian, then A is symmetric and B is skew symmetric.*
2. *C is Hermitian iff $\{C\}$ is symmetric.*
3. *If C is Hermitian, then* $\det(C)^2 = \det(\{C\})$.
4. *The eigenvalues of a Hermitian matrix are real.*
5. *The eigenvalues of a skew Hermitian matrix are imaginary.*
6. *Let C be Hermitian. The eigenvalues of C are all equal to zero iff $C = O$.*
7. *Let C be Hermitian. The eigenvalues of C are all equal to one iff $C = I_p$.*
8. *If C is Hermitian and* $\operatorname{rank}(C) = 1$, *then* $\det(I_p + C) = 1 + \operatorname{tr}(C)$.

A.8 Unitary Matrix

Definition A.15 Unitary
Let $C \in \mathbb{C}^{p \times p}$ be nonsingular. If $C^* = C^{-1}$, then C is unitary.

Notice that a complex matrix being unitary corresponds to a real matrix being orthogonal.

Theorem A.7 Rules for unitary matrices
For $C \in \mathbb{C}^{p \times p}$ the following properties hold.

1. *C is unitary iff $\{C\}$ is orthogonal.*

2. *If C is Hermitian, then it holds that there exists a unitary $p \times p$ matrix U such that $U^*CU = \text{diag}(\lambda_1, \lambda_2, \dots, \lambda_p)$, where λ_j for $j = 1, 2, \dots, p$ are the eigenvalues of C.*

A.9 Positive Semidefinite Complex Matrices

Definition A.16 Positive semidefinite complex matrix
Let $C \in \mathbb{C}_H^{p \times p}$. If $c^*Cc \geq 0$ for all $c \in \mathbb{C}^p$, then C is positive semidefinite. This is denoted by $C \geq O$.

The set of all $p \times p$ positive semidefinite complex matrices is denoted by $\mathbb{C}_S^{p \times p}$.

Theorem A.8 Rules for the positive semidefinite complex matrices
For $C \in \mathbb{C}_H^{p \times p}$ the following properties hold.

1. *$C \geq O$ with $\text{rank}(C) = r$ iff there exists a $D \in \mathbb{C}^{p \times p}$ with $\text{rank}(D) = r$ such that $C = D^*D$.*

2. *If $C \geq O$ and $D \in \mathbb{C}^{n \times p}$, then $DCD^* \geq O$.*

3. *$C \geq O$ iff $\{C\} \geq O$.*

4. *$C \geq O$ iff the eigenvalues of C are nonnegative.*

5. *If $C \geq O$, then $\text{tr}(C) \geq 0$.*

A.10 Positive Definite Complex Matrices

Definition A.17 Positive definite complex matrix
Let $C \in \mathbb{C}_H^{p \times p}$. If $c^*Cc > 0$ for all $c \in \mathbb{C}^p \setminus \{0\}$, then C is positive definite. This is denoted by $C > O$.

The set of all $p \times p$ positive definite complex matrices is denoted by $\mathbb{C}_+^{p \times p}$.

Theorem A.9 Rules for the positive definite complex matrices
For $\boldsymbol{C} \in \mathbb{C}_H^{p\times p}$ the following properties hold.

1. $\boldsymbol{C} \geq \boldsymbol{O}$ *and* $\boldsymbol{C}$ *is nonsingular iff* $\boldsymbol{C} > \boldsymbol{O}$.
2. $\boldsymbol{C} > \boldsymbol{O}$ *iff* $\boldsymbol{C}^{-1} > \boldsymbol{O}$.
3. $\boldsymbol{C} > \boldsymbol{O}$ *iff there exists a nonsingular complex matrix* $\boldsymbol{D} \in \mathbb{C}^{p\times p}$ *such that* $\boldsymbol{C} = \boldsymbol{D}\boldsymbol{D}^*$.
4. *If* $\boldsymbol{C} > \boldsymbol{O}$, *then there exists a complex matrix* $\boldsymbol{C}^{\frac{1}{2}} > \boldsymbol{O}$ *such that* $\boldsymbol{C} = \left(\boldsymbol{C}^{\frac{1}{2}}\right)^2$.
5. *If* $\boldsymbol{C} > \boldsymbol{O}$ *and* $\boldsymbol{D} \in \mathbb{C}^{n\times p}$ *with full rank, then* $\boldsymbol{D}\boldsymbol{C}\boldsymbol{D}^* > \boldsymbol{O}$.
6. $\boldsymbol{C} > \boldsymbol{O}$ *iff* $\{\boldsymbol{C}\} > \boldsymbol{O}$.
7. $\boldsymbol{C} > \boldsymbol{O}$ *iff the eigenvalues of* $\boldsymbol{C}$ *are positive.*
8. *If* $\boldsymbol{C} > \boldsymbol{O}$, *then* $\operatorname{tr}(\boldsymbol{C}) > 0$.
9. *If* $\boldsymbol{C} > \boldsymbol{O}$ *and* $a\boldsymbol{C} + b\boldsymbol{D} > \boldsymbol{O}$ *for all* $a, b \in \mathbb{R}_+$ *and* $\boldsymbol{D} \in \mathbb{C}^{p\times p}$, *then* $\boldsymbol{D} \geq \boldsymbol{O}$.
10. *If* $\boldsymbol{C} > \boldsymbol{O}$, $\boldsymbol{D} \in \mathbb{C}_+^{p\times p}$ *and* $a \in \overline{\mathbb{R}}_+$, *then* $\boldsymbol{C} + a\boldsymbol{D} > \boldsymbol{O}$.

A.11 Direct Product

Definition A.18 The direct product of matrices
Let $\boldsymbol{C} = (c_{jk})$ and $\boldsymbol{D} = (d_{rs})$ be $n \times p$ and $m \times q$ complex matrices, respectively. The direct product $\boldsymbol{C} \otimes \boldsymbol{D}$ is the $nm \times pq$ matrix with elements given as

$$(\boldsymbol{C} \otimes \boldsymbol{D})_{jr,ks} = c_{jk}\overline{d}_{rs} \,.$$

Theorem A.10 Rules for the direct product
For suitable complex matrices the direct product satisfies the following properties.

1. $\boldsymbol{O} \otimes \boldsymbol{C} = \boldsymbol{C} \otimes \boldsymbol{O} = \boldsymbol{O}$.
2. $\boldsymbol{C} \otimes (\boldsymbol{D} + \boldsymbol{E}) = \boldsymbol{C} \otimes \boldsymbol{D} + \boldsymbol{C} \otimes \boldsymbol{E}$.
3. $(\boldsymbol{C} + \boldsymbol{D}) \otimes \boldsymbol{E} = \boldsymbol{C} \otimes \boldsymbol{E} + \boldsymbol{D} \otimes \boldsymbol{E}$.
4. $(c\boldsymbol{C}) \otimes (d\boldsymbol{D}) = \left(c\overline{d}\right)(\boldsymbol{C} \otimes \boldsymbol{D})$, *where* $c, d \in \mathbb{C}$.
5. $(\boldsymbol{C} \otimes \boldsymbol{D})^* = \boldsymbol{C}^* \otimes \boldsymbol{D}^*$.
6. $(\boldsymbol{C} \otimes \boldsymbol{D})(\boldsymbol{E} \otimes \boldsymbol{F}) = (\boldsymbol{C}\boldsymbol{E}) \otimes (\boldsymbol{D}\boldsymbol{F})$.
7. *If the inverse complex matrices* $\boldsymbol{C}^{-1}$ *and* $\boldsymbol{D}^{-1}$ *exist, then* $(\boldsymbol{C} \otimes \boldsymbol{D})^{-1} = \boldsymbol{C}^{-1} \otimes \boldsymbol{D}^{-1}$.

8. *If* C *and* D *are square matrices, then* $\operatorname{tr}(C \otimes D) = \operatorname{tr}(C)\,\overline{\operatorname{tr}(D)}$.

9. *If* $C \in \mathbb{C}^{n\times n}$ *and* $D \in \mathbb{C}^{p\times p}$*, then* $\det(C \otimes D) = \det(C)^p\,\overline{\det(D)^n}$.

10. *If* C *and* D *both are positive (semi)definite matrices, then so is* $C \otimes D$.

11. *If* C *and* D *are orthogonal projections, then so is* $C \otimes D$.

A.12 Partitioned Complex Matrices

Theorem A.11
Let $C \in \mathbb{C}^{p\times p}$ *be partitioned as*

$$C = \begin{pmatrix} C_{11} & C_{12} \\ C_{21} & C_{22} \end{pmatrix},$$

where C_{jk} *has dimension* $p_j \times p_k$ *for* $j, k = 1, 2$*, and* $p_1 + p_2 = p$*. Further let* $C_{11\cdot 2} = C_{11} - C_{12}C_{22}^{-1}C_{21}$ *and* $C_{22\cdot 1} = C_{22} - C_{21}C_{11}^{-1}C_{12}$.

1. *If all the inverses exist, then*

$$C^{-1} = \begin{pmatrix} C_{11\cdot 2}^{-1} & -C_{11}^{-1}C_{12}C_{22\cdot 1}^{-1} \\ -C_{22}^{-1}C_{21}C_{11\cdot 2}^{-1} & C_{22\cdot 1}^{-1} \end{pmatrix}$$

or

$$C^{-1} = \begin{pmatrix} C_{11\cdot 2}^{-1} & -C_{11\cdot 2}^{-1}C_{12}C_{22}^{-1} \\ -C_{22}^{-1}C_{21}C_{11\cdot 2}^{-1} & C_{22}^{-1} + C_{22}^{-1}C_{21}C_{11\cdot 2}^{-1}C_{12}C_{22}^{-1} \end{pmatrix}.$$

2. *If* $p = 2$*, then*

$$C^{-1} = \frac{1}{\det(C)} \begin{pmatrix} C_{22} & -C_{12} \\ -C_{21} & C_{11} \end{pmatrix}.$$

3. *If* C_{11} *is nonsingular, then*

$$\det(C) = \det(C_{22\cdot 1})\det(C_{11}).$$

4. *If* C_{22} *is nonsingular, then*

$$\det(C) = \det(C_{11\cdot 2})\det(C_{22}).$$

5. *If* C_{11} *is nonsingular and* $C_{21} = C_{12}^*$*, then*

$$C \geq O \text{ iff } C_{11} \geq O \text{ and } C_{22\cdot 1} \geq O.$$

6. *If* C_{22} *is nonsingular and* $C_{21} = C_{12}^*$, *then*

$$C \geq O \text{ iff } C_{22} \geq O \text{ and } C_{11\cdot 2} \geq O\,.$$

7. *If* $C_{21} = C_{12}^*$, *then*

$$C > O \text{ iff } C_{11} > O \text{ and } C_{22\cdot 1} > O\,.$$

8. *If* $C_{21} = C_{12}^*$, *then*

$$C > O \text{ iff } C_{22} > O \text{ and } C_{11\cdot 2} > O\,.$$

9. *If* $C \geq O$, *then* $C_{11} \geq O$ *and* $C_{22} \geq O$.

10. *If* $C > O$, *then* $C_{11} > O$ *and* $C_{22} > O$.

B

Orthogonal Projections

This appendix contains results on orthogonal projections. The proofs are omitted but can be found in Eaton (1983) and MacDuffee (1956).

Definition B.1 Orthogonal projection
Let N and $N^{\perp}$ be subspaces in $\mathbb{C}^n$ such that $N \oplus N^{\perp} = \mathbb{C}^n$ and $N^{\perp}$ is the orthogonal complement of N w.r.t. the inner product on $\mathbb{C}^n$. If $\boldsymbol{x} = \boldsymbol{y} + \boldsymbol{z} \in \mathbb{C}^n$ with $\boldsymbol{y} \in N$ and $\boldsymbol{z} \in N^{\perp}$, then $\boldsymbol{y}$ is called the orthogonal projection of $\boldsymbol{x}$ onto N and $\boldsymbol{z}$ is called the orthogonal projection of $\boldsymbol{x}$ onto $N^{\perp}$.

Theorem B.1
For $\boldsymbol{P}_N \in \mathbb{C}^{n \times n}$ the following properties hold.

1. *Let $\boldsymbol{x} = \boldsymbol{y} + \boldsymbol{z} \in \mathbb{C}^n$ with $\boldsymbol{y} \in N$ and $\boldsymbol{z} \in N^{\perp}$. If $\boldsymbol{P}_N$ is a complex matrix such that $\boldsymbol{y} = \boldsymbol{P}_N \boldsymbol{x}$, then $\boldsymbol{P}_N$ represents the orthogonal projection of $\mathbb{C}^n$ onto N. The complex matrix $\boldsymbol{P}_N$ is also called a projection matrix.*
2. *Let $\boldsymbol{P}_N$ represents the orthogonal projection of $\mathbb{C}^n$ onto N. Then $\boldsymbol{I}_n - \boldsymbol{P}_N$ represents the orthogonal projection of $\mathbb{C}^n$ onto $N^{\perp}$. The complex matrix $\boldsymbol{I}_n - \boldsymbol{P}_N$ is also denoted by $\boldsymbol{P}_N^{\perp}$.*
3. *If $\boldsymbol{P}_N$ is an idempotent Hermitian matrix, i.e. $\boldsymbol{P}_N = \boldsymbol{P}_N^2 = \boldsymbol{P}_N^*$, then $\boldsymbol{P}_N$ represents the orthogonal projection of $\mathbb{C}^n$ onto $\mathcal{R}[\boldsymbol{P}_N]$.*
4. *If $\boldsymbol{P}_N$ is a projection matrix, then* $\text{rank}\,(\boldsymbol{P}_N) = \text{tr}\,(\boldsymbol{P}_N)$.
5. *Let $\boldsymbol{P}_N$ represent the orthogonal projection of $\mathbb{C}^n$ onto N. Then $\mathcal{R}[\boldsymbol{P}_N] = N$ and the dimension of N is* $\text{tr}\,(\boldsymbol{P}_N)$.
6. *Let $\boldsymbol{P}_N \in \mathbb{C}^{n \times n}$ be Hermitian. The complex matrix $\boldsymbol{P}_N$ is idempotent of rank r iff it has r eigenvalues equal to one and $n - r$ equal to zero.*
7. *If $\boldsymbol{Z} \in \mathbb{C}^{n \times p}$, $\mathcal{R}[\boldsymbol{Z}] = N$ and $\boldsymbol{Z}$ has full rank p, then a matrix representing the orthogonal projection of $\mathbb{C}^n$ onto N is given by $\boldsymbol{P}_N = \boldsymbol{Z}\left(\boldsymbol{Z}^*\boldsymbol{Z}\right)^{-1}\boldsymbol{Z}^*$.*
8. *If $\boldsymbol{Z} \in \mathbb{C}^{n \times p}, \mathcal{R}\left[\boldsymbol{Z} \otimes \boldsymbol{I}_p\right] = M$ and $\boldsymbol{Z}$ has full rank p, then a matrix representing the orthogonal projection of $\mathbb{C}^{n \times p}$ onto M is given by $\boldsymbol{P}_M = \boldsymbol{Z}\left(\boldsymbol{Z}^*\boldsymbol{Z}\right)^{-1}\boldsymbol{Z}^* \otimes \boldsymbol{I}_p$.*
9. *The vector space N_0 is a subspace of N iff $\boldsymbol{P}_N \boldsymbol{P}_{N_0} = \boldsymbol{P}_{N_0} \boldsymbol{P}_N = \boldsymbol{P}_{N_0}$.*

Index

Notation

The list given below contains symbols used in the book. For each symbol a short explanation is stated and if necessary pagereferences for further information are given in brackets.

Sets

$\mathbb{R}$	field of real numbers.
$\overline{\mathbb{R}}_+$	set of nonnegative real numbers.
$\mathbb{R}^p$	real vector space of p-dimensional real vectors.
$\mathbb{R}^{n\times p}$	real vector space of $n \times p$ real matrices.
$\mathbb{C}$	field of complex numbers.
$\mathbb{C}^p$	complex vector space of p-dimensional complex vectors.
$\mathbb{C}^{n\times p}$	complex vector space of $n \times p$ complex matrices.
$\mathbb{C}_H^{p\times p}$	complex vector space of $p \times p$ Hermitian matrices.
$\mathbb{C}_S^{p\times p}$	set of $p \times p$ positive semidefinite complex matrices.
$\mathbb{C}_+^{p\times p}$	set of $p \times p$ positive definite complex matrices.
$\mathbb{C}_H(\mathcal{G})$	set of $\|V\| \times \|V\|$ Hermitian matrices, which contains zero entries according to missing edges in $\mathcal{G} = (V, E)$, (125).
$\mathbb{C}_S(\mathcal{G})$	set of $\|V\| \times \|V\|$ positive semidefinite complex matrices, which contains zero entries according to missing edges in $\mathcal{G} = (V, E)$, (129).
$\mathbb{C}_+(\mathcal{G})$	set of $\|V\| \times \|V\|$ positive definite complex matrices, which contains zero entries according to missing edges in $\mathcal{G} = (V, E)$, (118).
$\mathcal{L}_2(\cdot)$	vector space over $\mathbb{C}$ of complex random "variables" with finite second moment, (5,8).

Complex Numbers

i	imaginary unit.
$\bar{c}$	complex conjugate of c.
$\mathrm{Re}(\cdot)$	real part.
$\mathrm{Im}(\cdot)$	imaginary part.
$\|\cdot\|$	absolute value.

Matrix Algebra

$\langle \cdot, \cdot \rangle$	inner product, (1).
$\{\cdot\}$	$2p \times 2p$ real matrix derived from a $p \times p$ complex matrix, (4).
$[\cdot]$	real vector space isomorphism between $\mathbb{C}^p$ and $\mathbb{R}^{2p}$, (3).
$[\![\cdot]\!]$	matrix obtained from a submatrix by filling in missing entries with zero entries, (116).
$\lVert \cdot \rVert$	length.
$\mathcal{R}[\cdot]$	range.
$\mathrm{tr}(\cdot)$	trace.
$\det(\cdot)$	determinant.
$\mathrm{rank}(\cdot)$	rank.
$\boldsymbol{A}^\top$	transpose of a real matrix $\boldsymbol{A}$.
$\boldsymbol{C}^*$	conjugate transpose of a complex matrix $\boldsymbol{C}$.
$\boldsymbol{C}^{-1}$	inverse of $\boldsymbol{C}$.
$\boldsymbol{C} \otimes \boldsymbol{D}$	direct product of two complex matrices, (2).
$\boldsymbol{I}_n$	$n \times n$ identity matrix.
$\mathbf{1}_n$	n-dimensional vector of ones.
$\boldsymbol{O}$	matrix of zeros.
$\mathrm{diag}(\cdots)$	diagonal matrix.
$\boldsymbol{C} \geq \boldsymbol{O}$	$\boldsymbol{C}$ is positive semidefinite.
$\boldsymbol{C} > \boldsymbol{O}$	$\boldsymbol{C}$ is positive definite.
$\boldsymbol{P}_N$	a matrix representing the orthogonal projection onto N.
$\boldsymbol{P}_N^\perp$	a matrix representing the orthogonal projection onto the orthogonal complement of N w.r.t. inner product, $N^\perp$.

Distributions

$\mathbb{E}(\cdot)$	expectation operator, (5, 8).
$\mathbb{V}(\cdot)$	variance operator, (6,9).
$\mathbb{C}(\cdot, \cdot)$	covariance operator, (5,9).
$\mathcal{L}(\cdot)$	distributional law.
$\mathcal{N}(\theta, \sigma^2)$	univariate real normal distribution with mean θ and variance σ^2.
$\mathcal{N}_p(\boldsymbol{\theta}, \boldsymbol{\Sigma})$	p-variate real normal distribution with mean $\boldsymbol{\theta}$ and variance matrix $\boldsymbol{\Sigma}$.
$\mathbb{C}\mathcal{N}(\theta, \sigma^2)$	univariate complex normal distribution with mean θ and variance σ^2.
$\mathbb{C}\mathcal{N}_p(\boldsymbol{\theta}, \boldsymbol{H})$	p-variate complex normal distribution with mean $\boldsymbol{\theta}$ and variance matrix $\boldsymbol{H}$.

$\mathbb{C}\mathcal{N}_{n\times p}(\Theta, \boldsymbol{J}\otimes\boldsymbol{H})$	$(n \times p)$-variate complex normal distribution with mean Θ and variance matrix $\boldsymbol{J}\otimes\boldsymbol{H}$.
$\mathbb{C}\mathcal{W}_p(\boldsymbol{H}, n)$	complex Wishart distribution with dimension p, n degrees of freedom and mean $n\boldsymbol{H}$.
$\mathbb{C}\mathcal{U}(p, m, n)$	complex U-distribution with parameters p, m and n.
χ^2_k	chi-square distribution with k degrees of freedom.
$\mathcal{B}(n, p)$	beta distribution with parameters n and p.
$\mathcal{F}_{n,p}$	F-distribution with parameters n and p.
P	distribution.
$\varphi_{\boldsymbol{X}}$	characteristic function of $\boldsymbol{X}$.
$f_{\boldsymbol{X}}$	density function of $\boldsymbol{X}$ w.r.t. Lebesgue measure.
$\exp(\cdot)$	exponential function.
$\boldsymbol{X}_1 \perp\!\!\!\perp \boldsymbol{X}_2$	$\boldsymbol{X}_1$ and $\boldsymbol{X}_2$ are independent.
$\boldsymbol{X}_1 \mid \boldsymbol{X}_2$	$\boldsymbol{X}_1$ given $\boldsymbol{X}_2$.

Simple Undirected Graphs

$\mathcal{G} = (V, E)$	simple undirected graph with vertex set V and edge set E.
$\mathcal{G}_A$	subgraph induced by A.
$\sim$	adjacent or neighbours.
$\not\sim$	nonadjacent.
$\alpha \sim_p \beta$	there exists a path between α and β.
$\alpha \not\sim_p \beta$	there exists no path between α and β.
$\mathrm{bd}(\cdot)$	boundary.
$\mathrm{cl}(\cdot)$	closure.
$\subset$	proper subset of.
$\subseteq$	subset o.f
$\cup$	union.
$\cap$	intersection.
$\dot\cup$	direct union.
$\mathcal{C}$	set of cliques in a simple undirected graph.
RIP	running intersection property, (92).
MCS	maximum cardinality search, (92).
$\lvert\cdot\rvert$	cardinality.
T_C	C-marginal adjusting operator, (130).
T	adjusting operator for all the cliques, (131).

Lecture Notes in Statistics

For information about Volumes 1 to 14
please contact Springer-Verlag

Vol. 15: K.R.W. Brewer and M. Hanif, Sampling with Unequal Probabilities. ix, 164 pages, 1983.

Vol. 16: J.P. Florens, M. Mouchart, J.P. Raoult, L. Simar, and A.F.M. Smith (Editors), Specifying Statistical Models: From Parametric to Non-Parametric, Using Bayesian or Non-Bayesian Approaches. xi, 204 pages, 1983.

Vol. 17: I.V. Basawa and D.J. Scott, Asymptotic Optimal Inference for Non-Ergodic Models. ix, 170 pages, 1983.

Vol. 18: W. Britton, Conjugate Duality and the Exponential Fourier Spectrum. v, 226 pages, 1983.

Vol. 19: L. Fernholz, von Mises Calculus For Statistical Functionals. viii, 124 pages, 1983.

Vol. 20: U. Herkenrath, D. Kalin, W. Vogel (Editors), Mathematical Learning Models — Theory and Algorithms: Proceedings of a Conference. xiv, 226 pages, 1983.

Vol. 21: H. Tong, Threshold Models in Non-linear Time Series Analysis. x, 323 pages, 1983.

Vol. 22: S. Johansen, Functional Relations, Random Coefficients and Nonlinear Regression with Application to Kinetic Data, viii, 126 pages, 1984.

Vol. 23: D.G. Saphire, Estimation of Victimization Prevalence Using Data from the National Crime Survey. v, 165 pages, 1984.

Vol. 24: T.S. Rao, M.M. Gabr, An Introduction to Bispectral Analysis and Bilinear Time Series Models. viii, 280 pages, 1984.

Vol. 25: E. Parzen (Editor), Time Series Analysis of Irregularly Observed Data. Proceedings, 1983. vii, 363 pages, 1984.

Vol. 26: J. Franke, W. Härdle and D. Martin (Editors), Robust and Nonlinear Time Series Analysis. Proceedings, 1983. ix, 286 pages, 1984.

Vol. 27: A. Janssen, H. Milbrodt, H. Strasser, Infinitely Divisible Statistical Experiments. vi, 163 pages, 1985.

Vol. 28: S. Amari, Differential-Geometrical Methods in Statistics. v, 290 pages, 1985.

Vol. 29: B.J.T. Morgan and P.M. North (Editors), Statistics in Ornithology. xxv, 418 pages, 1985.

Vol 30: J. Grandell, Stochastic Models of Air Pollutant Concentration. v, 110 pages, 1985.

Vol. 31: J. Pfanzagl, Asymptotic Expansions for General Statistical Models. vii, 505 pages, 1985.

Vol. 32: R. Gilchrist, B. Francis and J. Whittaker (Editors), Generalized Linear Models. Proceedings, 1985. vi, 178 pages, 1985.

Vol. 33: M. Csörgo, S. Csörgo, L. Horváth, An Asymptotic Theory for Empirical Reliability and Concentration Processes. v, 171 pages, 1986.

Vol. 34: D.E. Critchlow, Metric Methods for Analyzing Partially Ranked Data. x, 216 pages, 1985.

Vol. 35: T. Calinski and W. Klonecki (Editors), Linear Statistical Inference. Proceedings, 1984. vi, 318 pages, 1985.

Vol. 36: B. Matérn, Spatial Variation. Second Edition. 151 pages, 1986.

Vol. 37: R. Dykstra, T. Robertson and F.T. Wright (Editors), Advances in Order Restricted Statistical Inference. Proceedings, 1985. viii, 295 pages, 1986.

Vol. 38: R.W. Pearson and R.F. Boruch (Editors), Survey Research Designs: Towards a Better Understanding of Their Costs and Benefits. v, 129 pages, 1986.

Vol. 39: J.D. Malley, Optimal Unbiased Estimation of Variance Components. ix, 146 pages, 1986.

Vol. 40: H.R. Lerche, Boundary Crossing of Brownian Motion. v, 142 pages, 1986.

Vol. 41: F. Baccelli, P. Brémaud, Palm Probabilities and Stationary Queues. vii, 106 pages, 1987.

Vol. 42: S. Kullback, J.C. Keegel, J.H. Kullback, Topics in Statistical Information Theory. ix, 158 pages, 1987.

Vol. 43: B.C. Arnold, Majorization and the Lorenz Order: A Brief Introduction. vi, 122 pages, 1987.

Vol. 44: D.L. McLeish, Christopher G. Small, The Theory and Applications of Statistical Inference Functions. vi, 124 pages, 1987.

Vol. 45: J.K. Ghosh (Editor), Statistical Information and Likelihood. 384 pages, 1988.

Vol. 46: H.-G. Müller, Nonparametric Regression Analysis of Longitudinal Data. vi, 199 pages, 1988.

Vol. 47: A.J. Getson, F.C. Hsuan, {2}-Inverses and Their Statistical Application. viii, 110 pages, 1988.

Vol. 48: G.L. Bretthorst, Bayesian Spectrum Analysis and Parameter Estimation. xii, 209 pages, 1988.

Vol. 49: S.L. Lauritzen, Extremal Families and Systems of Sufficient Statistics. xv, 268 pages, 1988.

Vol. 50: O.E. Barndorff-Nielsen, Parametric Statistical Models and Likelihood. vii, 276 pages, 1988.

Vol. 51: J. Hüsler, R.-D. Reiss (Editors). Extreme Value Theory, Proceedings, 1987. x, 279 pages, 1989.

Vol. 52: P.K. Goel, T. Ramalingam, The Matching Methodology: Some Statistical Properties. viii, 152 pages, 1989.

Vol. 53: B.C. Arnold, N. Balakrishnan, Relations, Bounds and Approximations for Order Statistics. ix, 173 pages, 1989.

Vol. 54: K.R. Shah, B.K. Sinha, Theory of Optimal Designs. viii, 171 pages, 1989.

Vol. 55: L. McDonald, B. Manly, J. Lockwood, J. Logan (Editors), Estimation and Analysis of Insect Populations. Proceedings, 1988. xiv, 492 pages, 1989.

Vol. 56: J.K. Lindsey, The Analysis of Categorical Data Using GLIM. v, 168 pages, 1989.

Vol. 57: A. Decarli, B.J. Francis, R. Gilchrist, G.U.H. Seeber (Editors), Statistical Modelling. Proceedings, Trento 1989, ix, 343 pages, 1989.

Vol. 58: O.E. Barndorff-Nielsen, P. Blæsild, P.S. Eriksen, Decomposition and Invariance of Measures, and Statistical Transformation Models. v, 147 pages, 1989.

Vol. 59: S. Gupta, R. Mukerjee, A Calculus for Factorial Arrangements. vi, 126 pages, 1989.

Vol. 60: L. Györfi, W. Härdle, P. Sarda, Ph. Vieu, Nonparametric Curve Estimation from Time Series. viii, 153 pages, 1989.

Vol. 61: J. Breckling, The Analysis of Directional Time Series: Applications to Wind Speed and Direction. viii, 238 pages, 1989.

Vol. 62: J.C. Akkerboom, Testing Problems with Linear or Angular Inequality Constraints. xii, 291 pages, 1990.

Vol. 63: J. Pfanzagl, Estimation in Semiparametric Models: Some Recent Developments. iii, 112 pages, 1990.

Vol. 64: S. Gabler, Minimax Solutions in Sampling from Finite Populations. v, 132 pages, 1990.

Vol. 65: A. Janssen, D.M. Mason, Non-Standard Rank Tests. vi, 252 pages, 1990.

Vol 66: T. Wright, Exact Confidence Bounds when Sampling from Small Finite Universes. xvi, 431 pages, 1991.

Vol. 67: M.A. Tanner, Tools for Statistical Inference: Observed Data and Data Augmentation Methods. vi, 110 pages, 1991.

Vol. 68: M. Taniguchi, Higher Order Asymptotic Theory for Time Series Analysis. viii, 160 pages, 1991.

Vol. 69: N.J.D. Nagelkerke, Maximum Likelihood Estimation of Functional Relationships. V, 110 pages, 1992.

Vol. 70: K. Iida, Studies on the Optimal Search Plan. viii, 130 pages, 1992.

Vol. 71: E.M.R.A. Engel, A Road to Randomness in Physical Systems. ix, 155 pages, 1992.

Vol. 72: J.K. Lindsey, The Analysis of Stochastic Processes using GLIM. vi, 294 pages, 1992.

Vol. 73: B.C. Arnold, E. Castillo, J.-M. Sarabia, Conditionally Specified Distributions. xiii, 151 pages, 1992.

Vol. 74: P. Barone, A. Frigessi, M. Piccioni, Stochastic Models, Statistical Methods, and Algorithms in Image Analysis. vi, 258 pages, 1992.

Vol. 75: P.K. Goel, N.S. Iyengar (Eds.), Bayesian Analysis in Statistics and Econometrics. xi, 410 pages, 1992.

Vol. 76: L. Bondesson, Generalized Gamma Convolutions and Related Classes of Distributions and Densities. viii, 173 pages, 1992.

Vol. 77: E. Mammen, When Does Bootstrap Work? Asymptotic Results and Simulations. vi, 196 pages, 1992.

Vol. 78: L. Fahrmeir, B. Francis, R. Gilchrist, G. Tutz (Eds.), Advances in GLIM and Statistical Modelling: Proceedings of the GLIM92 Conference and the 7th International Workshop on Statistical Modelling, Munich, 13-17 July 1992. ix, 225 pages, 1992.

Vol. 79: N. Schmitz, Optimal Sequentially Planned Decision Procedures. xii, 209 pages, 1992.

Vol. 80: M. Fligner, J. Verducci (Eds.), Probability Models and Statistical Analyses for Ranking Data. xxii, 306 pages, 1992.

Vol. 81: P. Spirtes, C. Glymour, R. Scheines, Causation, Prediction, and Search. xxiii, 526 pages, 1993.

Vol. 82: A. Korostelev and A. Tsybakov, Minimax Theory of Image Reconstruction. xii, 268 pages, 1993.

Vol. 83: C. Gatsonis, J. Hodges, R. Kass, N. Singpurwalla (Editors), Case Studies in Bayesian Statistics. xii, 437 pages, 1993.

Vol. 84: S. Yamada, Pivotal Measures in Statistical Experiments and Sufficiency. vii, 129 pages, 1994.

Vol. 85: P. Doukhan, Mixing: Properties and Examples. xi, 142 pages, 1994.

Vol. 86: W. Vach, Logistic Regression with Missing Values in the Covariates. xi, 139 pages, 1994.

Vol. 87: J. Møller, Lectures on Random Voronoi Tessellations. vii, 134 pages, 1994.

Vol. 88: J. E. Kolassa, Series Approximation Methods in Statistics. viii, 150 pages, 1994.

Vol. 89: P. Cheeseman, R.W. Oldford (Editors), Selecting Models From Data: AI and Statistics IV. xii, 487 pages, 1994.

Vol. 90: A. Csenki, Dependability for Systems with a Partitioned State Space: Markov and Semi-Markov Theory and Computational Implementation. x, 241 pages, 1994.

Vol. 91: J.D. Malley, Statistical Applications of Jordan Algebras. viii, 101 pages, 1994.

Vol. 92: M. Eerola, Probabilistic Causality in Longitudinal Studies. vii, 133 pages, 1994.

Vol. 93: Bernard Van Cutsem (Editor), Classification and Dissimilarity Analysis. xiv, 238 pages, 1994.

Vol. 94: Jane F. Gentleman and G.A. Whitmore (Editors), Case Studies in Data Analysis. viii, 262 pages, 1994.

Vol. 95: Shelemyahu Zacks, Stochastic Visibility in Random Fields. x, 175 pages, 1994.

Vol. 96: Ibrahim Rahimov, Random Sums and Branching Stochastic Processes. viii, 195 pages, 1995.

Vol. 97: R. Szekli, Stochastic Ordering and Dependence in Applied Probablility, viii, 194 pages, 1995.

Vol. 98: Philippe Barbe and Patrice Bertail, The Weighted Bootstrap, viii, 230 pages, 1995.

Vol. 99: C.C. Heyde (Editor), Branching Processes: Proceedings of the First World Congress, viii, 185 pages, 1995.

Vol. 100: Wlodzimierz Bryc, The Normal Distribution: Characterizations with Applications, viii, 139 pages, 1995.

Vol. 101: H.H. Andersen, M. Højbjerre, D. Sørensen, P.S. Eriksen, Linear and Graphical Models: for the Multivariate complex Normal Distribution, x, 184 pages, 1995.